C000181085

Environmental Archaeology

Series Editors:

Co-ordinating Editor

John A. Matthews
Department of Geography, University of Wales, Swansea, UK

Editors

Raymond S. Bradley
Department of Geosciences, University of Massachusetts, Amherst, USA

Neil Roberts
Department of Geography, University of Plymouth, UK

Martin A. J. Williams
Mawson Graduate Centre for Environmental Studies, University of Adelaide, Australia

Titles in the series

Already published:

Cultural Landscapes and Environmental Change (Lesley Head, University of Wollongong, Australia)

Environmental Change in Mountains and Uplands (Martin Beniston, University of Fribourg, Switzerland)

Glaciers and Environmental Change (Atle Nesje and Svein Olaf Dahl, Bergen University, Norway)

Natural Hazards and Environmental Change (W.J. McGuire, C.R.J. Kilburn and M.A. Saunders, University College London, UK)

Environmental Archaeology

Theoretical and Practical Approaches

Nick Branch
Royal Holloway, University of London, Egham, Surrey UK

Matthew Canti
English Heritage, Fort Cumberland, Eastney, Portsmouth UK

Peter Clark
Canterbury Archaeological Trust, Canterbury, UK

Chris Turney
University of Wollongong, NSW, Australia

Hodder Arnold
A MEMBER OF THE HODDER HEADLINE GROUP

First published in Great Britain in 2005 by
Hodder Education, a member of the Hodder Headline Group,
338 Euston Road, London NW1 3BH

www.hoddereducation.co.uk

Distributed in the United States of America by
Oxford University Press Inc.
198 Madison Avenue, New York, NY10016

© 2005 Nick Branch, Matthew Canti, Peter Clark, Chris Turney

All rights reserved. No part of this publication may be reproduced or transmitted in any form or by any means, electronically or mechanically, including photocopying, recording or any information storage or retrieval system, without either prior permission in writing from the publisher or a licence permitting restricted copying. In the United Kingdom such licences are issued by the Copyright Licensing Agency: 90 Tottenham Court Road, London W1T 4LP.

Hodder Headline's policy is to use papers that are natural, renewable and recyclable products and made fromwood grown in sustainable forests. The logging and manufacturing processes are expected to conform to the environmental regulations of the country of origin.

The advice and information in this book are believed to be true and accurate at the date of going to press, but neither the editors, the authors nor the publisher can accept any legal responsibility or liability for any errors or omissions.

British Library Cataloguing in Publication Data
A catalogue record for this book is available from the British Library

Library of Congress Cataloging-in-Publication Data
A catalog record for this book is available from the Library of Congress

ISBN-10: 0 340 80871 3
ISBN-13: 978 0 340 80871 9

1 2 3 4 5 6 7 8 9 10

Typeset in Palatino by Servis Filmsetting Ltd, Manchester
Printed and bound in Great Britain by J.W. Arrowsmiths, Bristol.

What do you think about this book? Or any other Hodder Education title?
Please send your comments to the feedback section on www.hoddereducation.co.uk.

Contents

Figures

Tables

Boxes

Preface to the series

The study of environmental change is a major growth area of interdisciplinary science. Indeed, the intensity of current scientific activity in the field of environmental change may be viewed as the emergence of a new area of 'big science' alongside such recognized fields as nuclear physics, astronomy and biotechnology. The science of environmental change is fundamental science on a grand scale: rather different from nuclear physics but nevertheless no less important as a field of knowledge, and probably of more significance in terms of the continuing success of human societies in their occupation of the Earth's surface.

The need to establish the pattern and causes of recent climatic changes, to which human activities have contributed, is the main force behind the increasing scientific interest in environmental change. Only during the past few decades have the scale, intensity and permanence of human impacts on the environment been recognized and begun to be understood. A mere 5000 years ago, in the mid-Holocene, non-local human impacts were more or less negligible even on vegetation and soils. Today, however, pollutants have been detected in the Earth's most remote regions, and environmental processes, including those of the atmosphere and oceans, are being affected at a global scale.

Natural environmental change has, however, occurred throughout Earth's history. Large-scale natural events as abrupt as those associated with human environmental impacts are known to have occurred in the past. The future course of natural environmental change may in some cases exacerbate human-induced change; in other cases, such changes may neutralize the human effects. It is essential, therefore, to view current and future environmental changes, like global warming, in the context of the broader perspective of the past. This linking theme provides the distinctive focus of the series and is mentioned explicitly in many of the titles listed overleaf.

It is intended that each book in the series will be an authoritative, scholarly and accessible synthesis that will become known for advancing the conceptual framework of studies in environmental change. In particular we hope that each book will inform advanced undergraduates and be an inspiration to young research workers. To this end, all the invited authors are experts in their respective fields and are active at the research frontier. They are, moreover, broadly representative of the interdisciplinary and international nature of environmental change research today. Thus, the series as a whole aims to cover all the themes normally considered as key issues in environmental change even though individual books may take a particular viewpoint or approach.

John A. Matthews (Co-ordinating Editor)

Acknowledgements

Many stimulating and fruitful discussions with colleagues have helped us formulate ideas over the years. We would particularly like to thank Jen Heathcote, Peter Kershaw, John Lowe, Matt McGlone; the Interpreting Stratigraphy Group (especially Kate Steane, Andrew Hutcheson, Steve Roskams, Max Adams, Natasha Hutcheson, Ed Harris, Jez Reeve, Andrew Westman, Adrian Cox, Tim Williams, Rob Young, Friederike Hammer, Liz Popescu, and Steve Stead). Also, thanks to John Athersuch, Tony Barham, Jen Heathcote, Helen Keeley, John Lowe, Chris Lowe, Nick Merriman, Libby Mulqueeny, Deborah Pearsall, Richard Preece, James Rackham, Rob Scaife, Jane Sidell, Chris Thomas, Maurice Tucker, Rick Turner, Clare Twiddle, Sylvia Warman and John Whittaker, for contributions to diagrams.

Thanks to Harry Kenward, Mark Robinson, Jonathan Holmes, Nigel Cameron for help with identifications, and to Karon Branch for patience.

Preface

"For the mind of man is far from the nature of a clear and equal glass, wherein the beams of things should reflect according to their true incidence; nay, it is rather like an enchanted glass, full of superstition and imposture, if it be not delivered and reduced"

Francis Bacon
The Advancement of Learning

"See first, think later, then test. But always see first. Otherwise you will only see what you were expecting. Most scientists forget that"

Wonko the Sane
So Long and Thanks for All the Fish by Douglas Adams

Environmental change occurs across many scales in both space and time. Through the last few interglacials, variation has been perceived, reacted to, and affected by humans in a complex relationship that is not easily unravelled, even when working with modern data. Rather than simply being players on a stage, our part in the processes of change has become more and more an effective component. As human activity increases, the nature of that role adopts greater significance for the planning of global futures, so the development of an accurate representation of the whole dynamic picture becomes ever more important.

Environmental archaeology has played a major part in fostering our awareness both of these changes, and our interaction with them. In order to do this effectively, the subject has had to negotiate firstly the complexity of the human input to environmental systems, and secondly the difficulties of relating these developments to cultural histories based on the results of material excavations. The respective sides of this duality now have well-developed approaches built up from experience of detailed analysis within their subject boundaries during the last half century or so, but there has still been surprisingly little dialogue between them. They have existed in tandem, sometimes integrated at the publication stage, but frequently presenting separate results in all but name.

This book was inspired by the need to bring together these two parts of the archaeological spectrum. There are numerous examples of projects where excellent integration has occurred, and we know many inspiring colleagues whose interests seamlessly span the whole range of cultural and environmental change. Our aim is not, therefore, to throw out pre-existing approaches and try to start again with a clean sheet. We acknowledge and describe specialist activity traditional to environmental archaeology; we offer detailed discussions of the approaches that have contributed most to the subject; and where appropriate, we present introductions to the methodology. However, the underlying concept aims to position environmental archaeology more as an emphasis within the broad framework of archaeology as a whole, rather than viewing it as a narrow specialist activity with its own world view and preoccupations. The simple act of co-operation between four authors from different parts of the archaeological spectrum in itself led to difficulties of parity and variations in meaning which are only partly resolved. We hope, nevertheless, that the exercise proves worthwhile to the reader in terms not only of the information contained here, but also of building a more integrated view of this fascinating subject.

Safety Note

This book contains recipes for extraction and examination of materials involving the use of chemicals and procedures which could be injurious to health if safety protocols are not properly observed. These operations should only be carried out after due consideration of *all* the safety issues. As a minimum, the work should be done by a qualified operative, fully protected from contact with dangerous chemicals by protective clothing, and from all vapours by completing the work in a fume cupboard. However, full assessment and minimisation of risk can only be achieved by a detailed examination of the recommended approaches and safeguards. Further information can be found at http://www.coshh-essentials.org.uk/ http://www.hse.gov.uk/pubns/indg136.pdf http://www.hse.gov.uk/hthdir/noframes/coshh/

1

Introduction to environmental archaeology

1.0 Chapter summary

This introductory chapter presents a detailed discussion of the conceptual framework of environmental archaeology. In the opening section, we make an assessment of the relationship between archaeology and science with the aim of putting environmental archaeology into a clearer context, and ultimately arriving at a better understanding of its role. This is followed by a broader discussion of the concept of time in environmental archaeology, with particular emphasis on the importance of understanding how the biotic and abiotic components of our environment change. We examine the four main spatial scales (micro, meso, macro and mega) of importance to environmental archaeology, and discuss their relative significance with respect to a range of investigations.

1.1 Defining environmental archaeology

1.1.1 Introduction

Environmental archaeology has emerged as a named discipline only in the last 30 years. It has rapidly grown in significance and is now an integral part of most excavation projects. It is taught as a standard course component in many university archaeology departments, and has its own infrastructure of academics, societies and journals. It has developed with little or no theoretical discourse, but has relied chiefly on a general pre-existing uniformitarian view taken by the geologists and biologists who were the first to analyse environmental remains on archaeological sites. During this same period, cultural archaeology has undergone considerable changes of theoretical basis, which have continued almost without inclusion of the treatment of environmental evidence. This apparent independence between two such closely interwoven subjects belies a relationship of some complexity. We therefore need first to examine environmental archaeology as a component of archaeology more generally, before defining it as a subject in itself.

1.1.2 Archaeology and science

It is a truism that there are as many definitions of archaeology as there are archaeologists, but here we shall identify it as *'the study of people and their relationship with the environment through time'*. This is a broad church, and encompasses many sub-disciplines, philosophical stances and individual ambitions. Crucially, though, archaeology concerns people's lives, from the individual decisions that led to a particular feature being created, up through the social frameworks engendered to structure and regulate their relationship with the world; how these evolved and changed in response to social and environmental factors; and on through the ancient tragedies of warfare and plague, to the whole sweep of history from humanity's earliest origins to the global billions of today.

Archaeology is an immense field, and unsurprisingly there is a wide range of different types of data that can be employed. The scope includes the traditional objects of archaeological study: ancient structures, artefacts and stratification, as well as human remains, economic and environmental indicators such as plant remains, molluscs, animal bones and insects. Equally pertinent are the study of artistic and symbolic forms – historic documents and iconography, religious and ritual expression, social theory and much more. Archaeology is a discipline where scholars from many different fields can make a large contribution, ranging from the schools of artistic and literary criticism employed in understanding historic texts, through to the hard science of chemistry and physics underlying absolute dating techniques. Clearly this range of different disciplines brings with it a host of different views on what is the most appropriate approach to archaeological research, and a great deal of ink has been used in the debate between the various camps (see Hodder, 1986; Trigger, 1989; Renfrew and Bahn, 2000). Much of this debate ultimately revolves around the issue of what we can actually say about the past, and what validity such statements have. Is there a 'true past' that our multidisciplinary teams can explore, or is it all a fantasy of our own making, to be rewritten whenever political or academic fashions change? Hand-in-hand with this question go methodological issues: can archaeology ever be truly objective, or must it, by its very nature, always remain a subjective discipline? Does archaeology belong with the humanities or the sciences?

The answer must surely be that archaeology encompasses both. The great sweep of human history (as well as the actions of a single individual) cannot be explained by the application of universal, formulaic laws of behaviour (though many have tried), but neither can it be thought of as a complete fabrication. Most people accept that there are traces of the past all around us, from buried artefacts to the earthworks of ancient field systems. Although these traces may have undergone many changes or transformations through time (Schiffer, 1987), and may be interpreted in different ways, they are *facts* that prescribe the formulation of hypotheses about the nature of the past.

First, though, we should return to the question of whether there is a 'true past' that we can collectively explore, subject to our fragmentary evidence and its interpretation. It is inevitable that all history is a human construction. Any attempt to explain or derive a narrative from the evidence of the past is a subjective, selective activity. What two individuals would explain the modern world in the same way today, even with all the evidence to hand and an understanding of the cultural references of that material? They would surely have different assessments of what issues are pertinent, and different perspectives on why things happen in a certain way. Both explanations might be equally valid, but radically different.

When studying the evidence for the past, archaeologists must assign significance to their data and structure their research according to a hypothesis or questioning of that data. The questions we ask are modern constructs; they dictate what evidence we seek and how we analyse it. The study of the initial colonization of the Pacific (Irwin, 1994), for example, would require a very different approach to that articulating the sequence of events during the siege of Alésia in 52 BC (Reddé, 2003). We might decide, therefore, that while there is a 'true past' (as represented by archaeological data in its broadest sense), there is no 'true history'. That is not to say that our histories and explanations of the past are not valid, but rather that our data are mute until we breathe life into them with hypothesis.

This view differentiates archaeology from science, for in science there is a belief that there exist universal laws that await discovery. Many have argued over the years that there are similar universal laws (or 'covering laws') that govern human behaviour (Hempel, 1942; 1963; Binford, 1977). In archaeology this was perhaps most strongly represented by the 'New Archaeology' in the 1960s and 1970s, greatly influenced by the writings of Louis Binford

(1977) and David Clarke (1978). This view that human behaviour can be predicted or explained by universal laws has always had its critics (e.g. Mandelbaum, 1961). In the last century the philosopher Karl Popper stated: 'I have shown that, for strictly logical reasons, it is impossible for us to predict the future course of history' (whatever our starting point in time) (Popper, 1957: iv). In recent years the concept (and with it the 'New Archaeology') has fallen into disfavour with the rise of 'post-processualism' as the dominant paradigm in archaeological thinking (Hodder, 1982; Miller and Tilley, 1984). Whatever the swings and roundabouts of academic fashion, we cannot ignore the effects of cultural perception on the archaeological record, and thus on our understanding of past societies. Human communities do not appear to follow patterns of behaviour predicated by a set of universal rules, or even to work towards optimal survival strategies. 'There are countless examples of the willingness of human beings to ignore the deleterious effects of their decisions for perceived benefits' (Conlin, 1998: 12). But if we are to argue that archaeology is not a science, can there be such a thing as archaeological science?

At the heart of the argument stands environmental archaeology. This branch of the discipline, with its foundations in the biological and earth sciences, is rooted in scientific procedure and scientific philosophy. Ultimately, its results are based on the concept of uniformitarianism, the idea that natural processes observed today also occurred in the past and produced the same physical effects. Thus, if we observe a particular erosion pattern of overbank deposits following river flooding, and a similar pattern occurs in archaeological deposits, we may infer that the archaeological deposits were subject to erosion by flood waters (Brown, 1997: 76, plate 3.10). Similar analogies can be made in the biological sphere. For example, we may observe today that a particular species of mollusc (such as *Lymnaea truncatula*) inhabits exclusively marshland or wet, muddy areas. If we find shells of that species in an archaeological deposit, we can assume (allowing for taphonomy) that a similar habitat existed nearby at the time the deposit was formed.

This approach works well with processes and materials that obey universal laws, and within its own boundaries, environmental archaeology can properly be described as a scientific pursuit. However, the introduction of human agency limits the efficacy of uniformitarianism as a means of providing a completely satisfactory explanation for archaeological phenomena. One challenge facing environmental archaeologists today is to maintain scientific rigour and standards in their research and incorporate the powerful insights it provides into a post-processualist (non-scientific) agenda. This is by no means to suggest that environmental archaeology is necessarily more rigorous than other types of archaeological research. All archaeologists have a duty and responsibility to be as rigorous as possible within the terms of reference of the data they are studying, but it is easier for particular emphases to be placed on interpretation within a post-processualist framework. The authors' experience, including working in Scotland, Lebanon and elsewhere, has demonstrated that there are individuals who are eager to distort the interpretation of archaeological data to support divisive nationalistic and racist agendas. An extreme example of this was Heinrich Himmler's formation in 1935 of *Das Ahnenerbe* (the Ancestral Heritage Society), a branch of the SS devoted to using archaeological data to support the racist policies of the Nazi party. While we might create the questions we ask of our data, we must nevertheless remain servants to that data.

If, as we have suggested, there is no such thing as 'true history' (in contradistinction to the 'true past'), then is there any value in archaeology at all? We may contrast modern perceptions of the humanities with the sciences: science gives us 'facts' and explores universal truths; the humanities appear self-referential and offer little of practical value. This perception is perhaps reflected in the relative levels of funding available to scientific disciplines as opposed to the humanities; no wonder that some archaeologists are keen to present their subject as a scientific pursuit, with the potential for extra university posts and funding possibilities that would bring. However, archaeology does deal with a real, physical set of data that prescribes and inspires our understanding of the past and of ourselves. The sixteenth century philosopher Pierre Charron claimed that 'the true science and study of man is man'; if this is so, then 'the study of people and their relationship with the environment through time' is one of the most challenging and noble pursuits open to society.

1.1.3 Space, time and scale in environmental archaeology

The interaction between humans and their environment operates over a number of different spatial and temporal scales (Table 1.1). Long-term processes that lead to the evolution of plants and animals, the extinction of particular organisms and changes within biotic communities include climate (e.g. glacial-interglacial cycles), soil development (e.g. podzolization), succession (e.g. arctic-tundra to woodland) and human impact (e.g. since the inception of prehistoric agricultural practices to the present day). These processes have assisted the successful migration, colonization and adaptation of humans to new environments by providing suitable biological resources for food, clothing, medicine, construction and much more. They have also acted as natural barriers creating inhospitable and impenetrable environments unsuitable for human habitation or resource utilization. Short-term processes include natural events such as disease, e.g. the elm decline (see Section 3.4.2), but are mainly confined to human-induced events including deforestation, burning of vegetation, cultivation and animal husbandry, which together represent the long-term process of cultural modification of the natural environment (Butzer, 1982; Bell and Walker, 1992).

1.1.4 Main developments in environmental archaeology

Since the inception of modern archaeology during the 1930s, several individuals have made major contributions to the historical development of environmental archaeology. In the centuries before that, important foundations were laid by eighteenth and nineteenth century geologists looking at the principles of uniformitarianism and stratigraphy, e.g. Hutton (1726–97) and Lyell (1797–1875); by

TABLE 1.1 Spatial and temporal scales (Modified from Dincauze, 2000)

Spatial scales	Area (km^2)	Spatial units
Mega	5.1×10^8	Earth
	$<10^8$	Hemispheres, Continents
Macro	10^4–10^7	Region (e.g. country), Major physiographic unit (e.g. Mediterranean)
Meso	10^2–10^4 1–10^2	Site catchment (sub-region, e.g. Thames valley), Locality (e.g. London)
Micro	<1	Site, Activity area (e.g. house)
Temporal scales	**Duration (yr)**	
Mega	>1 million	e.g. evolution of plants and animals
Macro	10000 to 1 million	e.g. glacial-interglacial biome changes
Meso	100 to 10000	e.g. vegetation migration
Micro	<1 to 100	e.g. woodland clearance

biologists developing the theory of evolution, e.g. Darwin (1809–82); and by anthropologists or archaeologists looking at chronology and cultural development, e.g. Pitt-Rivers (1827–1900), Morgan (1818–81) and Taylor (1832–1917). These building blocks formed the basis for aggressive data gathering in archaeology between 1930 and 1960. Combined with major theoretical advances, e.g. cultural ecology by Steward (1902–72) and practical techniques, e.g. radiocarbon dating by Libby (1908–80), this resulted in a widespread recognition of the importance of understanding human-environment interactions in cultural change (Clark, 1952; Childe, 1957).

During the 1960s, the conceptual basis of archaeology changed, with the advent of the processual approach (Binford, 1968; Clarke, 1968), applying an inherently scientific rationale to research, and seeking to explain rather than simply describe cultural change (Renfrew and Bahn, 2000). Embedded in these developments was the need for multidisciplinary investigations, drawing on new practical approaches from the disciplines of physics, chemistry, geography, biology, sedimentology and pedology (Brothwell and Higgs, 1963), and new complex theoretical frameworks such as systems theory (Flannery, 1968). Environmental archaeology, incorporating the fields of geoarchaeology, archaeobotany and zooarchaeology epitomized these efforts, with its aim (on one level) to 'define the characteristics and processes of the biophysical environment that provide a matrix for and interact with socio-economic systems, as reflected, for example, in subsistence activities and settlement patterns' (Butzer, 1982: 6).

Several key developments have taken place in each of these fields that are worth briefly mentioning. In geoarchaeology, there has been improved understanding of landscape context (Gladfelter, 1977), stratigraphic context and site formation (Limbrey, 1975), and site modification and landscape modification (Vita-Finzi, 1969; Davidson and Shackley, 1976). New concepts have been introduced, such as formation processes and taphonomy (Schiffer, 1972), as well as new archives of information on palaeoclimatology, palaeohydrology and glacial history (Lowe and Walker, 1997), and new methods, such as soil micromorphology (Davidson and Simpson, 2001). Archaeobotany, with its emphasis on environmental reconstruction, and the inter-relationships between people and plants (Butzer, 1982), is an equally diverse field. Analytical research in pollen (Dimbleby, 1978, 1985), phytoliths (Pearsall, 1982; Piperno, 1988), diatoms (Mannion, 1987) and seeds (Hillman, 1973; Renfrew, 1973), coupled with new extraction techniques (flotation revolution; Jarman *et al.*, 1972) and experimentation in palaeoethnobotany (van Zeist and Casparie, 1984; Hastorf and Popper, 1989), has certainly improved understanding of natural and human-induced environmental change (Ucko and Dimbleby, 1969; Bell and Walker, 1992; Zohary and Hopf, 1993). Finally, zooarchaeology has always complemented archaeobotany in its main aim, which is to understand human subsistence practices (Butzer, 1982), but has developed major research themes including the study of domesticated and wild animal exploitation (Zeuner, 1963; Bender, 1975; Clutton-Brock, 1981; Clutton-Brock and Grigson, 1983), taphonomy (Behrensmeyer and Hill, 1980), quantification and anatomy (Legge and Rowley-Conwy, 1988) and the general reconstruction of the human environment (Buckland, 1976; Osbourne, 1976).

A closer integration between archaeology and science has manifested itself in several major developments within the discipline, of which three may be linked with the sub-discipline of environmental archaeology.

First, and perhaps arguably the most important, have been the advancements in geochronological methods. Modern archaeology is heavily reliant upon a range of 'absolute' dating methods to provide greater chronological precision than is available from 'relative' approaches (see Chapter 4). These include radiocarbon (especially using accelerator mass spectrometry; Harris, 1987), optically stimulated luminescence, thermo-luminescence, electron spin resonance and U series dating. Although these techniques are widely available through research and commercial laboratories, there remains a poor understanding of their strengths and weaknesses within the archaeological community generally. There is a clear need to support initiatives to refine these core methods and to develop newer approaches (e.g. single grain OSL dating), but most importantly we require an improved educational programme to ensure that archaeologists are fully equipped with the necessary theoretical and practical skills when choosing a method, selecting materials for dating and assessing the results (especially their statistical significance; see Bayliss, 1998). Nevertheless, these methods have permitted archaeologists to ask important questions about the timing and duration, and therefore the patterns, of human occupation (Rick, 1987; Housley *et al.*, 1997), and the relationships between the timing of changes in the natural environment (e.g. climate), cultural evolution and human activities (van Andel and Tzedakis, 1996; Binford *et al.*, 1997).

Secondly, we would include the various research areas concerned with human origins, evolution, worldwide human colonization and adaptation (e.g. Gamble, 1986; Aldenderfer, 1999). In Britain, for example, Palaeolithic archaeology received an exciting boost following the extraordinary discovery of human skeletal remains at Boxgrove, dated to c. 500,000 years. This study, as well as several others throughout the world, demonstrated the importance of an integrated approach to understanding the relationships between site formation processes, environmental change (e.g. climate change and vegetation history) and human activities (e.g. faunal exploitation and the utilization of flint resource) across a range of spatial and temporal scales (Gamble, 1993, 1995). Linked to these studies have been the developments in the use of ancient mtDNA (mitochondrial deoxyribonucleic acid; Jones, 2001), which have permitted human genetic history to be mapped, allowing archaeologists to reconstruct the patterns of human movement and their origins (Ammerman and Cavalli-Sforza, 1984; Cann *et al.*, 1987; Richards *et al.*, 1996; Sykes, 1998), as well as modelling of evolutionary advancements in cognition and language (Aiello and Dunbar, 1993). One concern, which has been articulated by Gamble (1998) and others, is the often imprecise nature of environmental reconstructions from terrestrial archives due to unsuitable geological or archaeological deposits, and/or poor taxonomic precision afforded by the range of biological proxies often used. This problem has led to considerable debate about the precise faunal and floral composition and structure of past ecosystems, and hence the availability (or not) of resources suitable to sustain human life over particular periods of time (e.g. during short-term climatic fluctuations) and in specific geographical areas. This is clearly one area of ongoing research that will continue to attract inter and multidisciplinary collaboration (e.g. the use of ancient plant and animal DNA).

The third research area is concerned with the study of plant and animal domestication and exploitation, and the origins of agriculture, throughout the Old and New world (Binford, 1968; Flannery, 1968; Higgs and Jarman, 1969; Ucko and Dimbleby, 1969). These pioneering studies marked an important transition in the way research was conducted around this central theme because they attempted to explain the move towards food production (agriculture), rather than food procurement (hunters and gatherers), as a continuum rather than an abrupt transition (Childe, 1957; Braidwood, 1960), and employed an ecological-evolutionary approach that considered the impact of these various processes on ecosystems. During the 1970s and 1980s, several authors continued to stress this approach, and introduced new terms (e.g. palaeoeconomy; Higgs, 1972) and concepts (Rindos, 1984), but fundamentally reinforced the idea that the domestication of plants and animals did not follow a series of exact developmental stages. Arguably, the best account of these processes was produced by Ford (1985), who clearly defined foraging in contrast to food production, and differentiated the latter into cultivation and domestication, being two separate processes having different levels of biological impact. Since then attention has continued to focus on three main areas of research (here specific to plants but they could equally apply to animals):

1 'to identify the areas of earliest cultivation and domestication of particular crops'

Plant exploitative activity	Ecological effects (selected examples)	Food-yielding system	Socio-economic trends	Time
Burning vegetation	Reduction of competition; accelerated recycling of mineral nutrients; stimulation of asexual reproduction; selection for annual or ephemeral habit; synchronization of fruiting	WILD PLANT-FOOD PROCUREMENT (Foraging)	Increasing sedentism (settlement size, density and duration of occupation) Increasing population density (local, regional and continental) Increasing social complexity (ranking → stratification → state formation)	↓
Gathering/collecting	Casual dispersal of propagules			
Protective tending	Reduction of competition; local soil disturbance			
Replacement planting/ sowing	Maintenance of plant population in the wild	WILD PLANT-FOOD PRODUCTION with minimal tillage		
Transplanting/sowing	Dispersal of propagules to new habitats			
Weeding	Reduction of competition; soil modification			
Harvesting	Selection of dispersal mechanisms: positive and negative			
Storage	Selection and redistribution of propagules			
Drainage/irrigation	Enhancement of productivity; soil modification			
Land clearance	Transformation of vegetation composition and structure	CULTIVATION with systematic tillage		
Systematic soil tillage	Modification of soil texture, structure and fertility			
Propagation of genotypic and phenotypic variants:	DOMESTICATION	AGRICULTURE (Farming)		
Cultivation of domesticated crops (cultivars)	Establishment of agroecosystems	Evolutionary differentiation of agricultural systems		

FIGURE 1.1 The evolutionary continuum of people-plant interaction. (Modified from Harris, 1989: 17)

(Harris and Hillman, 1989: 5), based upon locating the wild progenitors of these crops.

2 to establish why 'particular taxa came to be selected (whether deliberately or not) as cultigens' (Harris and Hillman, 1989: 6).
3 to understand the 'process of domestication', which led in many cases to either rapid or gradual changes in the genotype and/or phenotype of particular species (Harris and Hillman, 1989: 6).

Harris' model of 'an evolutionary continuum of people-plant interaction' (Harris, 1989) also continues to provide an excellent conceptual framework in which to address these research areas through field and laboratory investigations. The model (see Fig. 1.1), which is largely based on ethnographic and historical records of plant utilization, presents essentially two forms. The first does not assume that the sequence from food procurement to agriculture represents a stage-by-stage process over time, in that all of the tasks listed are still conducted today. The second does assume a sequence of stages, in that on a worldwide scale there has clearly been a reduction in wild food procurement and an expansion of agricultural activities. This is perhaps most clearly illustrated in south-west Asia (consult Harris and Hillman, 1989 for various papers on worldwide wild plant procurement and domestication), where bioarchaeological evidence indicates that following the initial stage of wild food procurement between 12,000 and 10,000 years ago, domestication of plants and animals was underway shortly after 10,000 years ago. This evidence indicates that the Near East was clearly a centre of domestication for many of the plant and animal species we utilize today (e.g. wheat, barley, rye, pea, lentil, sheep, goat) not only in Europe, but throughout the world. The cause of this 'revolution', in the words of Childe, has been intensively debated, and hypotheses include: sedentism, demographic pressure, social interaction, cognitive development and climate change.

During the course of the last 10,000 years many geographical areas that lie outside the main worldwide zones of domestication have therefore undergone a continuum of change from wild food procurement to agriculture (e.g. Mesolithic–Neolithic transition, e.g. parts of continental Europe and Britain). This has generated a plethora of fascinating research questions that have occupied environmental archaeologists for 30 years or more. In Britain we continue to be preoccupied with understanding the origins of agriculture, the nature of ancient land use, and prehistoric economy and diet (Fowler, 1983). Due to poor preservation of plant and animal remains in archaeological contexts, these issues remain to be resolved. New integrated approaches to archaeology are clearly required, employing a range of environmental archaeological methods in the fields of geoarchaeology, geochronology and bioarchaeology (e.g. mtDNA), to obtain the maximum amount of information from each site.

1.1.5 Definition of environmental archaeology

The scientific development of environmental archaeology described above is both the reason for its success, and the central focus for the difficulties of its integration with a post-processual paradigm. The position occupied by environmental archaeology is more complex now than it was during the ascendancy of the 'New Archaeology' (Wilkinson and Stevens, 2003: 242–268). One approach to understanding the relationship between environmental archaeology and archaeology generally might be through Stephen Gould's 'non-overlapping magisteria' (Gould, 2001). He argues (with regard to science and religion, though clearly pertinent here) that different domains of inquiry frame their own rules and admissible questions, and set their own criteria for judgement and resolution (the 'magisterium' or teaching authority of a particular philosophical approach). Thus different approaches operate within their own frame of reference, and there is no conflict to resolve as the issues they address do not overlap.

Perhaps, then, environmental archaeology can only be regarded as a distinct subunit of this broad church called archaeology? Certainly, relative to cultural data, the changes in the environment itself are less subject to the view that there is no true history. The causes of those changes, however, are very much open to a range of cultural and climatic interpretations. There is a distinct interplay between the two fields which appears to be more a matter of emphasis than of radical differences in need of reconciliation. If we define environmental archaeology as '*the study of the environment and its relationship with people through time*' we can reflect simply that difference of emphasis while mirroring our original definition of archaeology as a whole.

Ultimately, environmental archaeology is a part of archaeology, and the subjects do have overlapping roles, but this plurality of approach is a reflection of the richness of our research: 'no single magisterium can come close to encompassing all the troubling issues raised by any complex subject' (Gould, 2001: 53). In recognizing the need for a multidisciplinary approach to archaeology, involving so many techniques, philosophies and areas of expertise, we should celebrate both the complexity and the value of our chosen field of study.

1.2 Concepts of change through time

1.2.1 Changes in material culture

The material culture of past societies is represented by their alteration of resources from the natural environment. This is readily seen in the portable artefacts retrieved from archaeological excavations: flint nodules fashioned into tools; clay shaped and fired to form pots; metal ores mined, smelted and cast into palstaves and knives. It can also refer to non-portable phenomena: stone, wood, earth and other natural materials transformed to form dwellings, enclosures, burial mounds, field systems and religious monuments. Whole landscapes may be altered for practical or symbolic reasons, like the lines and geometric figures carved onto the plains of Nazca in Peru (Aveni, 1986). We should distinguish, however, between alterations to the natural environment brought about by deliberate action and those caused inadvertently; the clearance of natural vegetation and the establishment of farmland may be described as material culture, but resultant soil erosion or desertification should not (Hughes, 1996). Of course there are grey areas – should domesticated animals be viewed as 'natural', or are they artefacts, deliberately created by people?

Material culture provides a fundamental tool in human adaptive strategies. From a determinist viewpoint, the manipulation of the natural environment is a mechanism to ensure the provision of essentials such as food, water and shelter. Crucially, though, the form and nature of material culture is predicated as much (if not more) by cultural criteria as by pragmatic efficiency. A pottery vessel may be made to hold liquid, but its form and decoration are dictated by a host of other concerns. The form of material culture may thus reflect the world view of a particular society, as has been suggested for the Neolithic monuments of Orkney in the northern UK (Richards, 1993), or the architecture of Christian monuments in medieval Europe (Woodman, 1981; Hill, 1986). Vested interests in maintaining and controlling social structure can be supported by the symbolism worked into the physical form of artefacts, as in the eleventh century AD metalwork from Igbo-Ukwu in south-east Nigeria (Shaw, 1970; Ray, 1987). Similar uses of material culture to support political and social control can also be seen in the coin issues of the Roman Empire (Burnett, 1987).

Material culture thus spans both a utilitarian and perceptual view of archaeology. Through its study, we attempt to understand past societies and their relationship with the environment, by assuming a correlation between the constructs and the detail of people's lives (Hodder, 1982; Tilley, 1989). This is further complicated by introducing chronology into our purview. Not only do different communities and societies find different adaptive strategies and different forms of material expression, but the form and nature of material culture changes through time. We can see this clearly in the archaeological record; changes in pottery form and decoration, the adoption of different kinds of dwelling, metal tools and burial practices, etc. It is not immediately self-evident why they occur. If a particular strategy proves successful, then why adopt something different? A primary role for the archaeologist is to document these changes, but we also need to explain them. There is no *a priori* reason for a successful subsistence strategy to be abandoned or indeed a particular artistic or symbolic motif to undergo transformation.

A widespread philosophy in Western society today is the pursuit of optimization – the quest for improvement in all things, be it better grain yields, more efficient transport or more deadly weaponry. This attitude is not necessarily applicable to past societies. In the West, it is a post-Renaissance phenomenon; in other world views, the importance of tradition, religion and social structure can mitigate against changes in the status quo. The variation in timescales over which change occurs also needs to be considered. Although the perception of rapid change in the archaeological record demands explanation, it is just as important to understand the extended periods of time where there is little change in material culture; for example the religious iconography of Ancient Egypt, which

Time (Years BP)	Epoch	Geological Period	Archaeological Highlights	North America	South America	SE Europe: Greece	S Europe: Italy, France	N Europe: Britain	Environmental Change in North Europe
0					Inca				Little Ice Age
1000				Maize cultivation				Roman	
2000						Classical Greece	Rome	Iron Age	
3000				Rise of farming (Squash, Sunflower)				Late Bronze Age	Climatic deterioration
4000	HOLOCENE	QUATERNARY		ARCHAIC		Aegean Early Bronze Age	Copper Age mixed farming	Early Bronze Age, clearance and mixed farming	Lime decline
5000									Elm decline
6000					Maize cultivation			Early Neolithic (farming)	CLIMATIC OPTIMUM
7000							Early Neolithic foraging, mixed farming		
8000						Early Neolithic mixed farming			
9000									
10000						Early Mesolithic foraging		Mesolithic	
11000	Pleistocene		Upper, Middle and Lower Palaeolithic	Clovis					End of the last glacial stage
12000					Monte Verde				
13000									
14000									
1.7m		Tertiary							

FIGURE 1.2 Major Quaternary cultural and environmental changes.

apart from minor deviations, remained remarkably consistent over several millennia.

In the first instance, we must look to external factors that could instigate change. Environment is clearly one area where development might be stimulated. As ecological circumstances changed, so communities responded by adapting their material culture to their new conditions; in this scenario, material culture might be viewed as an 'extrasomatic means of adaptation' (Binford, 1965: 205). However, though this is clearly a logical approach, it is not predictive. The actual form of the response would be underwritten by a host of cultural and social references (Tilley, 1989). Another stimulus to change is through contact and social intercourse between communities with different cultures. The nature of influence between social groups is, of course, infinitely variable, from the wholesale imposition of new cultural expression through force, to more subtle influences such as religious conviction, social aspiration and economic gain. At the same time as reflecting and reinforcing social perceptions, material culture is itself dynamic in terms of the way it is perceived within different cultural contexts (Kopytoff, 1986); not only can material culture change, but its interpretation must be understood within the social context. For example, note the symbolic importance of prehistoric burial mounds in Anglo-Saxon England (e.g. Parfitt and Brugmann, 1997) or the way in which the Kula valuables of Melanesia acquire different meanings and significance as they are exchanged from person to person (Campbell, 1983).

Thus we are again faced with the complexity of archaeological interpretation. Living within the determinist forces of the natural environment, human societies leave behind material culture arising from strategies to provide the practical necessities of life, allow expression and support of a world view and help negotiate relationships. Changes in ecological, religious and social conditions are reflected in that material culture. Understanding the links between the environment and social dynamics, and the hows and whys of change through time is a major goal for archaeological study.

1.2.2 Environmental change

1.2.2.1 Geology, geomorphology and pedology

The changes that affect our environment do so across an infinite range of scales, from catastrophic sedimentary events taking minutes, to continental movements needing unimaginable time spans to take effect. The primary driving force at the geological scale is internal planetary circulation and cooling which produces convection currents and crustal movements. These affect the environment at more human timescales as well, since they produce both volcanic and seismic events often with dramatic consequences for individuals and for whole cultures. Crustal movements also cause mountain building where continents collide, and are thus a major contributor to sedimentation patterns.

Once geological materials have been uplifted and exposed, they become landforms and are immediately subject to erosion. Gravity and water act continuously to flatten them again at a range of temporal scales (Selby, 1982; Summerfield, 1997). For the most part, the less prominent landforms are lowering over 1 million and ten million year timescales, so appear as an unchanging background to human lives. However, their apparent stability can be the result of a finely tuned relationship between the erosive power of water, and the protection afforded by plant cover. If the gradients are steep enough, or storms powerful enough, then any change of land use may well be a critical factor in causing sudden sedimentation events (Cook and Doornkamp, 1990). Modern mudslides that sweep away cars and homes in areas of unrestrained economic development are the most extreme expression of this type of erosion, but less dramatic cases can be found in the archaeological record where lives were radically changed by sedimentary events. The northern European alluvial record, for example, seems to contain broadly similar increases in fine-grained sedimentation occurring wherever cultivation was starting to significantly affect catchments from the mid Bronze Age onwards (Bell, 1982; Brown and Barber, 1985; Salvador *et al.*, 1993). Although a clear link cannot be proven (Canti, 1992), the sediment yields attendant on cultivation can be compared to those measured in historical situations (see Section 2.5.3).

FIGURE 1.3 A natural profile showing clay translocation in the horizon below the topsoil.

Sediments that remain stable for tens of years start to undergo soil formation processes. This first becomes apparent when humus builds up either on the surface or intimately mixed with the mineral material. Underneath that humus-rich layer, a leached or otherwise altered layer will eventually develop in many situations depending on acidity and climate relationships (Fig. 1.3). This layer, and the layer immediately below it define the nature of the soil processes and thus the soil type (see Sections 1.3.2 and 1.3.3). There may be iron oxides and humus leaching downwards, or iron and clay, or just clay. There may be a changing water table redistributing iron and manganese oxides into mottles. Calcium carbonate or other salts may precipitate regularly if the climate is one where evaporation of soil water exceeds leaching (FitzPatrick, 1983). All of these different chemical redistributions are time-dependent and many are irreversible depending on the solubility of the precipitates.

Thus, although soil processes are part of the evidence for environmental change, the most recent effects are often intermingled with previous layers of evidence in a palimpsest of complexity that challenges simple analysis.

1.2.2.2 Ecology, ecosystems and environmental archaeology

Ecology (from the Greek word *Oikos*, meaning household) is the study of our biosphere, and of the complex processes, both natural and cultural, which determine the plant and animal composition of our ecosystems. Knowledge and understanding of ecological processes is essential for the interpretation of sub-fossil (or fossil) plant and animal (including human) remains (palaeoecology). The importance of this knowledge is particularly highlighted by the uniformitarian principles upon which interpretations rest. To make environmental archaeology work at all, we must be confident, for example, that plant and animal species from an early Holocene (c. 10,000 years ago) context are the same ones that we recognize in the environment today. It must be remembered that although the genetic content (genotype) and appearance (phenotype) of plants and animals is inherited, enabling them to survive and even adapt to new environmental conditions, alteration of both the genotype and phenotype does occur through time as a consequence of natural variation (e.g. mutation), natural selection (evolution) and human selectivity (e.g. domestication of plants and animals). Nevertheless, the fossil record has enabled the evolutionary pathway of plants and animals to be understood and arranged into a classificatory system (taxonomy) ranging from the simplest to the most complex species. In this system, species that are closely related are grouped into a genus, e.g. *Triticum spelta*, and then a family (e.g. Poaceae), order, class and phylum (see Table 1.2 for further non-taxonomic divisions).

By recording the major changes in speciation over time, it has been possible to elucidate whether these changes have occurred as a response to major environmental changes (e.g. climate) or human activity (e.g. extinction or selective breeding). However, environmental archaeology is not only concerned with studying the more complex plants and animals, such as the Mammalia (e.g. humans, apes, cows, sheep, goats) and Angiosperms (Spermatophyta: monocotyledonous (e.g. Poaceae) and dicotyledonous (e.g. Rosaceae) flowering plants), which, as we will see in later chapters, contribute much of the information about past human economies and diet, and the environmental context of these activities. There are also the Protista (unicellular algae, e.g. diatoms), Bryophyta (e.g. mosses), Pteridophyta (e.g. ferns), Gymnosperms (Spermatophyta: conifers), Coelenterates (e.g. corals), Platyhelminthes and Nematoda (e.g. parasites), Annelida (e.g. worms), Arthropoda (Crustacea, e.g. crabs; Insecta, e.g. flies), Mollusca (e.g. snails), Echinodermata (e.g. starfish) and Chordata (vertebrates, which includes Pisces (fish), Amphibia (e.g. frog), Reptilia (e.g. snakes), Aves (birds)), all of which provide useful information on the daily life of humans and the past environment. Further details of plant evolution, especially angiosperms, can be found in Roberts (1977), Cronquist (1981) and Jones and Luchsinger (1987) , and for animal evolution see Kerkut (1960) and Pough *et al.* (1989).

TABLE 1.2 Examples of non-taxonomic subdivisions of plants and animals (See Simmons, 1979)

Plants	Animals
1. Autotrophic (feeding by photosynthesis)	1. Heterotrophic (feeding on plants and animals)
2. Xerophytes	2. Herbivores
3. Hydrophytes	3. Carnivores
4. Mesophytes	4. Omnivores
	5. Saprovores
	6. Parasites

At any point in time, the distribution of plants and animals is determined by abiotic (e.g. temperature, water, light, humidity, wind, pH, salinity, nutrients, topography) and biotic (e.g. succession, competition, reproduction, dispersal, predation) factors, which may limit distribution to a narrow range (stenotopic) or enable colonization over a wide range of environmental conditions (eurytopic). At a global scale (megascale), plant and animal species occur in major world biomes determined principally by climate (Table 1.3) but also physiog-

TABLE 1.3 Major world biomes (See Simmons, 1979)

Biome	Details
1. Tropical rainforests	High and constant temperature and humidity High biodiversity E.g. Lowland areas of South America, Southeast Asia, Meso-America
2. Tropical seasonal forests	Pronounced wet and dry seasons Lower biodiversity than (1) E.g. India, Southeast Asia mainland, North Australia
3. Tropical savannas	One dry season Grassland dominated E.g. Africa, Australia, South America
4. Temperate sclerophyll woodland and scrub	Warm, dry summers and cool, wet winters Dominated by trees and shrubs E.g. Southeast and Southwest Australia (mallee scrub), Mediterranean basin (maquis), California (chaparral), Chile (matorral)
5. Temperate grasslands	Low rainfall E.g. North America (prairies), Asia (steppe), Africa (veldt), South America (pampas)
6. Deserts	Low rainfall Low biodiversity, high adaptation E.g. Chile (Atacama), California, Southwestern America (Mojave)
7. Deciduous forests of temperate climates	Pronounced hot and cold seasons, average rainfall E.g. Western and Central Europe, Northeastern America
8. Boreal forests	Long, cold winters Dominated by evergreen forests E.g. North Eurasia, North America
9. Tundra	Low temperature and rainfall Dominated by low shrubs and grassland E.g. Central Alaska, Greenland
10. Montane (mountains)	Complex altitudinal and latitudinal variation in climate Highly adapted biota
11. Islands	Low species diversity
12. Seas	
13. Freshwater	

raphy (e.g. mountains). Each biome will exhibit wide variations in biodiversity as a consequence of these factors, and in special circumstances may contain species with highly restricted distributions (endemic).

At finer scales, plants and animals occur in particular habitats, determined by specific biotic and abiotic factors (e.g. soil pH). The study of groups of species (synecology), as opposed to individual species (autecology), and their interaction with the abiotic environment (the ecosystem approach) has provided complex information on the processes that affect ecosystem formation and change through time over a range of spatial scales. The position of humans in this dynamic relationship cannot be underestimated, especially their impact on the food web, e.g. the reduction in species diversity as a consequence of woodland clearance and cultivation of crops. Prior to the end of the last cold stage and the adoption of food producing subsistence systems, food resources were gathered, hunted and fished creating a self-regulating system that managed the sustainability of the resources. Therefore the impact on ecosystems would have been in many instances negligible, although there are exceptions where fire was used extensively and where humans are thought to have led to the extinction of species.

As we briefly mentioned above, the domestication of plants and animals had a significant effect on genotype and phenotype evolution, as well as bringing about changes in worldwide ecosystems. The 'agricultural revolution' was a continuum of human-plant-animal interaction that varied both spatially and temporally, and in many respects it is still continuing today, e.g. genetic engineering. The road to domestication, as discussed in a previous section, was a gradual change rather than a series of general stages. This included the systematic exploitation of resources by humans probably based on a symbiotic relationship with plants and animals, followed by experimental propagation of plants involving seeds and/or root crops, and finally the establishment of multiple centres of domestication of wild progenitors throughout the Old and New Worlds. These processes would have had a profound effect on the dynamics of the ecosystems, in particular natural succession. There are two types of succession, one controlled by self-regulating internal factors, the other by external factors. For example, the gradual infilling of a small pond by vegetation is known as autogenic succession, while mature (climax) woodland clearance by humans followed by abandonment and the re-colonization of shrubland and woodland is known as allogenic succession.

The plant and animal population levels within an ecosystem are controlled by several factors, namely immigration, emigration, natality, mortality and density. Implicit within these factors is the idea of carrying capacity, which is a self-regulatory system that governs the total number of individuals living in the ecosystem because of the resources available for survival. These factors would have been especially important during periods of wild food procurement to provide humans with sustainable food resources. However, following the ecosystem changes that occurred after the inception of agriculture, the limiting factors for the successful reproduction of plants and animals in the newly created 'artificial' ecosystems would have been largely abiotic.

1.2.2.3 Climate change

As in the other environmental factors discussed above, climate change operates at a number of timescales. In the Quaternary as a whole (approximately the last 2 million years), there was a succession of global changes in ice-cap volume that can be loosely used as a proxy of overall climate. These changes were due to

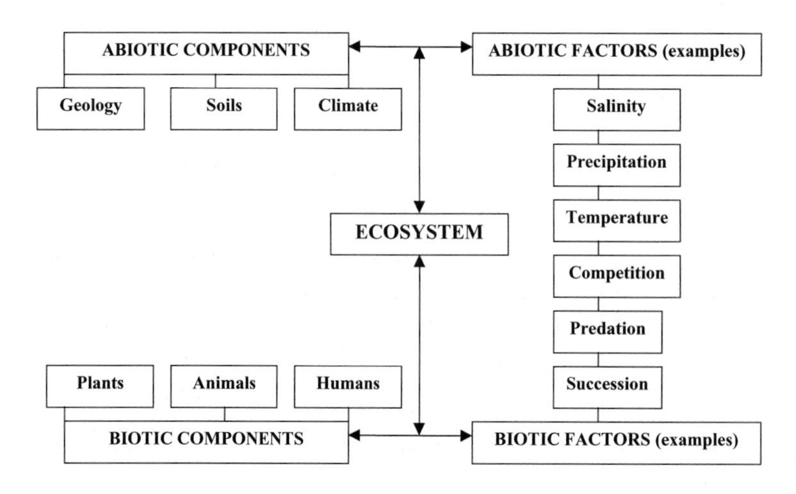

FIGURE 1.4 Simplified model of an ecosystem.

TABLE 1.4 A selection of cultivated plants originated in the Near East and Mediterranean (See Simmons, 1979)

Plants	Common name
Cereals	
Avena sativa	Oats
Hordeum vulgare	Barley
Secale cereale	Rye
Triticum aestivum	Bread wheat
Triticum dicoccum	Emmer
Triticum monococcum	Einkorn
Triticum turgidum	Tetraploid wheat
Oils, fruits, nuts and fibres	
Cannabis sativa	Hemp
Corylus	Hazelnut
Ficus carica	Fig
Juglans regia	Walnut
Linum usitatissimum	Flax, Linseed
Olea europaea	Olive
Pistacea vera	Pistachio
Vitis vinifera	Grape
Vegetables and pulses	
Allium cepa	Onion
Brassica oleracea	Cabbage, etc.
Cicer arietinum	Chickpea
Daucus carota	Carrot
Vicia faba	Broad bean

subtle variations in aspects of the Earth's orbit such as eccentricity, tilt and precession of the equinoxes, discovered by the Serbian Milutin Milankovitch (Milankovitch, 1941). They do not in themselves significantly alter the total amount of solar radiation entering the Earth's atmosphere, but they do influence the spatial distribution and seasonality of this input.

Milankovitch argued that the crucial changes were brought about by reduction in summer warmth, which allowed northern hemisphere snow to survive through the seasonal cycle to develop long-term into an ice mass, explaining troughs and peaks in global ice volume (the glacial-interglacial cycles) through the interplay of the different orbital components. These variations are insufficient to account for the large changes in ice volume required to induce a glaciation, suggesting that the shift of the global climate system into one state or the other is additionally driven by positive feedback mechanisms, such as ice albedo, atmospheric greenhouse gas content and/or intensification of the hydrological cycle (Berger *et al.*, 1999; Ruddiman, 2003; Kukla and Gavin, 2004).

Embedded within these large cycles are periods of rapid climate change that cannot be explained by orbital variations. Much work has focused on these changes in the important North Atlantic region. Here the late-Quaternary was characterized by a series of rapid (<100 years) and severe (up to 10°C) millennial-scale climatic oscillations referred to as Dansgaard-Oeschger events that have been widely identified in oceanic, ice and terrestrial records (Voelker *et al.*, 2002). These events can be bundled into regular, decreasing amplitude, cooling cycles (Bond *et al.*, 1993) that culminate in massive ice discharges into the North Atlantic (Heinrich Events). Their exact timing, magnitude and global extent are currently uncertain (Sarnthein *et al.*, 2002), and ideas on their origin have centred around reorganization of the ocean's thermohaline circulation and changes in the tropical atmosphere-ocean dynamics (Broecker, 2003; Turney *et al.*, 2004).

In contrast to the above, the present (Holocene) interglacial was long considered to be a period of exceptional climatic stability (Dansgaard *et al.*, 1993). However, changes in the North Atlantic sediments during this period have been shown to be quasi-cyclic, recording roughly 1500 year southward movements of cold, ice-bearing waters from the Labrador and Nordic Seas (Bond *et al.*, 2001). Similar quasi-cycles have been recorded throughout the Northern Hemisphere (O'Brien *et al.*, 1995; Bianchi and McCave, 1999; Hu *et al.*,

2003; Langdon *et al.*, 2003), consistent with Dansgaard-Oeschger events (Bond *et al.*, 1993) and suggesting a climate oscillation independent of glacial-interglacial cycles and anthropogenic effects. The presence of a 1500 year cycle persisting through the Holocene has recently been questioned, with some favouring shorter periodicities (Schulz and Paul, 2002). Potential mechanisms forcing such variability include relatively small changes in solar activity (Bond *et al.*, 2001) possibly amplified by other factors (such as stratospheric ozone concentration; Shindell *et al.*, 1999), changes in the North Atlantic circulation (Bianchi and McCave, 1999; Schulz and Paul, 2002), manifestation of a bipolar seesaw (Broecker, 2003) and distant changes in the El Niño-Southern Oscillation (Russell *et al.*, 2003; Turney *et al.*, 2004), though at present none of these are universally accepted by the scientific community.

1.3 Concepts of spatial organization

1.3.1 Microscale – contextual analysis

Underpinning most modern field archaeology is the concept of individual stratigraphic units (or contexts) and the relative chronological sequence derived from their stratigraphic relationships. Each context represents the physical manifestation of a past event of varying duration, typically a depositional event (e.g. a mosaic floor) or one of truncation (the terracing of a hillside or the cutting of a pit). Many field archaeologists go to great lengths to identify and demarcate context boundaries; uncertain or indistinct edges are anathema to good excavation technique and context recording practice. This emphasis is based on the need to unequivocally identify discrete stratigraphic units as the basis for the interpretation of the stratigraphic sequence (Harris, 1977; Pearson and Williams, 1993), and to increase confidence in the assemblages derived from the context (in the case of deposits). The approach has proved hugely successful over the years, but of course tends to emphasize a particular attribute set at the expense of others which may also be of great importance and interest to other archaeological researchers.

The idea of a single context representing a single event has some drawbacks when site formation processes are examined in detail. Soil processes continue to act on the deposits making up a context, and can affect its boundaries to a greater or lesser extent. (See Fig. 1.5.) Deposits undergoing redox processes involving wetting and drying are particularly prone in this regard, because of the soluble nature of reduced iron and manganese salts. Under anoxic conditions, iron and manganese enter solution and move around with groundwater only to redeposit as the familiar black or rust-coloured materials when oxygen is reintroduced. Luckily, the textural changes that commonly define context edges have a tendency to be picked out, rather than obscured by iron staining, so the process is not often an interpretative problem. More worrying can be the action of earthworms. These creatures have a variable range of effects including moving objects around on the surface and introducing extraneous material into their burrows (Canti, 2003). Their most important activity from the contextual point of view is a tendency to blur contexts, producing a gradual change between two deposits that started life having a distinct boundary (Stein, 1983).

The desire of field archaeologists to identify discrete stratigraphic units is emphasized at the local level by the common practice of refusing to correlate units as part of the primary record of the stratification. The site is broken down into its constituent parts (the contexts) for recording, and these may be grouped together in various combinations to allow interpretation (e.g. four walls might be grouped together to form a room). Grouping of contexts is considered to be a higher level of interpretation than the identification of the individual contexts, and therefore should not form part of the primary record of the site itself. This approach attempts to address the requirements of 'preservation by record', a theoretically dubious analogy between the *in situ* preservation of archaeological sites and their destruction by archaeological excavation. The importance of the identification and recording of individual

FIGURE 1.5 Pit fill associated with iron working in the Belgian loess. The fill has undergone post-depositional podzolization to produce a bleached horizon at the base. (Photo, Jen Heathcote)

contexts is paramount. This is something we shall return to in the next chapter, but in practice, it means that a field archaeologist will not conflate physically separate stratigraphic units. For example, if a ditch has been bisected by a later pit, each half of the ditch would be deemed a separate entity, and numbered and recorded independently; the decision to group the two halves together and describe them as a single ditch would be made during the analysis and interpretation of the site records. The approach is sometimes taken to extremes, but it is a response to the historical technique of producing phase plans as the primary record of an archaeological site, which allowed little if any opportunity for researchers to produce alternative interpretations to those arrived at during excavation. Such antiquated recording techniques have thankfully almost disappeared from modern archaeology.

A similar tension between objective description and the need for interpretation at the microscale has developed in thin section micromorphology. This method is used to examine soil or context fabrics microscopically in transmitted light, and requires a complex vocabulary to describe the soil particles, their arrangements and separations (Bullock *et al.*, 1985; Stoops, 2003) in such a way as to communicate a purely morphological description – theoretically repeatable and without interpretative bias. However, if non-interpretation is too rigorously enforced, the description ends up lacking any form of comprehensibility to the reader, and can easily become a ritual activity carried out in place of thought.

1.3.2 Mesoscale – local divisions

At the mesoscale, soil horizons are a striking division visible in any section of the earth's surface layers. Horizons are physicochemical zones produced as a result of the interactions of a range of variables among loose sediments or rock fragments close to the surface. The main variables are grouped under climate and parent material, with additional action from biological agents including humans (Bridges, 1997). Thus the humus-rich topsoil layer is a horizon produced by organic matter falling on the mineral soil and becoming incorporated (mull humus) or building up into a surface pad (mor humus). The leached layer often found beneath the humus layer is another horizon from which iron or clay may have been removed by percolating rainwater together with organic compounds from the plant matter above (Avery, 1990).

Material from a particular soil horizon may have distinct visual characteristics which can be traced across sites and even across landscapes. It can be a type of stratigraphic marker, at least in so far as it signals material formed under similar conditions, but cannot be taken as necessarily meaning that separate areas are actually contemporary. Many contexts are formed of materials derived from a particular

FIGURE 1.6 An earthwork showing the remains of the podzolised soil horizons from which it was built. (Photo: Helen Keeley)

soil horizon and redeposited as part of an activity. If the material is recognizable, then that horizon must have existed as part of the local soil profile when the activity took place (see Fig. 1.6).

Particular soils occupy particular positions in a given landscape (Young, 1976). Thus, for example, sandstone ridges intercut by river valleys may show podzolic soils (highly leached) on the tops, acidic brown earths on the valley sides and groundwater gleys (waterlogged) on the valley floor. Each of these zones would have offered resources and constraints for past habitation and economies, and the archaeology should thus tend to show similar activities in similar parts of the landscape. However, it is obvious that relationships between this natural resource base and the resulting human activity are often not linear in the way that early geographers foresaw. One of the central reasons for this seems to be that human societies have rarely developed to the point where they are obliged to maximize the use of available land. Without that constraint, people can still operate at both a social and personal level, with some individuals or groups acting differently, even counterproductively in terms of the common good.

It is important also to consider the way in which the landscape was perceived culturally. Basing an understanding of the socio-economic dynamics of a past society exclusively on environmental factors would be unsatisfactory. The social relationships between different kinship groups, or other aspects of the politics, can have a marked effect on the resources available to communities and their subsistence strategies; economic resources and routeways could be controlled, resulting in a different perceptual map of the landscape than might be at first apparent from a modern perspective. Transport technologies could also have a marked effect on how distance and the relative difficulty of travel across the landscape was viewed. Finally, we should also be aware of the influence of ritual and symbolism. It must have affected resource use markedly, not only in terms of the manipulation of the landscape (by the creation of monuments and so forth), but also in terms of the perception of natural places and the relationship of a society with the natural world generally. Such factors are difficult to identify and their impact hard to assess, but they cannot be ignored.

1.3.3 Macroscale – regional interpretation

Sediments are deposited by a characteristic process in a particular environment, for example colluvium on a hillslope or alluvium in a valley floor. Similar processes and environments produce comparable sediments, which may be traceable across large areas. As with soil horizons (see above), the similarity over a distance implies at least matching conditions, but it should not be assumed that this represents the same time period unless there is a clear physical link. Sediments of quite different ages can end up looking very similar, especially after post-depositional processes have modified the colours.

The physicochemical groupings represented by soil horizons (see Section 1.3.2) feed into higher category groupings known as soil types. There is a tendency for any given soil process to gradually become less significant over distance as conditions change. Often, another process will be starting to become more significant along the same gradient. Soil classification attempts to summarize those dynamic changes into individual soil types, usually by placing some requirement on the thickness of a given diagnostic horizon, or a threshold on its chemical characteristics. Soil types are thus found at regional or even global scales depending on the steepness of the environmental gradients which separate them, and the level of classification being employed.

Moving up to a regional scale presents significant problems of sampling and characterization for geoarchaeology. Dealing with soils as an example, a single soil type is unlikely to be found covering a particular region, so characterization will often require some description of the different landscape positions and the soil types found within them. Such a pedogeomorphological approach is harder for the non-specialist to grasp, and defies the kind of simplification that is needed for easy communication. Worse still, political boundaries often cut across landscape units, so summaries respecting this additional dimension will require descriptions not only for the different landscape positions, but for different topographies or geologies as well.

Analysis at a regional scale introduces interpretative constraints similar to the problems of soil-human relationships described for landscapes (see Section 1.3.2), with a tendency, if anything, to be even more pronounced. Political differences increase over distance, and regional variations of land use practice or agricultural tradition will significantly alter any expected patterns based on environmental factors. Furthermore, climatic variation starts to play a part in the equation when sites are hundreds of kilometres apart.

1.3.4 Megascale – global perspectives

Our spatial concepts at a global scale arise chiefly from the use of modern communications and technology. It is unlikely that any ancient peoples viewed themselves as part of a continent or a hemisphere. Although in some cases ancient astronomers had grasped the spherical nature of the Earth, in others, cultures capable of highly sophisticated astronomical predictions appear not to have questioned their own spatial position, or at least not to have written their views down (Aveni, 1996).

Continents are the only spatial units occurring at the megascale and their characteristics, in so far as they can be simplified, tend to be related to deep time geology and climatology rather than the brief episode of human history that we study. The immense stable land surfaces of Africa and Australia, for example, have acted as a backdrop for the drama of human and animal evolution, not so much by affecting its progress but by providing a record for us through their very stability. Conversely, the numerous glacial advances and retreats of the temperate and circumpolar regions have probably stimulated cultural development but have tended to wipe out most of the traces.

The boundaries of continents have little respect for crustal geology. Thus, some of the Scottish mountains are continued as the Appalachians of the eastern United States, and the similarities of lithology between West Africa and Brazil form one of the cornerstones of continental drift theory (Plummer *et al.*, 2001). The slowness of these processes precludes any direct effects on people. The Atlantic Ocean today is a few metres wider than it was when Christopher Columbus sailed across and a few tens of metres wider than when the Carlston Annis shell mound was accumulating, but these facts make no difference to the various cultures concerned. Paradoxically, the side effects of continental processes – volcanic eruptions and earthquakes – are among the most powerful mesoscale events found in environmental archaeology, being able to drastically change the landscape and destroy its cultures overnight.

REFERENCES

Aiello, L. and Dunbar, R. (1993) Neocortex size, group size, and the evolution of language. *Current Anthropology* **34**, 184–193.

Aldenderfer, M. (1999) The Pleistocene/Holocene transition in Peru and its effects upon the human use of the landscape. *Quaternary International* **53/54**, 11–19.

Ammerman, A.J. and Cavalli-Sforza, L.L. (1984) *The Neolithic Transition and the Genetics of Populations in Europe.* Princeton: Princeton University Press.

Aveni, A.F. (1996) Astronomy in the Americas. In (C. Walker, ed.) *Astronomy before the Telescope*, 269–303. London: British Museum Press.

Aveni, A. (1986) The Nazca lines: patterns in the desert. *Archaeology* **39**, 32–39.

Avery, B.W. (1990) *Soils of the British Isles.* Wallingford: CAB International.

Bayliss, A. (1998) Some thoughts on using scientific dating in English archaeology and building analysis for the next decade. In (J. Bayley, ed.) *Science in Archaeology, an Agenda for the Future*, 95–108. London: English Heritage.

Behrensmeyer, A.K. and Hill, A.P. (eds.) (1980) *Fossils in the Making: Vertebrate Taphonomy and Paleoecology.* Chicago: University of Chicago Press.

Bell, M. (1982) The effects of land-use and climate on valley sedimentation. In (A.F. Harding, ed.) *Climatic Change in later Prehistory*, 127–142. Edinburgh: Edinburgh University Press.

Bell, M. and Walker, M.J.C. (1992) *Late Quaternary Environmental Change.* Harlow: Longman.

Bender, B. (1975) *Farming in Prehistory: From Hunter-Gatherer to Food Producer.* New York: St Martin's Press.

Berger, A., Li, X.S. and Loutre, M.F. (1999) Modelling northern hemisphere ice volume over the last 3 Ma. *Quaternary Science Reviews* **18**, 1–11.

Bianchi, G.G. and McCave, I.N. (1999) Holocene periodicity in North Atlantic climate and deep-ocean flow south of Iceland. *Nature* **397**, 515–517.

Binford, L. (1965) Archaeological systematics and the study of cultural process. *American Antiquity* **31**, 203–210.

Binford, L.R. (1968) Post Pleistocene adaptations. In (R. Binford and L.R. Binford, eds.) *New Perspectives in Archaeology*, 313–341. Chicago: Aldine Press.

Binford, L. (ed.) (1977) *For theory building in archaeology: essays on faunal remains, aquatic resources, spatial analysis, and systemic modeling.* London: Academic Press.

Binford, M.W., Kolata, A.L., Brenner, M., Janusek, J.W., Sedden, M.T., Abbott, M. and Curtis, J. (1997) Climate variation and the rise and fall of an Andean civilisation. *Quaternary Research* **47**, 235–248.

Bond, G., Broecker, W., Johnsen, S., McManus, J., Labeyrie, L., Jouzel, J. and Bonani, G. (1993) Correlations between climate records from north Atlantic sediments and Greenland ice. *Nature* **365**, 143–147.

Bond, G., Kromer, B., Beer, J., Muscheler, R., Evans, M.N., Showers, W., Hoffman, S., Lotti-Bond, R., Hajdas, I. and Bonani, G. (2001) Persistent solar influence on North Atlantic climate during the Holocene. *Science* **294**, 2130–2136.

Braidwood, R.J. (1960) The agricultural revolution. *Scientific American* **203**, 130–148.

Bridges, E.M. (1997) *World Soils.* Cambridge: Cambridge University Press.

Broecker, W.S. (2003) Does the trigger for abrupt climate change reside in the ocean or in the atmosphere? *Science* **300**, 1519–1522.

Brothwell, D.R. and Higgs, E. (eds.) (1963) *Science in Archaeology.* London: Thames and Hudson.

Brown, A. (1997) *Alluvial geoarchaeology: Floodplain archaeology and environmental change.* Cambridge: Cambridge University Press.

Brown, A.G. and Barber, K.E. (1985) Late Holocene palaeoecology and sedimentary history of a small lowland catchment in central England. *Quaternary Research* **24**, 87–102.

Buckland, P.C. (1976) The use of insect remains in the interpretation of archaeological environments. In (D.A. Davidson and M.L. Shackley, eds.) *Geoarchaeology*, 369–396. London: Duckworth.

Bullock, P., Fedoroff, N., Jongerius, A., Stoops, G. and Tursina, T. (1985) *Handbook for Soil Thin Section Description*. Wolverhampton: Waine Research Publications.

Burnett, A. (1987) *Coinage in the Roman World*. Seaby.

Butzer, K.W. (1982) *Archaeology as Human Ecology*. Cambridge: Cambridge University Press.

Campbell, S. (1983) Attaining rank: a classification of shell valuables. In (J. Leach and E. Leach, eds.) *The Kula: New perspectives on Massim exchange*, 229–248. Cambridge: Cambridge University Press.

Cann, R., Stoneking, M. and Wilson, A. (1987) Mitochondrial DNA and human evolution. *Nature* **325**, 31–36.

Canti, M.G. (1992) Anthropogenic modification of the early Holocene soil environment. *SEESOIL, Journal of the South East England Soils Discussion Group* **8**, 29–42.

Canti, M.G. (2003) Earthworm activity and archaeological stratigraphy: a review of products and processes. *Journal of Archaeological Science* **30**, 135–148.

Childe, V.G. (1957) *The Dawn of European Civilisation*. London: Routledge.

Clarke, D.L. (1968) *Analytical Archaeology*. London: Methuen.

Clarke, D.L. (1978) *Analytical Archaeology* (2nd edition). London: Methuen.

Clark, J.G.D. (1952) *Prehistoric Europe: the Economic Basis*. London: Methuen.

Clutton-Brock, J. (1981) *Domesticated Animals from Early Times*. London: British Museum.

Clutton-Brock, J. and Grigson, C. (eds.) (1983) *Animals and Archaeology, Volume 1*. Oxford: British Archaeological Reports International Series, 163.

Conlin, D. (1998) Ship evolution, ship 'ecology', and the 'Masked Value Hypothesis'. *The International Journal of Nautical Archaeology* **27.1**, 3–15.

Cook, R.U. and Doornkamp, J.C. (1990) Geomorphology in environmental management. Oxford: Clarendon Press.

Cronquist, A. (1981) *An Integrated System of Classification of Flowering Plants*. New York: Columbia University Press.

Dansgaard, W., Johnsen, S.J., Clausen, H.B., Dahl-Jensen, D., Gundestrup, N.S., Hammer, C.U., Hvidberg, C.S., Steffensen, J.P., Sveinbjörnsdottir, A.E., Jouzel, J. and Bond, G. (1993) Evidence for general instability of past climate from a 250-kyr ice-core record. *Nature* **364**, 218–220.

Davidson, D.A. and Shackley, M.L. (1976) (eds.) *Geoarchaeology*. London: Duckworth.

Davidson, D.A. and Simpson, I.A. (2001) Archaeology and soil micromorphology. In (D. Brothwell and A.M. Pollard, eds.) *Handbook of Archaeological Sciences*, 167–177. Chichester: Wiley.

Dimbleby, G. (1978) *Plants and Archaeology*. London: Paladin.

Dimbleby, G. (1985) *The Palynology of Archaeological Sites*. London: Academic Press.

Dincauze, D.F. (2000) *Environmental Archaeology: Principles and Practice*. Cambridge: Cambridge University Press.

FitzPatrick, E.A. (1983) *Soils: their formation, classification and distribution*. London: Longman.

Flannery, K.V. (1968) Archaeological systems theory and early Mesoamerica. In (B.J. Meggers, ed.) *Anthropological Archaeology in the Americas*, 67–87. Washington: Anthropological Society of Washington.

Ford, R.I. (1985) The processes of plant food production in prehistoric North America. In (R.I. Ford, ed.) *Prehistoric Food Production in North America*, 1–18. University of Michigan. Museum of Anthropology.

Fowler, P. (1983) *The Farming of Prehistoric Britain*. Cambridge: Cambridge University Press.

Gamble, C.S. (1986) *The Palaeolithic Settlement of Europe*. Cambridge: Cambridge University Press.

Gamble, C.S. (1993) *Timewalkers: the Prehistory of Global Colonisation*. Stroud: Alan Sutton.

Gamble, C.S. (1995) The earliest occupation of Europe: the environmental background. In (W. Roebroeks and T. van Kolfschoten, eds.) *The Earliest Occupation of Europe*, 279–295. Leiden: University of Leiden.

Gamble, C.S. (1998) Questions for Palaeolithic science and science for Palaeolithic questions. In (J. Bayley, ed.) *Science in Archaeology, an Agenda for the Future*, 3–8. London: English Heritage.

Gladfelter, B. (1977) Geoarchaeology: the geomorphologist and archaeologist. *American Antiquity* **42**, 519–538.

Gould, S. (2001) *Rocks of Ages: Science and Religion in the Fullness of Life*. London: Jonathan Cape.

Harris, D.R. (1987) The impact on archaeology of radiocarbon dating by accelerator mass spectrometry. *Philosophical Transactions of the Royal Society of London Series* **A232**, 23–43.

Harris, D.R. (1989) An evolutionary continuum of people-plant interaction. In (D.R. Harris and G.C. Hillman, eds.) *Foraging and Farming, the Evolution of Plant Domestication*, 11–26. London: Unwin Hyman.

Harris, D.R. and Hillman, G.C. eds. (1989) *Foraging and Farming, the Evolution of Plant Domestication*. London: Unwin Hyman.

Harris, E. (1977) *Principles of archaeological stratigraphy*. London: Academic Press.

Hastorf, C.A. and Popper, V.S. (eds.) (1989) *Current*

Palaeoethnobotany: Analytical Methods and Cultural Interpretations of Archaeological Plant Remains. Chicago: University of Chicago Press.

Hempel, C. (1942) The function of general laws in history. *Journal of Philosophy* **39**, 35–48.

Hempel, C. (1963) Reasons and covering laws in historical explanation. In (S. Hook, ed.) *Philosophy and History: A Symposium*, 143–163. New York: New York University.

Higgs, E.S. ed. (1972) *Papers in Economic Prehistory*. Cambridge: Cambridge University Press.

Higgs, E.S. and Jarman, M.R. (1969) The origins of agriculture: a reconsideration. *Antiquity* **43**, 31–41.

Hill, D. (1986) *Canterbury Cathedral*. London: Bell and Hyman.

Hillman, G.C. (1973) Crop husbandry and food production: modern basis for the interpretation of plant remains, *Anatolian Studies* **23**, 241–244.

Hodder, I. (ed.) (1982) *Symbolic and Structural Archaeology*. Cambridge: Cambridge University Press.

Hodder, I. (1986) *Reading the past: Current approaches to interpretation in archaeology*. Cambridge: Cambridge University Press.

Housley, R.A., Gamble, C.S., Street, M. and Pettitt, P. (1997) Radiocarbon evidence for the Late glacial human recolonisation of northern Europe. *Proceedings of the Prehistoric Society* **63**, 25–54.

Hu, F.S., Kaufman, D., Yoneji, S., Nelson, D., Shemesh, A., Huang, Y., Tian, J., Bond, G., Clegg, B. and Brown, T. (2003) Cyclic variation and solar forcing of Holocene climate in the Alaskan subarctic. *Science* **301**, 1890–1893.

Hughes, D. (1996) *Pan's Travail: Environmental Problems of the Ancient Greeks and Romans*. 2nd edition. Baltimore: Johns Hopkins University Press.

Irwin, G. (1994) *The Prehistoric Exploration and Colonisation of the Pacific*. Cambridge: Cambridge University Press.

Jarman, H.N., Legge, A.J. and Charles, J.A. (1972) Retrieval of plant remains from archaeological sites by froth flotation. In (E.S. Higgs, ed.) *Papers in Economic Prehistory*, 39–48. Cambridge: Cambridge University Press.

Jones, M. (2001) *The Molecule Hunt, Archaeology and the Search for Ancient DNA*. London: Penguin.

Jones, S.B. and Luchsinger, A.E. (1987) *Plant Systematics*. Singapore: McGraw-Hill.

Kerkut, G.A. (1960) *Implications of Evolution*. Oxford: Pergamon Press.

Kopytoff, I. (1986) The cultural biography of things: Commoditization as process. In (A. Appadurai, ed.) *The Social Life of Things*, 64–91. Cambridge: Cambridge University Press.

Kukla, G. and Gavin, J. (2004) Milankovitch climate reinforcements. *Global and Planetary Change* **40**, 27–48.

Langdon, P.G., Barber, K.E. and Hughes, P.D.M. (2003) A 7500-year peat-based palaeoclimatic reconstruction and evidence for an 1100-year cyclicity in bog surface wetness from Temple Hill Moss, Pentland Hills, southeast Scotland. *Quaternary Science Reviews* **22**, 259–274.

Legge, A.J. and Rowley-Conwy, P.A. (1988) *Star Carr Revisited: A Reanalysis of the Large Mammals*. London: Birkbeck College.

Limbrey, S. (1975). *Soil Science and Archaeology*. London: Academic Press.

Lowe, J.J. and Walker, M.J.C. (1997) *Reconstructing Quaternary Environments*. Harlow: Longman.

Mandelbaum, M. (1961) The problem of 'covering laws'. *History and Theory* **1**, **3**, 229–242.

Mannion, A.M. (1987) Fossil diatoms and their significance in archaeological research. *Oxford Journal of Archaeology* **6**, 131–147.

Milankovitch, M. (1941) *Kanon der Erdbestrahlungen und seine Awendung auf das Eiszeitenproblem*. Belgrade.

Miller, D. and Tilley, C. (eds.) (1984) *Ideology, Power and Prehistory*. Cambridge: Cambridge University Press.

O'Brien, S.R., Mayewski, P.A., Meeker, L.D., Meese, D.A., Twickler, M.S. and Whitlow, S.I. (1995) Complexity of Holocene climate as reconstructed from a Greenland ice core. *Science* **270**, 1962–1964.

Osbourne, P.J. (1976) Evidence from the insects of climatic variation during the Flandrian period: a preliminary note. *World Archaeology* **8**, 150–158.

Parfitt, K. and Brugmann, B. (1997) *The Anglo-Saxon Cemetery on Mill Hill, Deal, Kent*. Society for Medieval Archaeology Monograph **14**. Leeds: The Society for Medieval Archaeology.

Pearsall, D.M. (1982) Phytolith analysis: applications of a new paleo-ethnobotanical technique in archaeology. *American Anthropologist* **84**, 862–871.

Pearson, N. and Williams, T. (1993) Single-context planning: its role in on-site recording procedures and in post-excavation analysis at York. In (E. Harris, R. Brown and G. Brown, eds.) *Practices of Archaeological Stratigraphy*, 89–103. London: Academic Press.

Piperno, D.R. (1988) *Phytolith Analysis: An Archaeological and Geological Perspective*. London: Academic Press.

Plummer, C.C., McGeary, D. and Carlson, D.H. (2001) *Physical Geology*. Boston: McGraw-Hill.

Popper, K. (1957) *The Poverty of Historicism*. Boston: The Beacon Press.

Pough, F.H., Heiser, J.B. and McFarland, W.N. (1989) *Vertebrate Life*. New York: Macmillan.

Ray, K. (1987) Material metaphor, social interaction and historical reconstructions: exploring patterns of association and symbolism in the Igbo-Ukwu corpus. In (I. Hodder, ed.) *The Archaeology of Contextual Meanings*, 66–77. Cambridge: Cambridge University Press.

Reddé, M. (2003) *Alésia*. Paris: Editions Errance.

Renfrew, C. and Bahn, P. (2000) *Archaeology: Theory, Methods and Practice*, (3rd edition). London: Thames and Hudson.

Renfrew, J. (1973) *Palaeoethnobotany*. London: Methuen.

Richards, C. (1993) Monumental Choreography: Architecture and spatial representation in Late Neolithic Orkney. In (C. Tilley, ed.) *Interpretive Archaeology*, 143–178. London: Berg.

Richards, M.R., Corte-Real, H., Forster, P., Macaulay, V., Wilkinson-Herbots, H., Demaine, A., Papiha, S., Hedges, R., Bandelt, H-J. and Sykes, B. (1996) Paleolithic and Neolithic lineages in the European mitochondrial gene pool. *American Journal of Human Genetics* **59**, 185–203.

Rick, J.W. (1987) Dates as data: an examination of the Peruvian preceramic radiocarbon record. *American Antiquity* **52**, 55–73.

Rindos, D. (1984) *The Origins of Agriculture: an Evolutionary Perspective*. New York: Academic Press.

Roberts, M.B.V. (1977) *Biology, a Functional Approach*. Middlesex: Nelson.

Ruddiman, W.F. (2003) Orbital insolation, ice volume, and greenhouse gases. *Quaternary Science Reviews* **22**, 1597–1629.

Russell, J.M., Johnson, T.C. and Talbot, M.R. (2003) A 725 yr cycle in the climate of central Africa during the late Holocene. *Geology* **31**, 677–680.

Salvador, P.G., Bravard, J.P., Vital, J., Lyon, J.L. and Voruz, J.L. (1993) Archaeological evidence for Holocene floodplain development in the Rhône valley, France. *Zeitschrift fur Geomorphologie* **88**, 81–95.

Sarnthein, M., Kennett, J.P., Allen, J.R.M., Berr, J., Grootes, P., Laj, C., McManus, J., Ramesh, R., SCOR-IMAGES Working Group 117 (2002) Decadal-to-millenial scale climate variability – chronology and mechanisms: summary and recommendations. *Quaternary Science Reviews* **21**, 1121–1128.

Schiffer, M.B. (1972) Archaeological context and systemic context, *American Antiquity* **37**, 156–165.

Schiffer, M. (1987) *Formation processes of the archaeological record*. Albuquerque: University of New Mexico Press.

Schulz, M. and Paul, A. (2002) Holocene climate variability on centennial-to-millennial time scales: 1. Climate records from the North-Atlantic realm. In (G. Wefer, W. Berger, K.-E. Behre and E. Jansen, eds.) *Climate Development and History of the North Atlantic Realm*, 41–54. Berlin: Springer-Verlag.

Shindell, D., Rind, D., Balachandran, N., Lean, J. and Lonergan, P. (1999) Solar cycle variability, ozone, and climate. *Science* **284**, 305–308.

Selby, M.J. (1982) *Hillslope materials and processes*. Oxford: Oxford University Press.

Shaw, T. (1970) *Igbo-Ukwu: An account of archaeological discoveries in Eastern Nigeria*. London: Faber.

Simmons, I.G. (1979) *Biogeography: Natural and Cultural*. London: Arnold.

Stein, J.K. (1983). Earthworm activity: a source of potential disturbance of archaeological sediments. *American Antiquity* **48**, 277–289.

Stoops, G. (2003) *Guidelines for Analysis and Description of Soil and Regolith Thin Sections*. Madison, Wisconsin: Soil Science Society of America.

Summerfield, M.A. (1997) *Global geomorphology*. London: Longman.

Sykes, B. (1998) Genetics and the Palaeolithic. In (J. Bayley, ed.) *Science in Archaeology, an Agenda for the Future*, 21–24. London: English Heritage.

Tilley, C. (1989) Interpreting material culture. In (I. Hodder, ed.) *The Meaning of Things*, 185–194. London: Unwin Hyman.

Trigger, B. (1989) *A History of Archaeological Thought*. Cambridge: Cambridge University Press.

Turney, C.S.M., Kershaw, P., Clemens, S., Branch, N., Moss, P. and Fifield, L.K. (2004) Millennial and orbital variations of El Niño/Southern Oscillation and high-latitude climate in the last glacial period. *Nature* **428**, 306–310.

Ucko, P.J. and Dimbleby, G.W. (eds.) (1969) *The Domestication and Exploitation of Plants and Animals*. London: Duckworth.

van Andel, T.H. and Tzedakis, P.C. (1996) Palaeolithic landscapes of Europe and environs, 150,000–25,000 years ago. *Quaternary Science Reviews* **15**, 481–500.

van Zeist, W. and Casparie, W.A. (eds.) (1984) *Plants and Ancient Man: Studies in Palaeoethnobotany*. Rotterdam: Balkema.

Vita-Finzi, C. (1969) *The Mediterranean Valleys*. Cambridge: Cambridge University Press.

Voelker, A.H.L. and workshop participants (2002) Global distribution of centennial-scale records for Marine Isotope Stage (MIS) 3: a database. *Quaternary Science Reviews* **21**, 1185–1212.

Wilkinson, K. and Stevens, C. (2003) *Environmental Archaeology: Approaches, Techniques and Applications*. Wiltshire: Tempus.

Woodman, F. (1981) *The Architectural History of Canterbury Cathedral*. London.

Young, A. (1976) *Tropical soils and soil survey*. Cambridge: Cambridge University Press.

Zeuner, F.E. (1963) *A History of Domesticated Animals*. London: Hutchinson.

Zohary, D. and Hopf, M. (1993) *Domestication of Plants in the Old World*. Oxford: Clarendon Press.

2

Defining the context: integrated approaches to stratigraphy

2.0 Chapter summary

The process of environmental archaeology starts with spatial recognition and organization. Both on site and off site remains need to be placed into ordered contexts and those contexts into understandable sequences. At the same time, cultural artefacts provide preliminary information on human activities during the period of site occupation or environmental change. This chapter looks at the concepts involved in stratigraphic recognition, description and interpretation. It examines the classic archaeological and geoarchaeological approaches, as well as the analytical methods which lead finally to a stratigraphic model into which subsequent environmental indicators are placed. Emphasis is continually placed on the importance of an integrated approach cutting across different specialisms to arrive at a holistic view of a site and its environment.

2.1 Geoarchaeology and stratigraphic analysis

On many archaeological projects, site formation processes are tackled from different directions by field archaeologists (usually including the site director) and geoarchaeologists, commonly viewed as specialists. While the stratigrapher places most emphasis on the context and how it fits into a site matrix driven by cultural imperatives, the geoarchaeologist concentrates more on sedimentary events, deposit alteration and soil formation episodes. The two views must, of course, be compatible if stratigraphy is being properly understood, but the fact that there are two views at all is a barrier to the more rounded perception of site formation and thus the realization of interpretative potential.

Is it possible to break down the barrier between these approaches? There have clearly got to be two pools of specialized knowledge available to any site, and it would never be possible for an individual to retain up-to-date details of both areas of study to the level required at complex sites. On the other hand, everyone involved should be able to deepen their (necessarily summarized) grasp of each other's subject area. This, then, is a practical aim of integrating geoarchaeological and standard stratigraphical approaches.

In trying to detail this type of integration, a difficulty arises from the wide range of inputs that geoarchaeology has to archaeological projects. This has, in turn, led to a variety of definitions of the word 'geoarchaeology' and given rise to different views of what should be included (Canti, 2001). However, it would probably be universally agreed that the dual activities of delineating a context and interpreting what's in it (activities that lie at the heart of archaeology) are both covered by some parts of the geoarchaeological spectrum.

In the sections below we discuss the fundamentals of the first part of that duality, examining the methodologies of spatial identification widely used both on and off site by field archaeologists and geoarchaeologists alike. Where possible, we try to assess the relationship between them, and the effect this has on the basic acquisition of data in both types of analysis. The discussion finishes with a number of examples showing archaeology and environmental interpretations based on spatial units operating at different scales.

2.2 Principles

2.2.1 Concepts of identity

In the previous chapter (see Section 1.3.1), we saw that contexts are the basic unit of spatial identification in archaeology. They are physical manifestations of past individual actions or processes, and the relationships between them help ascertain the chronology of the stratigraphy. The concept has close parallels with the sedimentological term 'facies'; both attempt to bring the complexity of the observed sediments towards a simplification that can be grasped by the stratigrapher and built into a sequence for analysis (see Section 2.2.4).

Deciding what constitutes a context is a subjective affair, but very important since the decisions made in the field will dictate all future analytical possibilities. The emphasis on process or action is critical to appreciating the detail of context definition. In the field, an archaeologist may not know what process(es) or event(s) caused the stratification; this may often only be ascertained by further study of samples of the deposit or its constituents after the excavation has been completed. In practice, therefore, stratigraphic units are usually identified by the physical characteristics or variations in those characteristics throughout the sequence. Although not normally the case, it is possible that a single entity can result from more than one event or process, and may therefore represent more than one stratigraphic unit. For example, a gravel deposit may have been deliberately laid to form a metalled trackway, and post-depositional processes may have acted on that gravel causing it to become cemented. The result is one layer, but two contexts.

The sediment facies is the classical way of describing and grouping sedimentary bodies. It was first used in the nineteenth century (Gressly, 1838) and has since come to have subtly different meanings for different groups of sedimentologists (Reading and Levell, 1996). However, it is essentially a type of sediment or suite of sediments with specific characteristics such as lithology, texture, sedimentary structure, fossils, colour etc. (Tucker, 1996) that can be recognized over a large area and grouped together as reflecting a particular depositional process or environment (Fig. 2.1). The limited extent to which particular sediments can be definitely ascribed to particular environments means that an element of interpretation is effectively brought to the facies model quite early in the environmental analysis. A particular observer chooses the important attributes, which must bring a series of preconceptions into the system. For some environments, schemes of facies distinction can be found in the literature (e.g. Ramos and Sopeña, 1983)

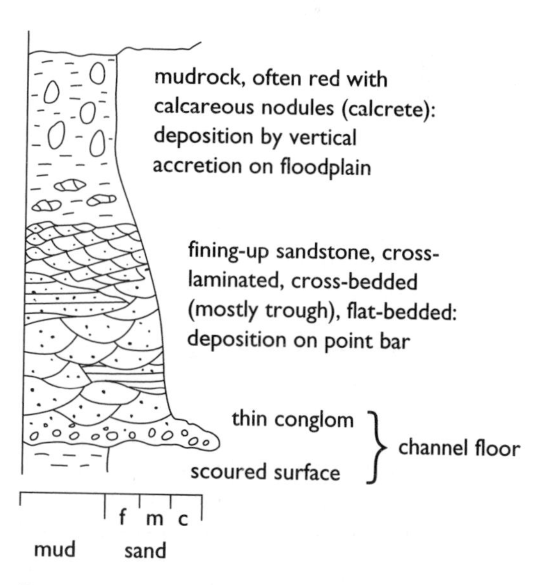

FIGURE 2.1 A facies unit produced by lateral migration of a channel. (From Tucker, 1996)

FIGURE 2.2 Calcium carbonate enriched ditch deposit from Godmanchester, UK.

and these promote a degree of unbiased interpretation. However, the natural variability of sedimentary systems makes the proscribed approach potentially dangerous in that it could force unnatural groupings onto attributes rather than allowing the development of those divisions on a local basis.

Different methodological emphases can give rise to different types of facies analysis, such as lithofacies – concentrating on physical characteristics; biofacies – concentrating on biological attributes; and microfacies – concentrating on microscopic features. Each of these approaches has particular strengths and weaknesses and needs to be deployed in situations where the benefits can be maximized. Environmental archaeology utilizes all three, with the main focus being a finely tuned approach to biofacies, detailed in Chapter 3.

Both types of identity (context and facies) are additionally affected by post-depositional alteration. Aside from the significance of unseen taphonomic changes to biological components, visible processes may act to form additional apparent contexts from a single original entity. This can occur through translocation of fine or soluble materials from the upper parts of a context and their re-deposition in a lower part. Variations of this type are caused by the core soil processes, such as podzolization, gleying and clay translocation. They can generate significant colour and even textural differences in materials originally deposited in a single event (Fig. 2.2). A field archaeologist must therefore be aware that events do not always follow a simple one-to-one relationship with the physical constituents of the context.

Both facies and context identification have a strong element of judgement interwoven with the observed physical differences. Although consistency is the goal, issues such as research objectives, the nature of the strata, time or financial constraints, and the experience of the observers inevitably have an influence on deciding what constitutes these units of identity in a particular field situation. For example, in well-preserved environments it may be possible to identify a series of very fine scale chronological events in the stratigraphic record, but a decision will have to be made on whether the recording of those events is worthwhile in terms of the research questions being asked of the sequence as a whole. At Rougier Street in York (Rankov *et al.*, 1982: 349), the clay sills of some Roman timber buildings had eroded through time, and the clean brickearth products of this erosion interdigitated with some of the occupational debris that accumulated on the floor of the building's interior. It was therefore possible to observe a sequence of stratigraphic units representing events that were in themselves presumably of short duration: erosion of clay sill, accumulation of occupation debris, further erosion, further accumulation, and so on. This sequence of contexts could be individually recognized, numbered and described; or alternatively the whole sequence could be recorded as a single stratigraphic unit, noting this sequence as an attribute of the context.

Other examples abound. Individual dumps may be visible in a thick deposit of mound or rampart material – should one record each individual dump or focus on the overall process of dumping and assign it a single entity? An apparently homogeneous, undifferentiated deposit may be arbitrarily divided into a number of stratigraphic units (by spitting, for example) in order to gain some control on artefact recovery during excavation. This decision-making process (termed 'lumping or splitting' by some field archaeologists) is a necessarily ubiquitous part of all archaeological recording, and has great importance for understanding the contextual nature of archaeology, even at the level of basic data collection. The field archaeologist must therefore be continually making value judgements about the significance of the stratigraphic phenomena perceived in the field in relation to the research questions that might be asked.

2.2.2 Concepts of boundary

The corollary of identifying depositional units, be they contexts or larger sediment bodies, is the delineation of their edges. Recognition of boundaries is predicated on the visible variation resulting from the changing process of deposition, and the attributes recorded (usually in section) are chosen to highlight that process or event. There are many types of boundary (or contact), though the two perhaps most relevant to environmental archaeology are the conformable contact and the unconformable contact (or 'unconformity'). A conformable contact is where the boundary is complete, with sediments representing the gradual change from one regime to another. This is an expected pattern in purely natural environments, indicating such changes as the slow movement of a river channel or a switch from channel to deltaic sediments. The concept can be a problem for many field archaeologists, who seek to clearly separate the physical identities of archaeological deposits in terms of a change in human actions. An unconformable contact is where the boundary between one deposit and another is very sharp, representing an episode of erosion or truncation. The sediments are incomplete, and missing information must be sought through other exposures.

On an archaeological site, not all depositional units have a physical identity. Units are generally classified as positive or negative, and this affects our view of their boundary. Positive contexts are self-explanatory: deposits, walls, timbers, stones – anything that has a physical presence. Negative features can be more problematic. The term derives from the realization that certain stratigraphic events can be recognized where there is evidence that strata have been removed by being cut away; hence the generic term 'cut' for negative contexts such as pits, post holes, ditches and grave pits. The action of (say) digging a pit is a stratigraphic event and is recognized as an individual stratigraphic unit; the only attribute that such a cut has is its surface, which can be identified during excavation by discontinuities in the positive strata.

Strictly speaking, the 'cut' or 'negative feature' is a sub-class of another class of stratigraphic unit, the 'interface' (Clark, 2000). The concept of the interface has proved extremely useful in recording and interpreting archaeological stratification. Any archaeological deposit consists of at least three stratigraphic subdivisions: its lower interface, i.e. the surface in contact with pre-existing stratification; the physical constituents of the deposit itself; and its upper interface, i.e. the surface in contact with overlying stratification. Attributes pertinent to the interfaces of a deposit may be different to those of its constituents; for example a deposit formed by the disposal of household refuse may have been subsequently trampled by horses, leaving the imprints of hooves on the upper surface (cf. the Bronze Age site at Krigshorm in Sweden; Neil Price, pers comm.). We can separate the evidence for the two events by recognizing and recording two stratigraphic units, namely the deposit and its upper interface. Moreover, an interface is not necessarily related to an individual deposit. For example, the mortar bedding deposits for a subsequently robbed stone floor at Canterbury Cathedral showed evidence of where the stone flags had once been (Blockley *et al.*, 1997: 25). This interface contained much important information, though it was not necessarily relevant to the physical constituents of each particular mortar spread, and required recording as a separate stratigraphic entity.

2.2.3 Concepts of description

Having identified individual stratigraphic units, they must be described in order to try and understand them and also record them for posterity. But what attributes are significant? The list of things one might record about a deposit seems endless: extent, thickness, mineral composition, particle size, acidity, colour, smell, compaction etc. Which of these should be recorded and in what detail?

It is worth distinguishing between stratigraphic units that have formed through natural processes and those that have been directly affected by human activity (though this differentiation is a crude convenience for discussion; a deposit of colluvium may have been brought about by natural processes, but ultimately triggered by human deforestation of a valley side). Naturally occurring strata may be understood by uniformitarian principles. Simply put, if we observe that process A produces the physical attribute B, then when we observe physical attribute B in the archaeological record we can assume process A has occurred.

Unfortunately, complications arise even in such apparently direct relationships. A given attribute may itself be derived from a complex overlay of processes in the same way as we saw with contexts. At Watertower Street, Chester, for example (LeQuesne, 1999), the textural composition of the Roman earth wall was a well-sorted sand, suggesting local water-borne deposits as a source. Only after mineralogical analysis and comparisons with local materials was it apparent that the wall was in fact built of a boulder clay that had inherited all its particle size characteristics from the underlying fluvially-deposited sandstone. Glacial action had scraped these materials up, and subsequent decay had left the sandstone's fluvial signature in the boulder clay soils from which the Romans cut their turves (Canti, 1999).

When describing anthropogenic strata, the introduction of an additional human component into site formation can further complicate the link between morphology and process. Context colours, for example, can present dualities or even multiplicities of interpretive options. Taking just one case, iron oxides in soils and sediments convert to the bright red mineral haematite on burning, but they also do this as a result of long-term weathering in hot desert environments. Thus, not only can burnt or desert materials be reddened in this way, but contexts made up of geologically-inherited desert materials (such as the northern European Permo-Triassic) are also frequently reddened. Attributes must therefore be linked to interpretation only with extreme care, and assumptions need to be continuously revisited.

In practice, the range of attributes designated as significant for recording anthropogenic and natural strata is derived from the earth sciences, particularly pedology and geology (see, for example, Canterbury Archaeological Trust, 2001; Spence, 1990; also Section 2.3.3.3). Although they do not usually contribute directly to interpretation, these attributes can be useful for informing decision making at the interpretive stage by identifying similarities (and differences) in the physical make-up of stratigraphic units (these issues are pertinent for both deposits and interfaces; Clark, 2000). However, although the immediate benefits are not always apparent on site, the precise and impartial recording of context characteristics must be an act of forbearance on the part of the practitioner. The detailed and sometimes tortuous conceptual breakdown of characteristics is the only way of guaranteeing a description for future generations.

2.2.4 Concepts of sequence

Critical to a discipline so concerned with issues of time is stratification. Without stratification, there can be no appreciation of change through time, and thus no environmental archaeology. Primarily, the relative chronology of a stratigraphic sequence may be understood by reference to the physical superposition of strata one upon the other, with the assumption that a layer overlying another must have been formed at a later date than that which it overlies: 'the principle . . . that (unless you stir them up) the discarded papers in your waste paper basket will have earlier dates near the bottom and later dates at the top' (Piggott, 1959: 34). This 'law of superposition' is the essential geological concept (Challinor, 1964: 238) underlying the different recording systems used in the various subjects that rely on information held in stratigraphy. Environmental archaeology borrows from a variety of these subjects, being commonly carried out in both off site and on site situations. Although underlying stratigraphic principles are essentially the same, the different approaches need to be understood by practitioners.

Similarities between the sedimentological and archaeological approaches (facies and contexts – see Section 2.2.1) continue to be apparent as we progress up the interpretive tree. Once facies are distinguished they can be grouped together into associations. This has a similar effect to multi-proxy approaches in bioarchaeology (see Chapter 3), where uncertainties in any given dataset are negated by concatenating them with information from other datasets. On the other hand, it also has the effect of reducing the detail inherent in each example. Some individual sedimentary environments can be interpreted confidently from single exposures because of the strong causal relationship between depositional style and sediment attributes. However, in other cases, a single sediment type could derive from one of a number of different environments of deposition. A cross-bedded sandstone, for example, could be deposited in fluvial, deltaic, lacustrine, shallow marine or even deep marine conditions (Tucker, 1996). Facies associations help to remove the uncertainties about environmental interpretation by confirming a number of different exposures with similar characteristics.

Facies analysis has many echoes in archaeological stratigraphy – the hierarchical nature of the grouping, the tension between the requirements for description and interpretation etc. Conceptually, the law of superposition is fundamental to both, along with other concepts inherent in archaeological stratigraphy, such as the 'law of horizontal deposition' (Harris, 1979b). However it became clear during the 1960s and 1970s that the study of archaeological stratification requires its own methodological and theoretical grammar (Brown III and Harris, 1993). Superposition in archaeological stratigraphy is not just relevant to deposits, but all the stratigraphic entities we have discussed above, including both positive and negative phenomena. Their relative chronological position can be articulated by reference to the law of superposition, though this is not always a simple physical relationship. While it can be expected that 'time's arrow' will run from the bottom of a stratigraphic sequence, establishing what is up and what is down may not always be straightforward (Thorpe, 1998). In Fig. 2.3, for example, we can see a stone-lined drain filled with sediment. Physically (both in section and in plan), the sediment (3) underlies the capping slab of the drain (2), yet our understanding of the deposit formation process indicates that the capping slab (2) was put into position prior to the build up of sediment (3). When built, the

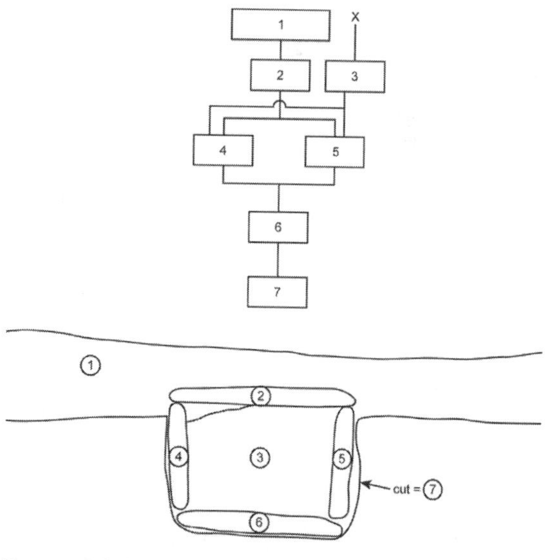

FIGURE 2.3 Time's arrow takes many routes: in this section of a stone-lined drain, the stone capping (2) is not stratigraphically later than the drain infill (3) even though it physically overlies it.

stone slabs (2), (4), (5) and (6) created a void through which waste water could pass. Sediment accumulated in this void, eventually rendering the drain ineffectual. Thus the sediment (3) is stratigraphically later than the drain base and side slabs (contexts 4, 5 and 6) but not stratigraphically earlier than the capping slab that overlies it. In this example we cannot ascertain what stratigraphic unit overlies the sediment (3).

The physical relationships between stratigraphic units can, furthermore, be very different to the stratigraphic relationships between them. In Fig. 2.4, for example, context 100 is physically in contact with contexts 200, 300 and 400, overlying all three and thus stratigraphically later. However, as context 200 overlies context 300, which in turn overlies 400, the stratigraphic relationship between context 100 and contexts 300 and 400 is said to be *redundant*, because if context 100 is stratigraphically later than context 200, it must evidently be later than 300 and 400.

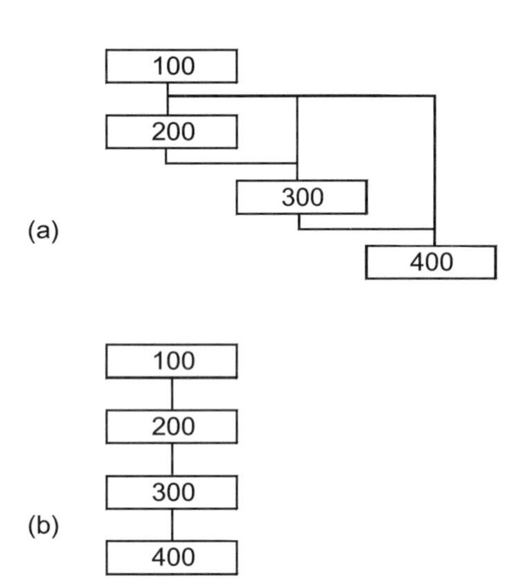

FIGURE 2.4 Different physical and stratigraphic relationships. Here context 100 is physically in contact with contexts 200, 300 and 400, overlying all three and thus stratigraphically later (a). However, as context 200 overlies context 300, which in turn overlies 400, the stratigraphic relationship between context 100 and contexts 300 and 400 are said to be redundant, because if context 100 is stratigraphically later than context 200, it must evidently be later than 300 and 400. The matrix may be drawn without these redundant relationships (b).

These detailed ideas of sequence resulted from development during the excavations at Winchester, UK, in the early 1970s (Harris, 1975; 1979a) and the principles of stratigraphic excavation, area excavation, the concept of interface, the single context plan and the site matrix are now axiomatic for most modern archaeological excavation (Harris *et al.*, 1993: 2–4). The site matrix (also known as the 'Harris-Winchester matrix' or the 'Harris matrix' after the eminent archaeologist Edward Harris, who has tirelessly promoted its benefits around the world for the last thirty years; Harris, 1975; 1979a; 1979b; 1983; 1989a; 1989b; 1991; 1992; 1995) shows 'in what relation one layer stood to others in the sequence . . . it assumes that any two units of stratification have either no stratigraphic connections, or they lie in superposition . . . These simple assumptions are of course the essence of notions of relative time' (Harris, 1979a: 117–118). The matrix is simply a visual representation of the relationships between individual stratigraphic units in an archaeological sequence: 'in its series of linked boxes, each of which shows defined context in relation to its immediate neighbours, is a formalized picture of the stratigraphic sequence of the site' (Hammond, 1993: 139–140).

The value of the matrix is discussed further below, along with its utility in post-excavation studies and hypothesis testing. Here, perhaps, it is worth emphasizing what the matrix is not; much of the critique of such matrices has stemmed from a failure to grasp the powerful, yet ultimately simple aims of such diagrams. First, as it is based on the 'laws of archaeological stratigraphy' (Harris, 1979b), which are generally held to be axiomatic among archaeologists, the relative chronology of stratigraphic units presented on the matrix is non-negotiable in terms of theoretical or interpretive hypotheses; in essence, the matrix is *true*. This is its strength; no interpretive hypothesis can contradict the sequence. However, the matrix makes no reference to the physicality or type of stratigraphic unit, nor does it address issues of chronological duration. Thus a single number on the matrix may represent an event lasting only a few seconds (such as the collapse of a masonry wall) or decades or centuries (such as a deposit formed by colluviation). Similarly, a masonry wall may be stratigraphically earlier than the first floor deposit laid abutting its face, but might remain an upstanding part of the physical landscape for hundreds of years. These aspects of stratification are not reflected in the formal presentation of the stratigraphic sequence on the Harris matrix, and are perhaps best addressed through a holistic consideration of all the data retrieved from an archaeological sequence.

2.3 Data acquisition

We move on now to the processes of acquiring data. Before applying the concepts described above, we must first look at the generalized physical context of a site. Many characteristics of archaeology from single layers to whole cultures have some roots based in the geological and depositional environment. The relationships can be complex, and of variable significance, but the physical basis needs to be understood if an integrated approach to environmental archaeology is to be achieved.

2.3.1 The site and its setting

2.3.1.1 Geological and geomorphological context

Archaeological sites are found in as many different situations as there are landforms to contain them. The geological basis of these landforms is critical to both preservation conditions and to stratigraphic build-up, but the differences actually found on sites are often modified by superficial processes in the overlying soil or sediment. These are of course part of the 'geology' as well, but the term is used most commonly to describe the solid underlying rock or drift deposits found on geological maps. Geology contributes much to geomorphology, by virtue of characteristics such as hardness and erodibility which are fundamental parts of the landforms equation. However, its most important archaeological contribution is probably the particle size and chemistry of the deposits from which stratigraphy is formed. Considerable differences in the agricultural potential of soils depend on particle size, at least at the extreme ends of the spectrum. Some geological materials can only deliver coarse sand, for example, to the surface soils, limiting the cropping range or promoting soil ameliorations on the part of inhabitants that can radically affect the archaeological record (Simpson, 1997).

Particle size is also critical to drainage and gaseous exchange, deciding (to a large extent) the fate of artefacts preserved by waterlogging and anaerobic conditions. The elemental make-up of particles ultimately decides the chemical characteristics of the deposit, affecting preservation through pH characteristics and biological activity. The most obvious example of this variation is between calcium carbonate rocks which yield alkaline stratigraphy, and silica which is usually acidic. Less apparent to the non-specialist are subtler issues which apply to all parent rocks. What, for example, is the content of mica, a mineral with both alkaline and siliceous components, delivering clay and cations to soils formed from it?

The relationship between archaeology and geology is mediated through geomorphology. Again, both cultural and taphonomic effects are fed through to the stratigraphic record. Firstly, a site can only be found if people were active in a particular setting. They must have found a living from the soil or through manufacturing derived from local materials; or they carried out social or ritual functions, perhaps perceiving special characteristics in landforms. Secondly, geomorphology has an intricate relationship with the preservation processes. The underlying question to be asked of any site is: why are cultural remains preserved at all? Why didn't they simply stay at the surface to be destroyed by the weather? Geomorphology is the setting for the delivery of sediment and water regimes that define decay pathways.

2.3.1.2 Depositional context

The parameters provided by geological and geomorphological factors are further modified in the detail of the depositional environment. People are active in many forms of sedimentary system, and their activities can also be covered or destroyed by sedimentary layers. The main groupings are:

Alluvial sediments

Alluvium is a general term for water-borne sediments and covers a wide range of materials regardless of particle size. However, the large part of the alluvium studied for archaeology is actually fine-grained sediment deposited in lowland river systems. In glaciated regions, this will frequently overlie a coarse gravel or boulder component from glacial and periglacial deposition that preceded the finer Holocene deposits which usually cover or contain the archaeological remains. Elsewhere, fine-grained deposits may lie directly over bedrock, with gravel deposits existing only as channel fills or outwash fans (van der Schriek *et al.*, 2003; Wilkinson and Pope, 2003), depending on the type of depositional environment. Channel dynamics examined at a detailed scale are extremely important when understanding both the depositional record, and the biological remains contained within it (Brown, 2003).

Colluvial sediments

Colluvium builds up where soil has moved downslope due to gravity, cultivation, animal burrowing etc. It is essentially an unsorted deposit, whose particle size characteristics frequently encompass the whole of the source material's range rather than concentrating a single portion as happens with many other processes. People commonly cause or exacerbate colluviation through activities on slopes that disturb the soil and therefore move it inevitably downhill (Govers *et al.*, 1996; Quine *et al.*, 1997; Wilkinson, 2003).

Windblown sediments

Windblown sands are found wherever expanses of bare sandy soils or sediments are open to wind action and a suitable area also exists to deposit the sediment load when velocity drops. Coastal dunes are the most widespread form of windblown sands, but inland deposits also occur commonly in desert environments and in drier temperate zones where vegetation is sparse. Deposition tends to be episodic, and can be very rapid, even building up significantly during the period of occupation (Loveluck, forthcoming).

At the finer end of the particle size spectrum, loess is the windblown silt found in various parts of the world especially China, Argentina, northern Europe and parts of the USA. Enormous thicknesses of loess are found in some of these areas; for example, in Romania, some loess layers are believed to be 80 m deep (Collard, 1988). The main deposits probably have a periglacial origin, and Palaeolithic sites are therefore the ones most commonly found in loess (Wymer, 1976). However, later Holocene events have deposited significant amounts on top of cultural remains and agricultural soils in parts of China (Huang *et al.*, 2002).

Marine sediments

Offshore sediments obviously cover archaeology mainly in situations where sea level has risen significantly since the period of occupation. For most passive continental margins, this will be in an area of continental shelf under 20 to 550 m of water and extending outwards between 2 and 1500 km (Leeder, 1982). Although there is great variation in these dimensions, they have considerable significance for the volume of undersea archaeology. Sea level has generally risen around 120 m in the last 15,000 years (Emery, 1969) with the result that large areas of archaeologically rich continental shelf must have become inundated.

Intertidal sediments are less straightforward, as they may be preserving evidence either for shoreline activity such as fish traps, or dry land sites which later became submerged (see Bell *et al.*, 2000).

Glacial sediments

Glaciers deposit boulder clay, an unsorted mass of (literally) boulders and clay. In northern Europe this is commonly Devensian or earlier, and thus does not often cover or contain archaeological remains. However, the periglacial area which moves across the landscape as glaciers retreat produces a number of artefact-bearing sediments. These are typically gravel bodies or solifluction deposits. The finds are mostly individual tools, and their appearance must define the point where sedimentary energy has been reduced enough for people to move safely around in the depositional environment, either ephemerally or living by hunting. Re-deposition is evident on many of the finds, testifying to the instability which characterized the periglacial landscape (French, 1996; Hosfield, 2001).

Volcanic sediments
Volcanoes produce large amounts of sediment, but deposits thick enough to contain or bury sites are usually found in a relatively small area of the globe. The lightness, fine particle size and recognizability of volcanic tephra make it one of the most widespread marker sediments of all. For example, Wastegård *et al.*, 2000 were able to trace the Icelandic Vedde ash (from an eruption around 10,300 BP) as far as parts of northwestern Russia (see also Section 2.5.4). In general, the destructiveness of the sedimentation (mainly due to heat and bombardment) means that preservation potential is low. Paradoxically, an example of unusually good preservation is found in the rapid all-engulfing pyroclastic flows of ash that buried Pompeii (Sigurðsson *et al.*, 1982), but this is a rare situation where neither of those destructive forces was active. The overwhelming speed of sedimentation was the main agent of preservation at that site.

Organic sediments
Organic matter will sometimes form the main contextual material of an archaeological site and where it builds up naturally (peat), it can provide some of the best sequenced archives of biological remains in datable contexts (e.g Clapham and Clapham, 1939; Pennington, 1970). However, there are many variations on the classical vertical accretion model of peat formation. Peat can be deposited in sedimentary environments, sometimes interleaved with mineral sediments. Where it grows on raised topography, the high water content and consequent weight of the material leads to slumping and re-deposition with exposure of underlying mineral materials, erosion and further mineral deposition. Although the site formation processes can thus be difficult to unravel, the preservation state of organic remains in peat based sites can be remarkable (Coles and Coles, 1986; Raftery, 1992).

Cave sediments
Caves enclose their own unique sedimentary systems involving many of the processes listed above in a small, frequently self-contained area. They are much more significant archaeologically than their total area would suggest, partly because of a strong tendency towards good stratification, and partly because of their considerable attraction for human activity. Caves also provide unusually low leaching conditions in which complexes of relatively soluble minerals can develop or be modified from concentrated human inputs such as ash and organic wastes (Boschian, 1997; Karkanas *et al.*, 2002; Weiner *et al.*, 2002).

2.3.2 Archaeological excavation techniques

2.3.2.1 Introduction

There are many approaches to excavation, and it should be emphasized from the outset that there is no right way to deal with a site, although there are many wrong ways (Wheeler, 1954: 1). It is axiomatic that any excavation requires the destruction of part of the archaeological resource: 'a fragile, finite, irreplaceable and rapidly disappearing source of information about the past' (Raab, 1984a: 57). Good field methodology should, therefore, be flexible enough to satisfy the principle of 'preservation by record' (ODPM, 2000; see also Lipe, 1974), while also coping with a wide range of archaeological phenomena by allowing the deployment of different excavation and recording techniques in a coherent fashion (Dunnell, 1984; Raab, 1984b; Clark, 1988a; Shay, 1989).

Early systematic recording developed the use of numbered layers, relying primarily on the sections revealed in the trench sides to record sequences, as well as a written description of the position and main characteristics of deposits (Wheeler, 1947; Kenyon, 1952; Webster, 1963). However, the limitations of sections became increasingly apparent during the late 1960s and 1970s: 'arguments of chronology or of the sequence of a complex stratigraphic situation based on sectional analysis must be suspect or completely fallacious' (Harris, 1975: 110). The introduction of open area excavation allowed the identification of archaeological phenomena not easily recorded in section (e.g. Barker, 1969, 1977; Biddle and Kjolbye-Biddle, 1969; Biddle, 1975), raising broader questions about the relationship between excavation methodology and interpretation. The issue crystallized around this central question: to what extent should archaeological stratification be interpreted in the field, and how should such field interpretation influence the recording of that stratification?

2.3.2.2 The single context system

The single context system was developed to provide a less interpretive form of recording, and has proved highly effective, being widely used by archaeologists around the world (e.g. Macleod *et al.*, 1988; Bibby, 1993; Brown and Muraca, 1993; Stuck, 1993). It is predicated on the simple principle that each and every stratigraphic unit is uniquely identified and recorded (Spence, 1993). Open area excavation is carried out in strict stratigraphic order, i.e. the uppermost stratigraphic unit is exposed in its entirety, recorded and removed (retaining artefacts and samples) before the next one down is excavated, and so on. The attributes to be recorded and the manner in which the record is compiled are usually dictated by prescriptive recording manuals (e.g. Scottish Urban Archaeological Trust, 1988; Spence, 1990; Canterbury Archaeological Trust, 2001).

Each context has a plan drawn of its topographical boundary, including details of surface morphology, areas of discoloration etc. The plan has a primary role in the single context system. Indeed, the idea of 'single context planning', suggested by the English archaeologist Laurence Keen, was first widely employed in London during the 1970s and has led to the term being synonymous with the recording system as a whole (Pearson and Williams, 1993;

Box 2.1 Open area excavation: the baths basilica at Wroxeter

As discussed elsewhere in this book, open area excavation allied to the single context system has provided an effective technique for understanding and recording complex stratigraphic sequences. Unsurprisingly, open area excavation also allows the observation of archaeological phenomena not easily perceived in narrow trench or box excavation. A fine example of this was Philip Barker's campaign at the Roman town of Wroxeter between 1966 and 1974 (Barker, 1975) (see Fig. 2.5). Here an area of some 4000 m^2 was stripped of topsoil and the underlying rubble cleaned by hand. The resultant surface was recorded by taking over 1500 vertical photographs, which were assembled into a photomosaic of the rubble spread. On examination, it became apparent that the superficially random rubble was in fact highly structured. Rectangular plots, occasionally associated with post pads, post holes and short lengths of timber slot, were laid out in rows to the north and south of an east-west aligned metalled street. These were interpreted as the platforms for timber framed buildings, one a very large rectangular structure with a southern facade that possessed a central portico with symmetrical flanking wings. These buildings demonstrate a major period of redevelopment with a planned complex of timber framed buildings, some of classical design and very large in size, dating to the second half of the fourth century AD. The vestigial traces of such timber framed buildings would be very difficult (if not impossible) to understand in section or in narrow box trenches. Indeed, prior to Barker's work, the site of the baths basilica at Wroxeter had been extensively excavated since at least 1859, with no suggestion of this important phase of rebuilding and redevelopment at the end of the Roman occupation of Britain.

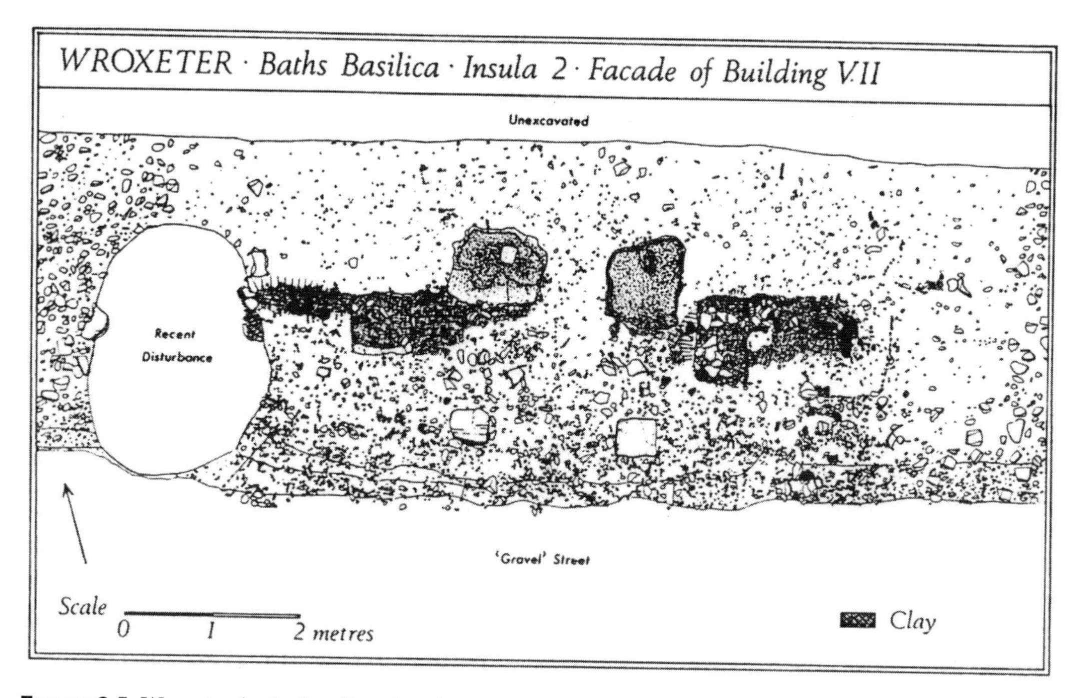

FIGURE 2.5 Wroxeter baths basilica: Insula 2: Facade of Building VII (from Barker, 1975).

Hammer, 2002). Individual plans, usually drawn on semi-transparent drafting film or digitized into computer CAD systems, can be overlain and examined to see if they overlap. If they do, by reference to the law of superposition, it may be demonstrated that a stratigraphic relationship exists between them. Because the plan record is comprehensive across the excavation area, all stratigraphic relationships can be identified. This may not be the case if relying on vertical sections, since not all stratigraphic units might appear in a particular section.

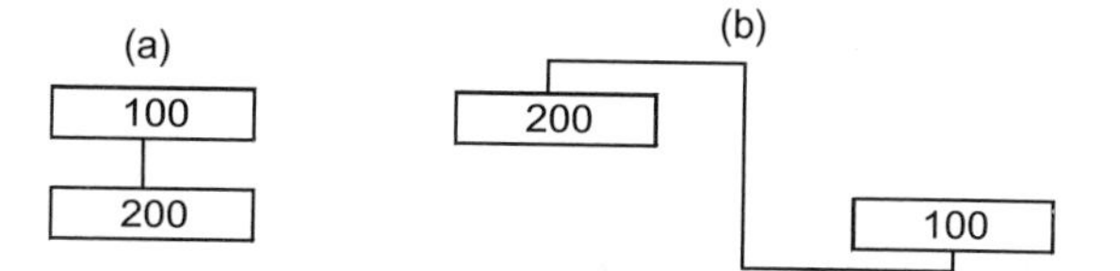

FIGURE 2.6 A stratigraphic relationship as depicted on a site matrix: in (a) context 100 can be seen to be stratigraphically later than context 200 by virtue of the line connecting the two boxes. The physical positioning of the numbered boxes on the matrix is immaterial; the same relationship is portrayed in (b).

The information collected by single context planning is gradually brought together into a diagram showing the stratigraphic relationships (and thus the relative chronological relationships) between individual contexts (Fig. 2.6). This is the matrix. It is a simple, but powerful tool that underpins most archaeological stratigraphy today, particularly in complex stratigraphic situations.

The matrix represents the site in such a way that an archaeologist can easily understand and interrogate the relative chronology. It is 'the unique testing pattern for all later (artefact, ecofact, chronological and topographical) analysis of a site' (Harris, 1998).

Sections remains a useful tool for presenting certain attributes of a stratigraphic sequence, particularly vertical relationships between units such as tip lines or episodes of erosion in cuts. By showing the full sequence in a particular location, and the boundaries between contexts, sections can inform certain sorts of sampling (e.g. by using monoliths). However, the fact of recording several contexts at once fits

Box 2.2 Using the single context system at Whitefriars, Canterbury

Between 1995 and 2003, large scale redevelopment within the city walls of the historic city of Canterbury occasioned a major programme of archaeological excavation covering more than 2 ha of the city centre (over 8 per cent of the total intramural area). This work had to be fitted around a complicated programme of construction, moving field teams onto sites as buildings were demolished, resulting in over forty archaeological interventions, ranging from minor watching briefs to numerous large scale open area excavations (see Fig. 2.7). The resultant site archives therefore represent a huge mosaic of deeply stratified urban stratigraphic sequences. A challenge for the post-excavation team is to integrate these myriad independent studies into a united site archive that will allow the whole sequence to be interrogated in a holistic fashion.

From the outset of the project, the excavation and recording methodology employed was predicated on the issue of integrating the site records. A strictly stratigraphic approach was adopted, using the recording methodology employed on excavations carried out in London during the 1970s and early 1980s, with some minor modifications (Canterbury Archaeological Trust, 2001). Based on the single context system, the individual context plans are compiled using the 'grid area' technique. This approach divides each site into a grid of 5 m squares; each context is planned in relation to the grid squares so that if a context crosses a boundary between two squares it is drawn on two plans, one for each grid square. This approach is extremely useful for compiling and verifying complex stratigraphic sequences over large areas, as well as facilitating the correlation of stratigraphy from different interventions via the production of plan matrices for each grid square. At Whitefriars, the single context plan approach is pivotal in articulating the chronological development of a major part of the city from the earliest times to the present day.

uncomfortably with the theoretical aspects of the single context system and their use is rarely encouraged, except in particular circumstances.

In an ideal world, the comprehensive nature of the single context record allows the site to be 're-excavated' from different theoretical perspectives or research agendas. By basing the system on explicit stratigraphic criteria, coupled with standardized and documented attributes for recording, comparability between sites is improved, as is the training and involvement of all members of the excavation and post-excavation team. However, if deployed unthinkingly, this powerful tool can become mechanistic, and transform itself into a liability – a shackle for the imagination.

2.3.2.3 The planum method

Distinct from the single context system, there is a non-stratigraphic approach to excavation and recording that has been used to great effect. This system excavates archaeological deposits in a series of spits, planing off sediment at a predetermined depth to an arbitrary horizon (usually by machine), which is then drawn, and the process repeated through the archaeological sequence. Environmental evidence and artefacts are collected, and assigned to the spit from which they were recovered. If discrete archaeological features are encountered, such as masonry walls or graves, these are recorded separately. Otherwise stratification is ignored, even where complex or tipping (Fig. 2.8).

Although not popular in the UK, the planum method is still commonly employed in continental Europe and has achieved impressive results. Its perceived advantages are that it is relatively cheap and that large areas can be excavated quickly with a comparatively unskilled labour force. It eschews sophisticated chronological understanding of the site and detailed analysis of environmental data and

FIGURE 2.7 One of the large open area excavations that formed part of the archaeological investigations at Whitefriars, Canterbury. (Canterbury Archaeological Trust)

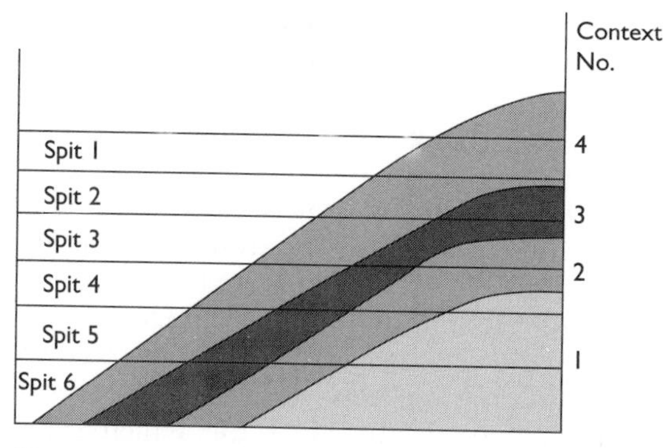

FIGURE 2.8 Illustration of the planum method in sloping stratigraphy.

material culture to concentrate on examining spatial distribution over large areas at low cost. The system has been used to great effect at sites such as Wijster (van Es, 1967), Feddersen Wierde (Haarnagel, 1979) and Dorestad (van Es and Verwers, 1980), among many others.

This relatively crude stratigraphic recording and artefact retrieval does not, at first sight, square up to ambitions of 'preservation by record' (ODPM, 2000). However, there are situations where the system has great advantages, for example in the excavation of graves where the organic remains of the body and other graves goods have completely decayed, but have left three-dimensional silhouettes of differential staining in the grave fill (e.g. Longworth and Kinnes, 1980: 23 and fig. 23).

Box 2.3 Using the planum method at Sutton Hoo

It is possible to excavate and record soil silhouettes stratigraphically, but this presents substantial practical difficulty and requires an excavator of enormous skill and experience. An alternative is to employ the planum method, removing the grave fill in spits, plotting and recording changes in colour and so forth at the surface of each spit (Rahtz, 1989, 1996). This may be supplemented by stratigraphic features as they are revealed by the removal of arbitrary spits, an approach adopted by the team excavating the 'sand bodies' at Sutton Hoo (Hummler and Roe, 1996). The result is a series of slices through the stratification, which have the potential in post-excavation analysis to be reconstituted to realize the three-dimensional forms of the stratigraphic phenomena recorded in the successive spits, an approach akin to tomography (usually carried out on artefacts: Spriggs, 1992; Anderson and Fell, 1995).

2.3.3 Techniques for establishing the position and character of sediments and soils

2.3.3.1 What are sediments and soils?

Archaeological sites occur in landscapes composed of soils and sediments. Location and description of these natural or semi-natural entities is therefore a fundamental activity for many projects, particularly during exploratory phases. It is important to be clear from the outset on the meanings of the words 'soils' and 'sediments'. They are not two mutually exclusive sub-categories of earth, but have a strong causal link:

- *Sediments* are masses of mineral and organic material deposited by any process involving natural or human agencies.
- *Soils* are discrete bodies of altered mineral material (either sediment or *in situ* rock remains) occurring only at the earth's surface. They are commonly a metre to a few metres deep, and have different layers (horizons) which develop over time in response to climatic, geomorphological and chemical factors.

While the word 'sediments' has a fairly broad definition, 'soils' is quite precise. One of the central communication difficulties that archaeologists and soil scientists have traditionally encountered is founded in the difference between this precise definition and the common usage of the word 'soil' to describe dark topsoil or just deposits of earth in general (Fig. 2.9).

2.3.3.2 Characterizing soils and sediments

Soils and sediments are the backbone of most archaeological sites. Many features are cut into the lower horizons of the soil on which people lived, and have disturbed topsoil above them. Sediments are the stuff of contexts – the fills of features and the medium in which finds are held; they occur frequently as interdigitating or sealing layers as well. Knowledge of their extent and characteristics is therefore frequently needed, and a range of methodologies is applicable to provide this information.

The traditional approach to soil survey is digging a series of soil pits across a landscape. In order to fully appreciate the nuance and detail of spatial changes to soil characteristics, roughly metre square pits are dug right through to the parent material perhaps one or two metres deep (Dent and Young, 1981). The expenditure and effort required to cover an area with pits makes this a rarely used procedure in modern investigations, but the concept of purposeful sections dug to look at soil changes alone is an ideal for which other approaches act as surrogate. In reality, mixtures of existing soil data (in some countries), incidental observations (road cuttings, river banks) and borehole survey are used when soil spatial characteristics are needed in environmental archaeology. Where existing data is available, it usually comes in the form of reports from consultancies concerned with development or agricultural issues, and occasionally as academic studies produced for geological and pedological purposes. These sources usually require interpretation and extrapolation due to their being produced for projects with different aims or in different parts of the landscape. Incidental observations at roadsides and construction sites are widely used in geology, sedimentology and pedology for data collection at a landscape scale. Availabililty is a key issue, particularly in areas of soft sediments where vertical sections tend not to survive. Though valuable, making sense of incidental observations often requires considerable experience on the part of the practitioner.

The difficulties of obtaining primary soil and sediment data are often overcome by using borehole surveys. Boreholes provide stratigraphic descriptions of which the detail is sacrificed in favour of increased scale. The information is never as clear as that taken from a soil pit or trench section, but on the other hand, borehole transects can stretch for hundreds of metres across the landscape or down to impracticable levels in very deep stratigraphy.

(a)

(b)

FIGURE 2.9 Different meanings of the word 'soil': (a) the common meaning used by, for example, gardeners, (b) meanings vary within the archaeological community, but the dark layer in this picture is commonly called a soil by field archaeologists, (c) a soil scientist uses the term 'soil' to describe the whole profile from the dark topsoil down to the parent material. (Photograph (b) by Mike Shaw).

FIGURE 2.10 Small scale mechanical auger system for soil and sediment study: (a) drilling in a window sampler, (b) levering it out.

A number of tools can be used to obtain borehole information, ranging in size from the ordinary hand auger through to vehicle- or derrick-mounted drilling systems. In general, hand augers are only really suitable for the top metre or so of the stratigraphy (except in very wet sediments), because of the difficulty of pulling them out from deeper holes. Powered systems are preferred because of the greater depth range and their capability of producing much larger intact cores. Vehicle- or derrick-mounted machines cost a great deal and usually give fairly coarse information. There are some modern exceptions to this rule, but the shell and auger systems widely used by geotechnical companies lose significant amounts of stratigraphy during normal operation. In general, however, the larger machines do have the advantage that they can be operated down to depths of many metres. Information from these systems is sometimes freely available as a result of preliminary soundings taken for geotechnical reasons on development projects. If so, this can provide an overview of the sediments but may lack the detail needed for the stratigraphic survey.

Smaller hand-held mechanical systems can be more practical for archaeological use, as they offer precisely positioned window samples (intact cores examined in the field through an open-sided corer) down to depths of a few metres (see Canti and Meddens, 1998) using equipment that can fit into an estate car or utility vehicle (Fig. 2.10). Window samplers are particularly valuable for preliminary scoping work on environmental remains in projects where the state of preservation is unknown. Although the stratigraphy may not be fully understood until full excavation, a series of samples can be taken at well-defined depths during an early phase of the work and estimates made of the likely future commitment for analysis.

2.3.3.3 Soil and sediment description

As for the context description criteria described above (see Section 2.2.3), the characterization of soils and sediments requires the use of descriptive systems with precise meanings rather than relying on generalized understandings between people, which are notoriously unreliable. The difficulty with this stricture is that the details inherited from the pure disciplines (pedology, sedimentology and geology) can often overtake the need for communication, leading to unwieldy or incomprehensible terminology. As a minimum, deposits need to have colour, texture and inclusions described according to one of the systems currently in use. Additional information such as soil structure (with terminology coming from pedology) or sedimentary structures (coming from sedimentology) may be needed in some cases, but the similarity of the basic recording systems means that there is already a good deal of interchangeability with context descriptions. Ultimately, it is probably more important that one of the systems has been used correctly, rather than worrying about which particular system it was.

The need for the defined meanings of commonly used words is nowhere clearer than in colour description, which contains all the typical elements of the larger problem. Although colours are widely perceived in a similar way, the detail of what shade they are, how dark or light they are etc., is variable both between observers and in different lighting conditions. To overcome this bias, the Munsell chart and its equivalents were created by defining colours according to exact chemical recipes that produce the same result whenever they are manufactured. The resultant materials are cut into squares and given names according to a system of definitions. Without this precision, the variation between observers would be highly significant. Even with a controlled system of words but no colour chart, there are problems of perceptual differences between people. Worse still would be the old-fashioned uncontrolled system where words like 'buff' were used, and meanings became so variable that comparisons with other studies are not possible.

2.3.4 Sampling techniques

Sampling is an attempt to represent reality with a few data points. It may be employed in the collection of finds for cultural analysis, environmental remains for laboratory study, or even existing records for the testing of dynamic trends. Whatever the data being sampled, the principles remain the same. The sample must provide an adequate cover of the range of types found, and must represent them in similar proportions to those of the whole data body.

The most obviously rigorous approach to this problem is the random sample. Here, in theory, the complete lack of bias must automatically lead to an accurate reflection of the whole. In fact, random sampling will only be securely unbiased if the dataset is relatively large, owing to the possibility that points will, by chance, not fall into certain important groups. This can be overcome by using the stratified random sample, which is based on the idea of randomness, but ensures representation by deliberately assigning a proportion of the points to predetermined groupings.

Replication is another critical area for scientific approaches to sampling and presents conceptual difficulties in many real life archaeological situations. Archaeology is not science, and at the big end of interpretation neither the site nor many of the sections and context relationships are likely to be in any sense replicable. As one moves down the interpretive tree, however, the opportunity for replication becomes more likely and the need more apparent. Within a particular context, for example, a feature may occur once in a micromorphological slide which just happened to be taken at that spot. Confidence can only be placed in a context-wide interpretation of that feature if other slides show it as well.

Once a sampling plan has been decided, sample collection is typically carried out using bags or buckets for loose sediment material or objects. This applies to situations where the positioning or orientation are insignificant information. Whole blocks of sediment or soil may be taken in such situations, if there is a need for detailed examination of fine stratigraphy, or where small interval sampling is more conveniently pursued indoors. The deliberate collection of intact blocks is also carried out where micromorphology is required. In this technique (see Section 2.4.3.3 below) a small slice of the soil or sediment is viewed intact microscopically, with stratigraphic relationships preserved by resin impregnation.

2.4 Interpreting stratigraphic sequences

2.4.1 Principles of stratigraphic analysis

Environmental archaeology is carried out through the analysis of both off site and on site sequences. Of course, in any given project, soils and sediments studied off site may well be influenced by the anthropogenic activity on site or in the landscape more generally. Although the division is thus arbitrary, it represents as much a convenient way to distinguish types of stratigraphic analysis as it does a distinction of formation processes.

Off site sequences will be analysed in a broadly facies-based system derived from sedimentology (e.g. Reading and Levell, 1996). This is not usually a detailed attempt to understand all of a sedimentary system, since the main focus is the provision of datable contexts with evidence from sedimentary characteristics and biological remains providing the interpretation of environmental change. The focus on usable sections can be detrimental if attention is not given to the details of stratigraphic build-up. Peat and alluvium both display non-sequential characteristics in some circumstances (see Section 2.3.1.2) and it is essential for the purposes of environmental analysis that clear sequences are isolated from sections with these types of problem. However, the relative simplicity of off site sequences means that a basic understanding of superposition, fundamental sedimentation processes, taphonomy and dating potential is often enough to allow useful material to be selected and sampled.

For on site work, the potential complexity of anthropogenic stratigraphy requires a systematized approach operating with considerable organizational and descriptive discipline. The analysis has to be hierarchical, dealing with information in an ordered and repeatable way with staged outcomes. In the first instance, interpretation of the records of the site sequence is based purely on stratigraphic criteria, taking no account of chronological or other information derived from other data types. This is not always possible on sites with little stratigraphic depth (for example, truncated rural sites; Clark, 1992), and here preliminary dating information derived from artefacts (spot dating) is often fed into the stratigraphic model to be subsequently tested against the results of detailed artefact studies.

During the analysis stage, contexts are grouped into higher level stratigraphic entities to which we can attribute interpretation and explanation. These groupings may themselves be amalgamated into larger entities, forming a hierarchy of stratigraphic groups with the individual context at its base. Hierarchical grouping is a key element of the single context system, and indeed many aspects of the excavation and recording procedure are predicated on its subsequent deployment.

The primary level of grouping, termed the 'set' (also, sometimes 'sub-group' or 'block'; Clark, 1993), brings together contexts with a high level of stratigraphic association. Thus a foundation cut and the masonry footings it contains might be grouped as a set, or a sequence of destruction deposits of similar composition, or a pit and its primary fill(s). At the set level, one is seeking a degree of grouping whose identification will remain robust when the results of analysis of other datasets are introduced, so it may only contain a single context. The sets may then be combined to form higher level 'groups' (though again different terms have been used). For example, a number of sets, variously interpreted as structural post holes, an eavesdrip gulley, a hearth and some storage pits, might (if the matrix allows) be brought together into a single group interpreted as a timber roundhouse. Groups may themselves be brought together into higher level interpretive entities and so on. The number of hierarchical levels and their nature is determined by the stratification and the research questions applied to it; in some circumstances the construction, use and demolition of a structure might be represented by three groups, while on another occasion these three elements might be subsumed into one.

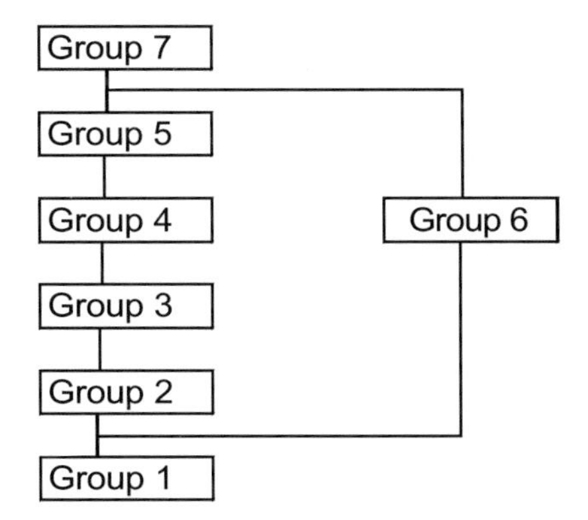

FIGURE 2.11 A sequence of groups set out in a Harris matrix. Group 6 is stratigraphically earlier than Group 7 and stratigraphically later than Group 1, but otherwise 'floats' between these two groups, potentially relating to any of groups 2–5, or none, either being later than Group 5 or earlier than Group 2.

By concentrating on stratigraphic criteria and temporarily ignoring the dating information available through artefact studies or absolute dating, the Hierarchical Grouping System is focused on relative chronology and interpretive (functional) aspects of the archaeology. Hierarchical grouping allows interpretive sequences to be described without reference to absolute chronology or cultural labels, and allows events to 'float' in relationship to each other. For example, in Fig. 2.11, a sequence of groups has been set out as a Harris matrix. From the bottom upwards, Group 1 represents a buried topsoil; this was overlain by the construction of a building (Group 2); within the structure a series of occupation deposits built up (Group 3), which were overlain by a sequence of destruction deposits (Group 4) relating to the demolition of the building. These were cut through by a series of refuse pits (Group 5), which were finally infilled and covered by an extensive gravel metalling (Group 7). Group 6, a human inhumation, can be seen to be later than the topsoil Group 1 and earlier than the metalled surface Group 7, but otherwise 'floats' between these two groups, potentially relating to any of Groups 2–5, or none, either being earlier than Group 2 or later than Group 5.

Stratigraphic analysis produces a narrative of site development based on the incontrovertible evidence of the sequence, forming the primary model to be tested, amended (if necessary) and enhanced by the study of other datasets. Thus, pottery and other artefacts, constructional materials, plant and animal remains, micromorphological slides, sediment samples and so on are studied within their own methodological frameworks, but the stratigraphic model provides meaningful provenance and allows resources to be targeted on problematic or important parts of the sequence.

Geoarchaeological studies will sometimes play a part in deciding the nature of the contextual relationships, particularly where an interplay of natural and human processes is at work in the stratigraphy. At Banwell Moor on the UK Somerset Levels, for example, periods of still-stand and soil formation occur as dark layers in alluvial stratigraphy. Soil micromorphology was used to determine the relationship between successive additions of alluvium which were visually indistinguishable from each other, and periods of soil formation and land use which left behind evidence of soil processes as well as additions of microscopic charcoal (Heathcote, 2000).

Inserting chronological data may also suggest new groupings and associations between stratification not immediately suggested or supported by stratigraphic criteria (though such groupings cannot, of course, contradict the site matrix). Similarly, analysis of different datasets can assess the validity of stratigraphic interpretation, and in most cases greatly enhance the understanding of past activities and processes represented by the site grouping.

Finally, the results of the different analyses may be brought together and amalgamated to form a consensus of interpretation and understanding of the archaeological sequence. This can (and should) be a dynamic process, as the results may be mutually supportive, contradictory, or suggestive of new or enhanced interpretations when considered holistically. The stratigraphic narrative may have to be revised, and it is one of the great strengths of the Hierarchical Grouping System (and ultimately the single context approach) that groupings can easily be broken down and rebuilt in different combinations; something impossible in excavation systems where field recording is selectively based on on site interpretation. (The suggestion made by a senior British archaeologist at a conference in Bradford in the early 1980s that the need to revise on site interpretations in the light of post-excavation analysis of material recovered from a site was a sign of archaeological incompetence may be dismissed as an example of both foolishness and hubris.)

2.4.2 The process of post-excavation analysis

2.4.2.1 Introduction

Practical guidance on the process of archaeological excavation has been produced by many distinguished authors over the years, particularly following World War II and during the archaeological boom in the UK during the 1970s (Wheeler, 1947, 1954; Kenyon, 1952; Webster, 1963; Barker, 1969, 1977; Hirst, 1976; Jefferies, 1977; Boddington, 1978). However, it was only from the late 1980s onwards that a number of authors began to produce detailed procedural post-excavation manuals, largely based on the single context system. The manuals were mostly unpublished documents for particular organizations (e.g. Williams, 1987; Clark, 1988b; Westman, 1990; Hammer, 2002), with only shorter papers appearing in the archaeological literature (e.g. Clark, 1993; Pearson and Williams, 1993; Shepherd, 1993; Hammer, 2000). Apart from these few documents, the actual process of post-excavation analysis has been surprisingly little discussed or debated among field archaeologists, despite its crucial importance.

2.4.2.2 Preparation

The compilation of a comprehensive concordance is a critical first step, cross-referencing all the site records, artefacts, ecofacts, samples etc. so that the stratigraphic provenance and associations between this material can be examined. Computers greatly facilitate this process: concordances can be prepared on a spreadsheet or (preferably) a relational database, context record sheets and photographs stored as digital images, plans and section digitized as CAD drawings, all of which can then be readily accessed by the post-excavation team. A number of bespoke archaeological database systems have been designed to do just this, such as the Integrated Archaeological Database (IADB), developed in Scotland during the mid 1980s (Rains, 1985; Stead, 1988).

2.4.2.3 The matrix

Ideally, the site matrix will have been compiled during the course of excavation. It must be checked against the plan and section record, recording sheets and master concordance, removing redundant relationships (Fig. 2.12) and potential ambiguities or errors (Fig. 2.13). This is because the matrix is a primary tool for the analysis and interpretation of stratigraphic sequences, particularly in deeply stratified situations, and it is vital that one is confident of its accuracy and consistency before proceeding. In particularly complex situations it may be appropriate to redraw the matrix to allow longest strand analysis (see Section 2.4.2.4), but on shallow sites where stratification is largely confined to cut features with little intercutting this is less necessary.

2.4.2.4 Longest strand analysis

In this procedure, the first step is to identify the longest strand (or 'primary route'; Pearson and Williams, 1993: 96) through the matrix. This is a route from bottom to top consisting of the maximum number of contexts having a stratigraphic relationship (Fig. 2.14), irrespective of their type, character or interpretation. As a first step, the longest strand is studied, interpreted and grouped at the set level. All subsidiary strands can then be integrated into the layout, resulting in a 'set matrix', showing the primary interpretive grouping of the stratification and the stratigraphic relationships between them. On particularly complex matrices, it may be appropriate to identify 'nodal points' (Clark, 1988b: 34), which helps subdivide the matrix into more manageable blocks and facilitate the integration and primary grouping of subsidiary

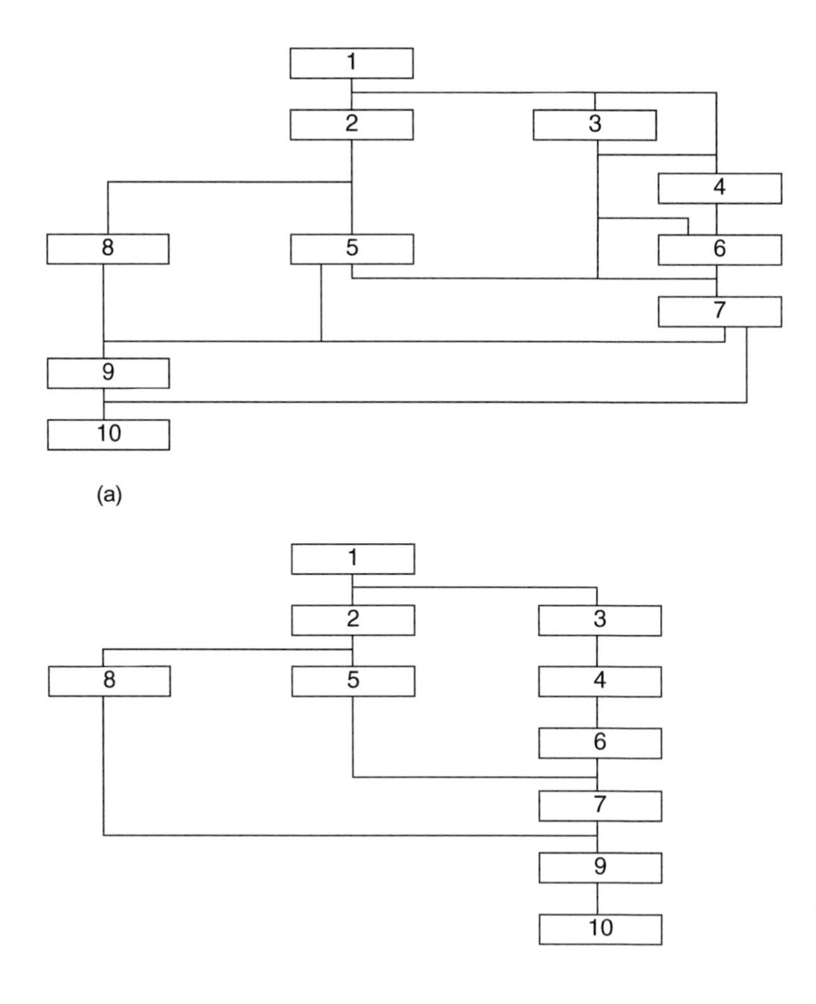

Figure 2.12 The redundant relationships portrayed in matrix (a) may be removed. The resultant matrix (b) shows the relative chronological relationships of the stratigraphic units with far greater clarity.

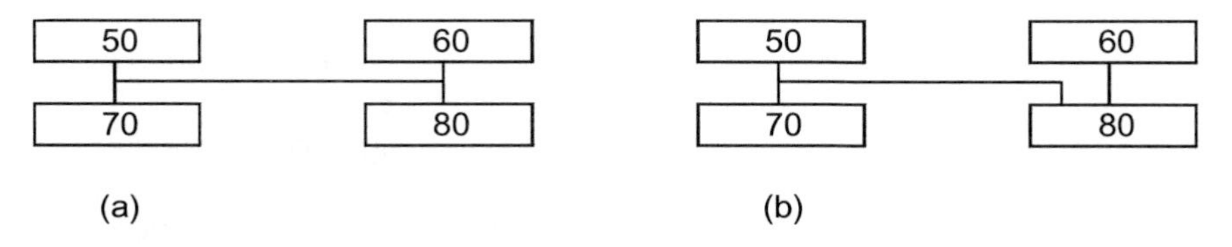

Figure 2.13 An 'H' relationship. Matrix (a) tells us that both context 50 and context 60 are stratigraphically later than contexts 70 and 80, and that they have no relationship with each other. Though this scenario is certainly possible, it is also rather rare stratigraphically. The appearance of H relationships on matrices is usually due to drawing errors, where separate lines showing relationships have been conflated, so that the correct set of relationships (as shown in matrix (b), where 50 overlies 70 and 80, but 60 overlies only 80) are misrepresented.

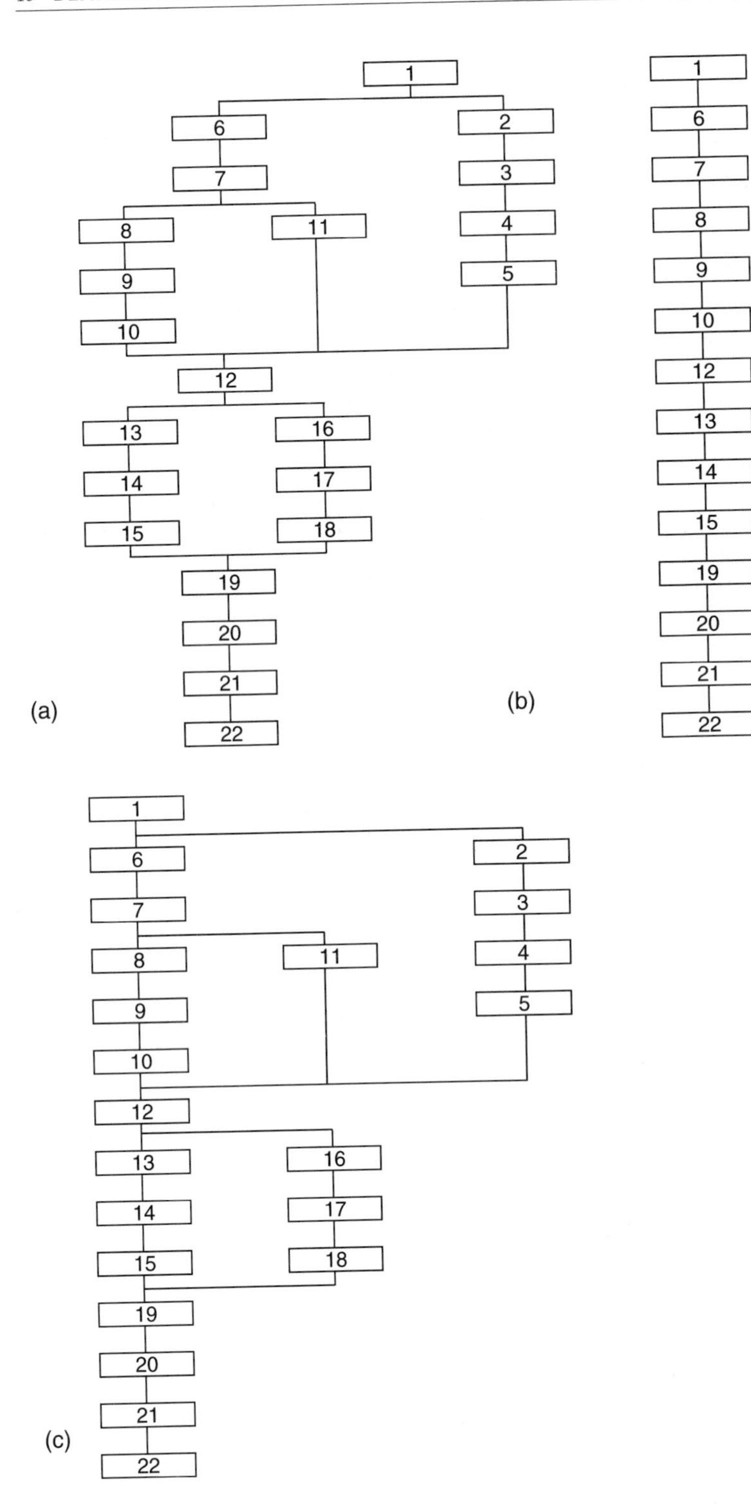

FIGURE 2.14 Longest strand analysis. From the original site matrix (a), the longest route through it can be isolated and drawn as a vertical unilinear sequence (b). Subsidiary strands can then be re-integrated (c).

strands. A nodal point is a single context which all of the stratification in a particular area either pre- or post-dates. In Fig. 2.15, context 200 forms a nodal point (but note also the use of 'limited nodal points'; Pearson and Williams, 1993: 98).

2.4.2.5 Interpretation and grouping

There are few stratigraphic attributes relating to anthropogenic strata that allow truly predictive interpretive statements to be made. Interpretation of individual contexts and their primary grouping therefore depends on the judgement of the analyst, assessing all of the available stratigraphic data. Similarities and contrasts in the composition of deposits, morphological attributes, the nature of inclusions, and so forth all help inform the grouping process. However, it is important to note that the results of analysing other datasets are not yet introduced, as they include interpretive factors outside the control of physical and stratigraphic data. Thus the stratigraphic analyst must decide which attributes recorded about a context are significant during analysis (a process already undertaken during primary recording) and deploy these in an argument justifying the association and interpretation of individual stratigraphic units. While this is a methodologically sound procedure, and avoids circular arguments (one (unpublished) example known to the authors was where pottery spot dating was employed to establish the stratigraphic sequence; the sequence was then used to analyse and confirm the dating of the pottery), it can limit the potential for interpretation at this stage. Contexts may be grouped together (say) on the basis of similarities of their mineralogical constituents, but not be open to satisfactory interpretation on stratigraphic grounds alone. This is to be expected in a methodologically rigorous post-excavation procedure; the integration of other datasets into the stratigraphic model may enhance, support or refute that model, or suggest totally new interpretations.

The initial grouping process, then, results in a set matrix, identifying base levels of association between contexts. The same process may then be repeated, identifying higher level stratigraphic entities (groups) often correlating contexts on multilinear stratigraphic sequences. The resulting group matrix can then be further analysed, and the process repeated according to the number of grouping hierarchy levels deemed appropriate.

2.4.3 Analysing and testing different datasets

2.4.3.1 Introduction

At this stage of post-excavation analysis, we may have samples taken from off site and on site stratigraphy containing biological and mineral information. Off site samples are already

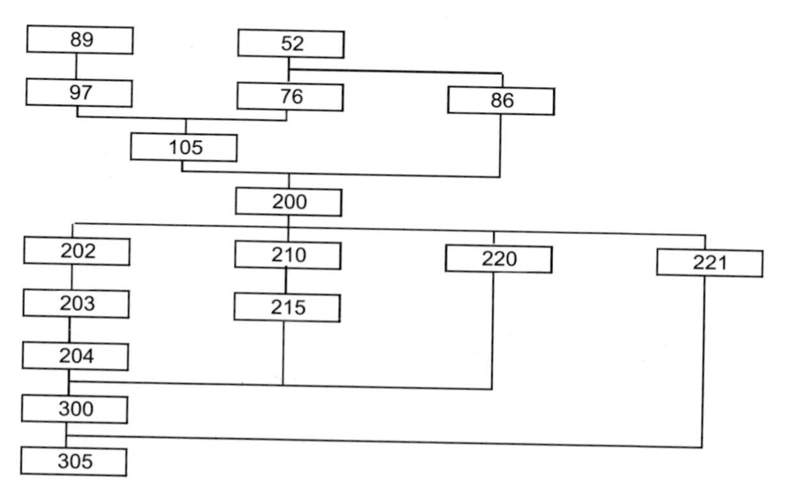

FIGURE 2.15 In this matrix, context 200 forms a 'nodal point'.

contextualized, having been taken from selected stratigraphy such as an alluvial plain around the site, a paleochannel, or a nearby peat deposit. Dating is probably still awaited on these deposits, but there is good reason to believe that they will cover similar periods to the preliminary evidence from the cultural stratigraphy. For the on site samples, we have an interpretive model based on physical and stratigraphic criteria, accompanied by matrices of the context, set and group sequences (together with any other hierarchical grouping levels deemed appropriate). This is the primary hypothesis that is ready to be tested and refined using other datasets retrieved during fieldwork. Analysis of both the environmental and cultural remains cannot sensibly be carried out without a proper understanding of their stratigraphic provenance and preferably an initial interpretive model to work with. Following the analysis of all the information derived from a stratigraphic sequence, the results are therefore brought together, compared, and the initial interpretive hypothesis tested against the new information.

2.4.3.2 Cultural datasets

At the primary stage of hypothesis testing, one focus is upon the cultural material derived from the stratigraphic units. This analysis requires the consideration of a range of attributes, described below, which can all help test and enhance the interpretations contained within the stratigraphic model.

Material constituents
At a basic level, artefacts may be classified by the material they are made of: organic matter (wood, bone, horn etc.) or inorganic matter (clay, stone, metal) – but note objects can be made of more than one material type, such as a metal knife with a bone handle. However, closer analysis can greatly enhance the opportunity for understanding the research potential of the object itself. Analysis of the clay used in creating a pottery vessel, for example, may indicate the geographical location of natural resources exploited. Study of the composition and structure of metal objects can offer clues to the technology employed in their manufacture and possible sources of raw materials. Post-depositional changes can also be studied by analysis of the constituent material of artefacts, for example by studying the compression of tree rays and rings in wooden objects or the mineralized remains of organic materials in metallic corrosion products.

Form
In the first instance, the form of an object is predicated by function – a cup must hold liquid, a knife must cut and a boat must float. Beyond this basic functionality, however, form can be dictated by a host of cultural imperatives influencing the shape and decoration of objects (in addition to the function of the object itself). This aspect of material culture offers an almost infinite variability in its expression, and provides a sensitive and critically important dataset in understanding past societies. Because of this sensitivity to cultural expression, material remains offer a rich insight into variability within and between social groups, as well as documenting change through time and space. For example, pottery used for different purposes may be of different shapes and have different decoration; this may vary between social groups and may change through individual innovation or in response to broad social change.

Decoration can give us insights into the world view and social practices of past societies, whether that be in the form of figurative decoration, such as the scenes of ceremony carved on the tomb slabs at Kivik in Sweden (Randsborg, 1993) or the abstract patterns on the stones of New Grange, Ireland (O'Kelly, 1995). In literate societies, inscriptions or graffiti on essentially temporary artefacts can offer a new perception of society apart from the often formalized and self-conscious records found elsewhere, for example the potter's stamps on Terra Sigillata found in the Roman Empire (Dragendorff, 1895, 1896) or the ostraca of Deir-el-Medina (e.g. Gasse, 1990) or Masada (Cotton and Geiger, 1989).

Chronology
The conjunction of material, form and stratigraphy can also offer important information about

chronology. In an idealized world, we might perceive a simple stratigraphic sequence where the lower strata produced pottery (say) decorated with zones of horizontal stripes (Type A), the middle strata pottery with vertical stripes (Type B) and the upper strata pottery decorated with circular stamps (Type C). On another stratigraphic sequence we may find that Type C derives from the earliest strata, with different types being present higher in the sequence. By combining many sequences, we may thus construct elaborate relative chronologies between such types of artefact, and chart changes through time and space. Ultimately a relative chronological relationship can be tied to calendric dates, either by cross-referencing to sequences with historical fixed points or through association with scientific (absolute) dating techniques such as dendrochronology.

The reality, of course, is not so simple. Human beings have the sometimes irritating habit of not changing their material culture synchronously: new fashions, new ideas and new forms of material expression may take place at different rates in different places or in different sections of a community; objects may have a perceived value because of their antiquity and can be curated for generations, demanding a more sophisticated assessment of the cultural assemblage derived from a stratigraphic sequence.

Furthermore, archaeologists need to consider the deposit formation processes that brought about the assemblage recovered from the archaeological record. There is no simple equation between the date of the creation of an archaeological deposit and the artefacts contained within it. People may live in one place for extended periods of time, and may dig into earlier occupation layers, bringing to the surface objects created by previous generations and incorporating them into contemporary deposits. In this way a deposit may contain a cultural assemblage primarily consisting of material dating decades or even centuries earlier. The recognition of such residual material is of crucial importance in understanding the chronological relevance of material culture derived from archaeological sequences (Carver, 1983, 1985).

Use

Use, as with function, can show complexities over and above a simple deterministic model. Naturally, some objects can be used for more than one purpose, but in addition objects may be used in different ways at different stages in their life history (Kopytoff, 1986). A Roman amphora might be created to serve as a container for garum (fish sauce) but subsequently re-used to hold a human burial (Stead and Rigby, 1989: 115–116; Curtis, 1991; Martin-Kilcher, 1994: 469–474). This multiplicity of use extends beyond utilitarian function, in that objects may be perceived in very different ways at different stages of their existence. A quernstone may initially serve as a purely utilitarian object for grinding cereal grains (even this role perhaps having symbolic connotations), but later be transformed into a ritual propitiation by its inclusion into a symbolic 'placed deposit' (e.g. at Flag Fen; Buckley and Ingle, 2001). Thus our study of material culture requires that we go beyond the quantifiable attributes of composition and form, and take into account the social perception of the object itself.

2.4.3.3 Geoarchaeological datasets

The importance of hypothesis testing has been stressed above, and geoarchaeological methods will often play an integral role in that testing, utilizing laboratory techniques to enhance the stratigraphic model. Commonly used approaches are as follows:

Particle size analysis

Sediments and soils are made up of mineral particles ranging in size from rocks down to the finest clays. The percentages of these particles present in a sample can be tested in the laboratory using analytical sieves for the coarse material, and various forms of sedimentation test for the fines. The result is a curve showing the complete range of particles and their concentrations (Fig. 2.16). In many cases, the curve can be a fingerprint for either the geological input (e.g. sorted sand from a sandstone) or various processes that the soil or sediment has undergone (e.g. clay translocation from a topsoil horizon into subsoil horizons). Particle size

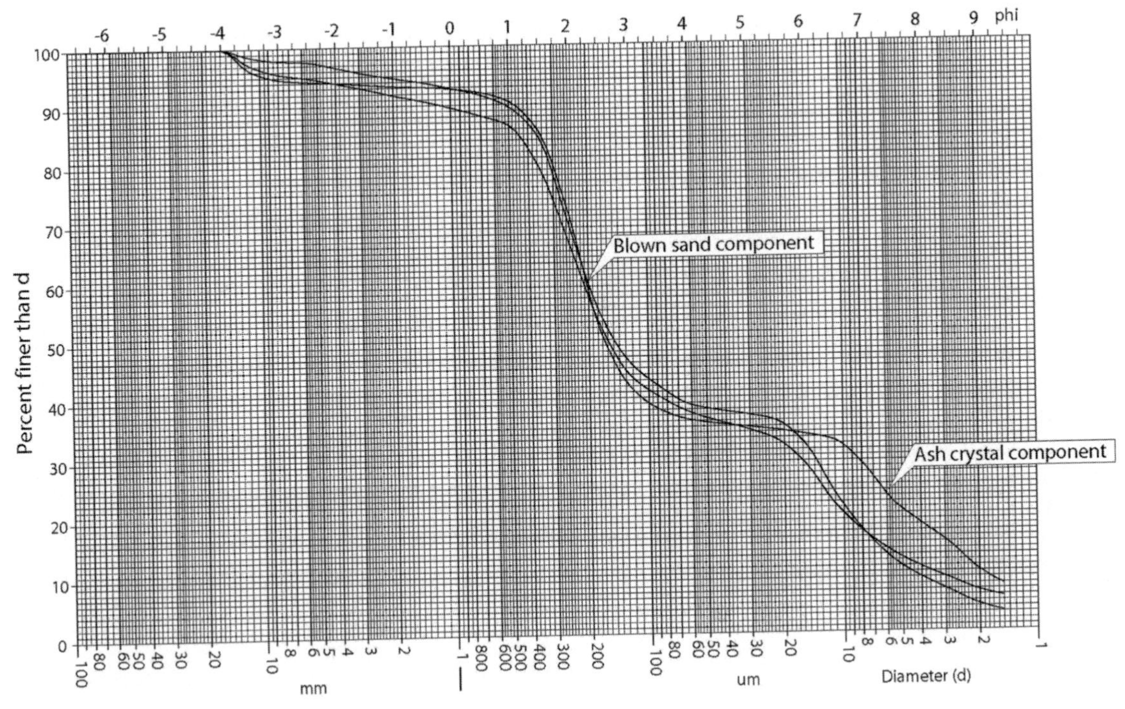

FIGURE 2.16 Particle size analysis of ash crystals (typically 5–20 μm) mixed with windblown sand (in this case 100–300 μm). The data is presented as a cumulative percentage curve, which can be read simply as large amounts of material where the curve is steep, and small amounts of material where the curve is flat. (from Canti, 2003a)

analysis can therefore be valuable for testing site formation hypotheses involving either sediment sourcing or alteration (Canti, 1991, 2003a).

Mineralogy

Soils and sediments contain a number of different minerals, often with varied origins. Primarily the mineral content is made up of grains inherited from the geological source. This will consist mostly of common quartz, but a small percentage of other minerals are usually present as well. These tend to be the more resistant types that have survived many geological cycles and therefore have little interpretive significance. However, in some cases the additional mineral content can be a sensitive indicator of origin, particularly where more than one possible source is available and results from particle size analysis are equivocal.

Minerals can also occur spontaneously in stratigraphy as a result of human or other biological processes. Calcium carbonate, in particular, and calcium phosphates to a lesser extent, precipitate in some soils associated with bacterial activity and high nutrient conditions, sometimes replacing organic ecofacts entirely (McCobb *et al.*, 2003). Iron compounds, particularly the phosphate vivianite (Gebhardt and Langohr, 1999) and the sulphide pyrite, are common neoformed minerals in waterlogged conditions. Variations in colour occur as a result of these mineral transformations, leading to significant differences in context delineation (Canti, 2003b).

Micromorphology

Micromorphology is a term used for a number of microscopic approaches to soil and sediment study, but refers chiefly to thin section analysis. This is carried out by removing intact blocks of stratigraphy, impregnating them with resin, then producing an extremely fine section which can be examined in polarized light. The method offers the potential for instant visual identification of minerals and inclusions, as well as detailed examination of context boundaries and recognition of typical fabrics. It is therefore well suited to the elucidation of anthropogenic effects on soil sequences (French and Whitelaw, 1999), explaining anomalous deposits (Carter, 2001), and detailing inclusions or residues from cultural layers (Simpson *et al.*, 2003).

Using micromorphology opens up a new dimension for the stratigrapher, with many different concepts and forms of language. It is important therefore to keep a clear understanding of the questions that are being asked, and a firm mental grasp on the microscopic scale at which it is carried out (see Section 2.3.4).

Magnetic susceptibility

Many soils and sediments can be magnetized by placing them in a magnetic field. The degree of response is called the magnetic susceptibility. It is a characteristic most strongly affected by burning the soil or the addition of heated materials, and to a lesser extent by normal soil development processes, some forms of bacterial activity, wetting and drying etc. (see Thompson and Oldfield, 1986; Dearing, 1999). This indication of generic alteration makes magnetic susceptibility a valuable prospection tool, but it can also play a part in locating soil development and human activity in stratigraphic sequences. At a site scale magnetic susceptibility is regularly used as part of a multi-proxy approach to the interpretation of activity areas (e.g. Smith *et al.*, 2001).

pH

The pH of a soil or sediment is a chemical shorthand for its hydrogen ion concentration which translates to acidity or alkalinity in everyday usage. It is one of the many stratigraphic characteristics that may be needed as part of the unravelling of site formation processes, but is often only required at a general level which can be estimated by an experienced observer from the geology, climate and vegetation. When a genuine measured value is called for, it must be carried out using a calibrated meter (e.g. Gale and Hoare, 1991) and the results considered in the knowledge of the environmental processes which produce them. It is no use, for example, allowing waterlogged deposits to oxidize in storage and then expecting their pH to reflect processes in the stratigraphy.

Organic matter

All sediments usually contain some organic matter, but significant amounts tend to occur only where plant remains have accumulated. In practical terms this means waste deposits or soil development (including peat growth). Measurement can take place in a relatively simple (if rather crude) way, by burning off the organic matter and recording the attendant weight loss. Various temperatures and burn times are favoured by different authors (e.g. Keeling, 1962; Davies, 1974), but the significant point is that above around 500°C, samples rich in calcium carbonate start to lose weight from CO_2 escape. For this reason the lower temperature burns are generally preferred.

Organic matter content is usually measured as part of a series of tests associated with understanding stratigraphic accretion. Stillstand events, if given enough time, become soil horizons, and the humus content and magnetic susceptibility will generally be higher in these layers than in stratigraphy above or below. The humus content may or may not be visible in sections, and the testing comes into its own where visual distinctions are unclear.

Chemical characteristics

A number of chemical characteristics can be important for understanding site formation processes.

Phosphorus in various forms is strongly adsorbed by many soil materials, so the phosphate content reflects a summary of inputs to the stratigraphy over time. Since nearly all biological wastes are relatively rich in phosphorus, the measurement acts as a broad proxy for human and animal activities in most cases. Phosphorus can be analysed as a method of areal survey or as part of the multivariate examination of stratigraphy (Crowther, 1997; Parnell *et al.*, 2002; Terry *et al.*, 2004).

Calcium carbonate is the common alkaline component of limestone and can derive from many other rock weathering situations, biological products, wood ash, mortars and numerous anthropogenic materials. Its concentration or depletion is a major control over stratigraphic pH, making it a lynchpin of preservation conditions. Calcium carbonate is slowly leached out of northern latitude soils and sediments by rainfall. In these situations, therefore, it is measured to assist in the understanding of soil development and stillstand phases in stratigraphic accretion.

A range of other elements have been successfully measured in soils and sediments as indicators of agricultural (Entwistle and Abrahams, 1998) and industrial processes (Parnell *et al.*, 2002), or use of space (Middleton and Price, 1996). Many human activities should, in theory, concentrate particular elements, and in some cases they will tend to remain in the stratigraphy due to adsorption onto clays or organic matter. Our understanding of both the concentration and the preservation processes is still limited, hampering interpretations of multi-element analytical results.

2.4.3.4 Integration and hypothesis testing

The individual approaches described above, and the datasets they produce all have particular strengths and weaknesses. Some have long traditions of use in certain areas and certain types of deposit; others are relatively new and are still developing methodologies and conceptual approaches. Whole conferences are held to study one approach, with detailed dissection and questioning of their value and efficacy. Can a single thread be run through them? How best can we elicit the interpretive returns they offer?

For the scientific approaches, the key to maximizing the benefit from analytical methods lies in hypothesis testing. Looking back over the last three decades, it has been a particular problem for geoarchaeology that the application of scientific methods has often been carried out without a clear testing framework. Numerous reports contain particle size or pH data on deposits, and sometimes micromorphological reports on buried soils, without a clear purpose or aim for the analysis. These sorts of laboratory techniques cannot regularly deliver worthwhile results outside such a framework, any more than randomly digging trenches on the earth's surface would deliver high quality archaeological sites.

Analysis of material remains has to consider many variables to realize its research potential. Constituents, form and decoration, chronology and social perception all contribute to developing an understanding of such material. Hypotheses can be tested within each specialist area, for example stylistic changes over time, or the use of different source materials in different areas. These approaches will impact on the site stratigraphy only at the early stages of analysis, enabling enhancement of the interpretive model at a site level. Further down the analytical road, they tend to become more isolated as specialist areas with the site informing the specialist subject rather than the subject informing the site.

The example presented in Section 2.5.1 below typifies the methodology and benefits of an integrated approach. The project's overall aims are strongly based on a hypothesis-testing framework; the excavation strategy has the clarity and flexibility to allow varying levels of incorporation of specialist data; the specialist data itself is thoroughly argued and tested; analyses are purposeful rather than routine, and results are presented from different angles. The final discussion actively involves all the information gathered, from documentary sources to micromorphological slides.

The full potential of any project is only realized by this kind of total assemblage analysis, encompassing material and environmental data, both at the level of the individual context and the various interpretive groupings proposed by the stratigraphic model. Ultimately, it is in the study of the combination of stratigraphic attributes, material culture and environmental evidence from a particular archaeological setting that the most interesting and satisfying interpretive narrative is found. The path of analytical deconstruction, independent analysis of different data types and a heterogeneous methodological and theoretical approach only truly fulfils its promise when brought together to consider all the results in a holistic fashion.

2.5 Examples

2.5.1 Microscale – interpreting the context: Papa Westray

The excavation at St Boniface Church, Papa Westray, Orkney represents a situation where site formation processes and the archaeological interpretation were inextricably linked. The central focus of the work (Lowe, 1998) was the analysis of a 'farm mound' to the north of the church, which comprised layers of dumped materials, and blown sand with an intervening buried soil. The mound overlay earlier sediments associated with extensive Iron Age activity mainly to the west of the church. (See Fig. 2.17.)

Prior to the excavations at St Boniface Church, interpretations of farm mounds had traditionally centred on two possible reasons for their accumulation. The first of these, the stockfish model, viewed the mounds as accumulations of farm waste (especially dung) that was no longer needed as fertilizer on the fields because of the economic effects of the stockfish trade in Europe. This trade in dried fish was of considerable significance in Medieval Europe, and allowed producers to buy grain rather than grow it, supposedly leading to the unwanted fertilizers accumulating in the farmyard. The second interpretation, the structure model, proposed that farm mounds accumulated as a result of the decay and replacement of turf buildings, along with minor additions of midden materials, manure and windblown sand.

FIGURE 2.17 The farm mound at St Boniface Church, Papa Westray, Orkney from the north. (Lowe, 1998)

For the farm mound itself, minimal cleaning up of the collapsing cliff face was carried out to provide a section. Single context recording was used, and stratigraphic groups or 'blocks' were identified, and then tested for their integrity as excavation progressed. The deposits were intensively sampled at four 0.5 m wide × 0.5 m deep columns through the whole mound (up to 3 m deep), selected to maximize the deposit variability. The Iron Age deposits, which naturally formed an eroded turf-covered slope, were dealt with by a mixture of vertical and horizontal excavation which produced the equivalent of a three-dimensional stratigraphic record (Lowe, 1998).

Two geoarchaeological methods (particle size analysis and micromorphology – see Section 2.4.3) helped to elucidate the detail of stratigraphic formation at Papa Westray. They were used together to compare the underlying till (boulder clay) with the soils developed over it in Block 108, 118 and 109. This showed clear evidence for the inclusion of decalcified blown sand. The sand had accumulated as a layer at least 0.5 m deep on the original soil, before being decalcified, eroded and finally incorporated.

The fabric of the 'farm mound' itself was noticeably layered at a field scale, corresponding to context boundaries, but also additionally at the microscopic scale, where bands of amorphous organic matter were interspersed with coarse mineral layers. Both carbonized and uncarbonized bands were present along with biogenic silica and mineral grains, consistent with ash being a major constituent of the mound. Whole carbonized peat fragments could also be found, indicating a possible source of the fuel. Since the mound materials were mostly mineral in nature, this interpretation would have to be based on the assumption that the organic part was largely lost during the burning process. Particle size analysis (carried out optically) of the mineral grains in the peat fragments showed that the farm mound sediments could be produced simply by removing the organic fraction. Additional evidence for burning as a major site formation process was provided by two other features of the slides. Firstly, much fine red material was apparent when the thin sections were viewed in oblique incident light. Secondly, ash melts were occasionally found composed of fused quartz grains set in a glassy matrix. These are characteristic features of higher temperature fires, and develop where small amounts of glass form due to the silica component melting in the presence of alkalines from the ash (Carter, 1998a).

The conclusion was that the size and prominence of the farm mound were directly attributable to the high ash content found in many of the available fuels. The pre-mound sediments, interpreted on site as some form of specialized midden formation, were directly shown by micromorphological analysis to be formed from ash dumps with intervening periods of soil formation.

Experimental work was carried out to help narrow down the exact origin of the ash layers. The carbonized and uncarbonized peat fragments that suggested the probable fuel type contained too much mineral material to be derived from a true blanket peat. Rather, they were likely to be from organic-rich sediments or soils, which served as lower grade fuels. Samples of combustible materials were collected from various parts of Orkney and burnt to provide data for their loss on ignition, ash percentage and ash bulk density. These materials included peats, turves and peaty alluvium, each of which has significantly different mineral contents depending on their origin. The results produced an interpretation of more than one source for the ash layers. In the Iron Age sediments underlying the mound they appeared to have derived from a terrestrial peat containing little mineral matter. The farm mound materials on the other hand, appear to have derived largely from an alluvial or lacustrine environment, where the best match of mineral material was found – particularly in the abundance of biogenic silica (Carter, 1998b).

Bringing all of the evidence together, including pollen and zooarchaeological results, the final interpretation of the mound was that it represented the remains of an early fish processing station, utilizing low grade fuel for activities such as the boiling of fish livers and production of oil. It had been built on an area of previously cultivated land, and had flourished for a short span of time, perhaps entirely within the twelfth century. The reason for its demise was not determined, but local conditions mean that catastrophic erosion of the landing site was a possibility.

2.5.2 Mesoscale – interpreting the feature: Neolithic causewayed enclosure, Ramsgate

Until recently, the only positive indicators of a Neolithic presence in Kent were represented by two groups of communal monuments: the earthen long barrows of the Stour valley and the megalithic monuments of the Medway valley (Clarke, 1982). However, the absence in Kent of causewayed enclosures, a common type site of the Neolithic in Western Europe, has long perplexed archaeologists. This has recently changed with the discovery of the first confirmed causewayed enclosure at Chalk Hill on the Isle of Thanet in east Kent, which has transformed our perceptions and understanding of the Neolithic in Kent.

A causewayed ditch was noted on an air photograph on the chalk cliffs above Pegwell Bay, near Ramsgate in 1989. Subsequent replotting in 1996 suggested the possibility of a causewayed enclosure typical of the interrupted ditch circuits that characterize these types of enclosures. However, a new road to Ramsgate Harbour was planned to pass through the site, and an excavation was therefore carried out in 1997–98 prior to its destruction. The site was stripped of topsoil and the archaeological features excavated by hand, with hand collection of artefacts and animal bone supplemented by bulk sampling of selected deposits. The complexity of the intercutting ditch segments making up the causewayed enclosure meant that the field team needed to maintain tight stratigraphic control of the excavation process, using techniques more commonly employed in complex urban situations (CAT, 2001). In tandem with this approach, environmental archaeologists intensively sampled the stratigraphic sequence, collecting numerous column samples, soil monoliths and samples for palaeoenvironmental analysis. The sampling techniques used in the field included particle size analysis, soil micromorphology, loss on ignition, calcium carbonate analysis, magnetic susceptibility and optically stimulated luminescence. The integration of these two approaches allowed the articulation of a sequence of events encompassing both natural erosion and deposition events and anthropic activity on the site.

The earliest feature encountered on the site was a steep sided gully, about 4 m deep, formed due to the action of water channelling and incising its way through the chalk landscape. Its lower 2.5 m was infilled with a succession of water-lain chalk and flint gravels, calcareous sands and silts, probably of pre-Devensian age, a date supported by the discovery of a struck flint flake at the base of the gulley, suggesting an origin no later than the Upper Palaeolithic.

An isolated group of post holes and stakeholes were sealed by an extensive deposit of silt, up to 1.5 m thick in places, which also filled the upper part of the early gulley. This was probably formed by the prevailing westerly wind moving fine material from the exposed loess at Pegwell Bay and depositing it upslope, where it was reworked by wind and surface water in a series of episodic events. It seems likely that the upper part of this deposit developed into a soil, which was subsequently removed by an episode of erosion or truncation prior to the establishment of the Neolithic causewayed enclosure. This enclosure measured about 150–170 m in diameter and consisted of three roughly concentric circuits of interrupted ditches, all of which underwent repeated recutting, particularly the outer circuit, which was the deepest and widest. However, the fills of the three circuits had significant differences. Following some initial natural erosion of the ditch sides, the fills of the inner circuit were characterized by cremated bone, often associated with pottery. The middle circuit contained mainly flints, noticeably scrapers and arrow-heads found at the butt-ends of the ditches. The outer circuit had evidence of possible feasting – animal bones and seashells in association with pottery sherds, and in one ditch segment parts of two human skulls were found, one placed in a re-cut above the other, with a third nearby suggesting repeated ritual deposition at specific locations within the ditch circuit. Also found in the outer circuit were a large number of 'placed' deposits, containing huge numbers of flint flakes, often in association with discrete piles of pottery or animal bone. (See Fig. 2.18.)

Cutting through the infilled ditches of the causewayed enclosure was a pair of parallel causewayed ditches, the fills of which also contained Neolithic pottery and flint assemblages. These were cut in turn by two parallel linear ditches, thought to represent a cursus. The alignment of the ditches heads off towards a barrow group and putative henge monument modified into a barrow a short distance away. A close parallel to this was discovered at Etton in Cambridgeshire, where the abandoned enclosure was cut through by a pair of cursus ditches that linked a henge monument some distance away (Pryor, 1998: 373). The fills of these ditches mostly contained Early Neolithic pottery, with a small amount of later prehistoric pottery also being found. Tempting as it is to see these ditches as a cursus, their interpretation must for now remain speculative.

Occupation and land use of the area continued in later prehistory. An Early Bronze Age barrow was built about 60 m to the south of the causewayed enclosure, containing a central burial accompanied by a fragmented pottery vessel and a shale object. During the later stages of the use of the causewayed camp, and continuing after its disuse, sealing the cursus and Early Bronze Age barrow was a further natural deposit of sandy silt, again primarily of windblown origin, clearly affected by soil forming processes. A later Bronze Age enclosure was built over this, which was in its turn overlain by a Late Bronze Age/Early Iron Age field system.

The evidence thus reveals a sequence of natural processes and secular and non-secular patterns of human activity throughout the Neolithic period and Bronze ages, and forms a small part of what is clearly a much larger multi-period prehistoric landscape. While research continues into the site, the adoption of an integrated stratigraphic and palaeoenvironmental approach has greatly enhanced the appreciation of the sequence of events and allowed an understanding of the site's development in the broader landscape over time.

FIGURE 2.18 A potential 'ritual' deposit in the outer circuit of the Neolithic causewayed enclosure at Ramsgate. (Canterbury Archaeological Trust.)

2.5.3 Macroscale – environmental change across a region: the alluviation of lowland Britain

Many of Britain's lowland river valleys contain large amounts of fine-grained alluvium overlying coarser gravels. A typical example (Fig. 2.19) of this sequence is at Yarnton on the flood plain of the river Thames in Oxfordshire (Hey, 2001). Here, the lowest lying flood plain area commonly supports 3 m thickness of silty clay and clay textured alluvium on top of limestone and chert gravels derived from the Jurassic bedrock to the west.

The basic stratigraphic pattern is widely ascribed to a periglacial phase of high-energy sedimentation corresponding to the glacial and late-glacial periods before about 12,000 BP. This was then followed by a transition to quieter Holocene conditions characterized by meandering rivers and episodic overbank flooding depositing silts and clays. While there can be no doubt about the very general climatic causes of this sequence, it wasn't until the 1970s that discoveries of archaeological material under the fine alluvium led slowly to the realization that UK lowland flood plains could represent a major source of Holocene environmental information. Much of this possible value depends, however, on an assessment of the relative human impact on the nature and extent of the upper fine materials.

Shotton (1978) reported a sequence in the Avon and Severn valleys, consisting of gravel layers overlain by various clayey and silty sediments representing overbank sedimentation and the remains of cutoffs. The variability of these layers was in clear contrast to the uppermost fine reddish clay or silty clay found everywhere at the surface. These latter deposits were taken to be eroded soils formed of the characteristic pinkish Permo-Triassic mudstones of the region. The dating evidence suggested the Bronze Age for the onset of this erosion.

Robinson and Lambrick (1984) proposed an alluvial history for the mid-Holocene onwards in the Upper Thames. It consisted of a Late Bronze Age and Iron Age rise in water table being essentially due to climatic change and not yielding significant quantities of alluvium, followed by a large increase in flooding and alluviation due to soil erosion in the Late Iron Age and Roman period. Although flooding continued thereafter through into post-Medieval times, it was only believed to have produced further substantial alluvium during the late Saxon to early Medieval periods.

Brown and Keough (1992) argued that there was a significant change in Midlands flood plains and river channels in the mid to late Holocene (4500 to 2500 B.P). This led to accelerated vertical accretion, a fall in channel width/depth ratios, and a reduction in flood plain relative relief, ultimately producing the profiles that are typically seen today in development along the rivers Soar and Nene (Fig. 2.20).

Echoing Robinson and Lambrick (1984), they indicated that there was a period of hydrological change before this major metamorphosis occurred. However, they placed the greatest emphasis on a large increase in sediment supply leading to a disequilibrium between channel bed and flood plain aggradation rates which produced a period of relative incision.

FIGURE 2.19 A thick section of alluvium from Yarnton on the flood plain of the river Thames in Oxfordshire.

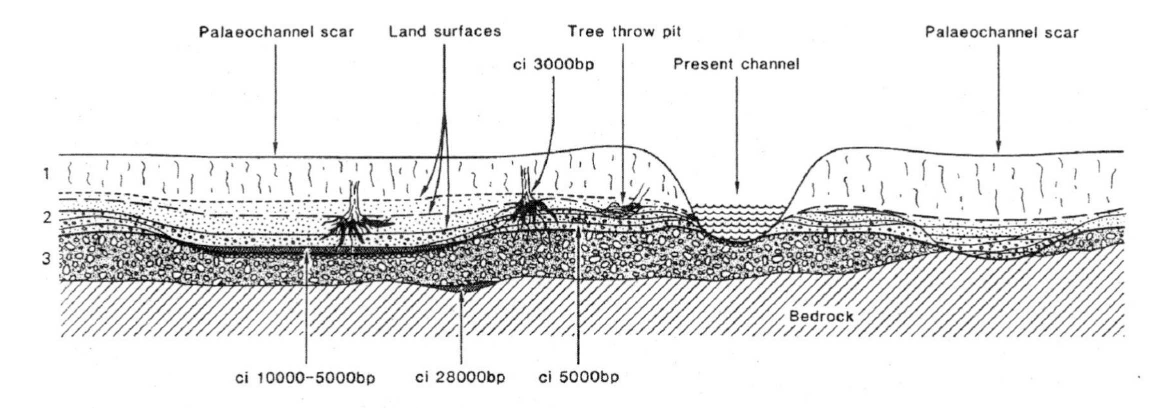

FIGURE 2.20 Stratigraphic model of the Soar and Nene flood plains. (From Brown and Keough, 1992)

The heightened sediment supply was ascribed to anthropogenic changes, particularly soil erosion attendant on cultivation.

Numerous papers have now been published which show similar trends, although sometimes from wider time periods. Needham (1992), for example, found evidence for both Neolithic and Bronze Age sedimentation at Runnymede Bridge in Surrey; Godwin and Vishnu-Mittre (1975) found Bronze Age clay layers in raised bogs in Huntingdonshire; and Keevill (1992) found Neolithic and Bronze Age archaeology sealed by alluvium in the Nene valley. There is evidence that similar events were occurring in continental Europe as well. Klimek (2002, 2003), for example, found periods of alluviation associated with both Bronze Age forest clearance around 1600–1300 BC, and also with early Medieval settlement in the eastern Sudety mountains of Poland.

The overall pattern of significant alluvial build-up led Brown and Barber (1985) to suggest that the expansion of 3000–2000 BP and the associated alluvium could be compared to the effects of the nineteenth century European colonization of the US midwest. This time period can be characterized by very large spikes in sediment delivery, as much as thirty times the normal rates in one Michigan lake (Davis, 1976). When trees are removed, the water reaching the ground surface increases by up to 50 per cent, and can stimulate an increase of up to 2000 fold in the sediment entering rivers (Leyton *et al.*, 1967; Limbrey, 1978; Robinson, 1981).

There are, however, complications within this basic approach. Burrin and Scaife (1988) found a greater degree of variation in the valley fills they studied in Sussex. Layers of peat, and complex interbedded sands, marls and tufa did not match the simplicity of the straightforward anthropogenic view of valley fills. Furthermore, their sites often showed significant differences of stratigraphy between one location and another in the same valley. The importance of this complexity is now grasped by most authors, and it is widely understood that anthropogenic effects have to be disentangled from the individual site-based evidence that may appear to fit the basic erosion model. Parker and Robinson (2003), for example, found significant increases of sedimentation after 5200 BP at Dorney on the river Thames, but cautioned that much of this was in fact calcium carbonate that had entered the system in solution.

Brown (2003) emphasized the importance of recognizing biogeomorphological feedback in any understanding of alluvial deposition and its relation to human activity. This could lead to indirect human impact on flood plains before any deforestation had occurred. Combined with the variations of scale at which sedimentary and palaeoecological changes occur, the identification of cause and effect in the relationship between flood plains and humans remains a considerable challenge for palaeoecologists and geoarchaeologists.

2.5.4 Megascale – volcanic eruptions, humans and the global environment

Active volcanoes are widespread throughout the world and the outputs of many of these volcanic centres can be traced over the length of the geological record. In many areas, there has therefore been a long, complex history of eruptions and deposition of volcanic sediments that have potential for geoarchaeological and environmental studies. Here we look at volcanic sedimentary units and the impact of associated eruptions on humans and their environment.

The mechanisms that form volcanic deposits are numerous, including air-fall, landslides, lava and pyroclastic flows, which in areas proximal to the eruption can form massive deposits (sometimes of the order of tens of metres thick). Considerable research has been carried out investigating the climatic and social repercussions of historical volcanic eruptions (Zielinksi *et al.*, 1995; Sadler and Grattan, 1999), partially because the ejecta can have a significant effect on humans and the environment at a wide range of distances from the source. Numerous mechanisms have been proposed for how eruptions can drive global climate in the short, medium and long term. Although inherently dangerous during eruption, the deposits generated by volcanoes can have considerable benefits to human populations, generating nutrient-rich agricultural soils (Cronin *et al.*, 1998; de Boer and Sanders, 2002).

Typically, lava flows in themselves do not form a major hazard to human populations. Although often large in area, they are also usually slow moving (of the order of tens of metres an hour), though the environmental impacts can often be felt for some considerable time following the eruption. For instance, in Iceland the Laki eruption (also known as the Skaftár Fires) of 1783–84 AD (Thordarson and Self, 1993) generated large lava flows from a succession of fissures associated with the Grimsvötn system that devastated what was already a poor agricultural colony of Denmark (Scarth, 1999). The eruptions did not kill anyone directly, though the devastating effects of 8 million tonnes of fluorine and 150 million tonnes of sulphuric acid led to the destruction of almost half the livestock in Iceland and subsequently resulted in the deaths of 10,500 people through starvation. The blue fog generated by the eruption was noted over large areas of Europe and the Middle East, and appears to have significantly impacted on some North American communities, such as the Inuit of Alaska (Jacoby *et al.*, 1999), though the environmental effects of the eruption have recently been contested (Zielinski *et al.*, 1995). The fissures and lava flows are still clearly seen but the eruption itself only destroyed 21 farming hamlets (of which only three were subsequently abandoned).

Larger magnitude eruptions can form sedimentary deposits over significant areas, many of which are associated with archaeological contexts. One of the best examples is the eruption of Santorini in the Cycladean Archipelago of the southern Aegean Ocean sometime around 1650 BC. On Santorini, exposures throughout the island allow three easily identifiable units to be described for this eruption (Friedrich, 2000). The lowermost pumice unit, typical of air-fall deposits has been identified as far afield as the Nile Delta (Stanley and Sheng, 1986). The middle unit is generally not very thick and largely consists of ash, representing a base surge. At this stage, the eruptive mechanism went through internal changes and the interaction with sea water led to a violent (phreatomagmatic) phase. The magma was torn into small particles that were surrounded by a thin layer of expanding steam. Clouds of ash suspended in steam spread outwards from the eruption centre in explosive rings. Closer investigation of this sedimentary unit shows that in the transition between the different types of eruption both fallout and base surges alternated before the latter dominated. The third and final phase is recognizable by the apparent homogeneity and the presence of dark fragments that are remnant blocks from earlier volcanic activity. In this final unit, the column did not reach the height of the first phase and the sediments were deposited as a pyroclastic flow, eroding some of the earlier layers.

Within these deposits, numerous archaeological finds have been made associated with the Minoan culture. In the southern part of Santorini, a major Minoan settlement called Akrotiri is being excavated out from under the volcanic deposits (Cioli *et al.*, 2000), only 8 km from the vent source. The preservation of Akrotiri is remarkable, especially considering this event happened over 2000 years before the eruption of Vesuvius in AD 79 that buried Pompeii and Herculaneum. Numerous frescoes have been identified in the site, with many two- and three-storey buildings being found at Akrotiri. No bodies have been discovered, almost certainly because there was significant earthquake activity prior to the eruption which led to the wholesale evacuation of the town. In one building, there was sufficient time to remove a set of three beds from the ruins of the earthquake and place them one on top of the other. Even in the first phase of the eruption, pumice covered the town, in places reaching depths of 2 m (Cioli *et al.*, 2000). As the eruption progressed, blocks of lava were flung far from the vent; some of these were sufficiently large to crush buildings and walls. Further afield, tsunami deposits including volcanic ash and pumice associated with the eruption, have been identified in coastal settings throughout the eastern Mediterranean, including archaeological contexts in central and eastern Crete (Marinatos, 1939; Minoura *et al.*, 2000) indicating a significant impact on Minoan settlements in the region.

Of all the eruptive products on the global scale, however, the generation of fine ash and sulphate aerosols are considered to be the most significant. Sulphate aerosols are composed of particles typically 500 nm in diameter, and if sufficiently powerful, an eruption may eject large quantities of these particles as far as the stratosphere (10–30 km). In this region, they have a residence time of several years (Rampino and Self, 1982; Devine *et al.*, 1984), and are transported around the planet in the Junge layer (Legrand and Delmas, 1987) reflecting incoming solar radiation and cooling global climate. The ignimbrite eruption of Toba (Sumatra, Indonesia) around 74,000 years ago was the largest known late Quaternary explosive volcanic eruption, generating between 2500 and 3000 km^3 of dense rock equivalent and covering several million square kilometres. Based on scaling up from known historic events, temperatures in the Northern Hemisphere are estimated to have reduced by 3–5°C (Rampino and Self, 1992).

Approximately at this time (100,000–50,000 BP), human mitochondrial DNA analyses suggest a genetic bottleneck, with a maximum of only 10,000 reproductive females present in the Old World (Rampino and Self, 1993; Ambrose, 1998), perhaps indicating that Toba caused a dangerous population decline in the human species.

REFERENCES

Ambrose, S.H. (1998) Late Pleistocene human population bottlenecks, volcanic winter, and differentiation of modern humans. *Journal of Human Evolution* **34**, 623–651.

Anderson, T. and Fell, C. (1995) Analysis of Roman cremation vessels by computerised tomography. *Journal of Archaeological Science* **22**, 609–617.

Barker, P. (1969) Some aspects of the excavation of timber buildings. *World Archaeology* **1**, 220–223.

Barker, P. (1975) Excavations on the site of the Baths Basilica at Wroxeter 1966–1974: An interim report. *Britannia* **6**, 106–117.

Barker, P. (1974) The origins and development of RESCUE. In (P. Rahtz, ed.) *Rescue Archaeology*. Penguin Books, 280–285.

Barker, P. (1977) *Techniques of Archaeological Excavation*. London: Batsford.

Bell, M., Caseldine, A. and Neumann, H. (2000) *Prehistoric Intertidal Archaeology in the Welsh Severn Estuary*. CBA Research Report, 120, 269–270.

Bibby, D. (1993) Building Stratigraphic Sequences in Excavations: An Example from Konstanz, Germany. In (E. Harris, M. Brown III and G. Brown, eds.) *Practices of*

Archaeological Stratigraphy, 104–121. New York: Academic Press.

Biddle, M. (1975) Winchester Cathedral cemetery. *World Archaeology* **7**, 86–108.

Biddle, M. and Kjolbye-Biddle, B. (1969) Metres, areas and robbing. *World Archaeology* **1**, 208–218.

Bishop, S. (1976) The methodology of post-excavation work. *Science and Archaeology* **18**, 15–19.

Blockley, K., Sparks, M. and Tatton-Brown, T. (1997) *Canterbury Cathedral Nave: Archaeology, history and architecture.* The archaeology of Canterbury (new series), **I**, Whitstable: Dean and Chapter of Canterbury Cathedral and Canterbury Archaeological Trust.

Boddington, A. (1978) *The Excavation Record: Part 1 Stratification*. Northamptonshire County Council.

Boschian, G. (1997) Sedimentology and soil micromorphology of the late Pleistocene and early Holocene deposits of Grotta dell'Edera (Trieste Karst, NE Italy), *Geoarchaeology*, **12**, 227–249.

Brown III, M. and Harris, E. (1993) Interfaces in archaeological stratigraphy. In (E. Harris, M. Brown III and G. Brown, eds.) *Practices of Archaeological Stratigraphy*, 7–20. New York: Academic Press.

Brown, A.G. (2003) Time, space and causality in floodplain palaeoecology. In (A. Howard, M.G. Macklin and D.G. Passmore, eds.) *Alluvial Archaeology in Europe*, 15–24. Lisse: A.A. Balkema.

Brown, A.G. and Barber, K.E. (1985) Late Holocene palaeoecology and sedimentary history of a small lowland catchment in central England. *Quaternary Research* **24**, 87–102.

Brown, A.G. and Keough, M.K. (1992) Holocene floodplain metamorphosis in the Midlands, United Kingdom. *Geomorphology* **4**, 433–466.

Brown, G. and Muraca, D. (1993) Phasing stratigraphic sequences at Colonial Williamsburg. In (E. Harris, M. Brown III and G. Brown, eds.) *Practices of Archaeological Stratigraphy*, 155–166. New York: Academic Press.

Buckley, D. and Ingle, C. (2001) The saddle querns from Flag Fen. In F. Pryor, *The Flag Fen basin: Archaeology and environment of a Fenland landscape*, 322–328. London: English Heritage.

Burrin, P.J. and Scaife, R.G. (1988) Environmental thresholds, catastrophe theory and landscape sensitivity: their relevance to the impact of man on valley alluviations. In (J.L. Bintliff, D.A. Davidson and E.G. Grant, eds.) *Conceptual Issues in Environmental Archaeology*, 211–232. Edinburgh: Edinburgh University Press.

CAT (2001) *Guidelines for site recording*. Canterbury: Canterbury Archaeological Trust.

Canti, M. (1991) *Soil particle size analysis: a revised interpretative guide for excavators.* English Heritage Ancient Monuments Laboratory Reports, **1/91**.

Canti, M.G. (1999) Soil Materials, pp. 77–87. In LeQuesne, C. (1999) *Excavations at Chester: The Roman and Later Defences Part 1: Investigations 1978–1990*. Chester City Council/Gifford and Partners: Chester Archaeology and Survey Report 11.

Canti, M.G. (2001) What is geoarchaeology? Re-examining the relationship between archaeology and earth-sciences. In (U. Albarella, ed.) *Environmental Archaeology: Meaning and Purpose*, 103–112. Dordrecht: Kluwer Academic.

Canti, M.G. (2003a) Aspects of the chemical and microscopic characteristics of plant ashes found in archaeological soils. *Catena* **54**, 339–361.

Canti, M.G. (2003b) Aspects of sediment diagenesis and taphonomy. In (M. Brennand and M. Taylor, eds.) The survey and excavation of a Bronze Age timber circle at Holme-next-the-Sea, Norfolk, 1998–9, 41–42. *Proceedings of the Prehistoric Society* **69**, 1–84.

Canti, M.G. and Meddens, F.M. (1998) Mechanical coring as an aid to archaeological projects. *Journal of Field Archaeology* **25**, 97–105.

Carter, S. (1998a) Soil micromorphology. In (C. Lowe, ed.) *Coastal erosion and the archaeological assessment of an eroding shoreline at St Boniface Church, Papa Westray, Orkney*, 172–186. Edinburgh: Historic Scotland.

Carter, S. (1998b) 'Farm Mounds' and the formation of the Area 1 sediments. In Lowe, C. (1998) *Coastal erosion and the archaeological assessment of an eroding shoreline at St Boniface Church, Papa Westray, Orkney*, 187–198. Edinburgh: Historic Scotland.

Carter, S. (2001) A reassessment of the origins of the St Andrews 'garden soil'. *Tayside and Fife Archaeological Journal* **7**, 87–92.

Carver, M. (1983) Theory and practice in urban pottery seriation. *Science and Archaeology* **21**, 3–14.

Carver, M. (1985) Theory and practice in urban pottery seriation. *Journal of Archaeological Science* **12**, 353–366.

Carver, M. (1987) *Underneath English Towns*. London: Batsford.

Challinor, J. (1964) *A Dictionary of Geology*. Cardiff.

Cioli, R., Gurioli, L., Sbrana, A. and Vougioukalakis, G. (2000) Precursory phenomena and destructive events related to the Late Bronze Age Minoan (Thera, Greece) and AD 79 (Vesuvius, Italy) plinian eruptions; inferences from the stratigraphy in the archaeological areas. In (W.G. McGuire, D.R. Griffiths, P.L. Hancock and I.S. Stewart, eds.) *The Archaeology of Geological Catastrophes*, 123–141. London: Geological Society Special Publications 171.

Clapham, A.R. and Clapham, B.N. (1939) The valley fen at Cothill, Berkshire. *New Phytologist* **38**, 167–174.

Clark, P. (ed.) (2004) *The Dover Bronze Age Boat*. London: English Heritage.

Clark, P. (1988a) Studying deposit formation processes on rescue sites: The toolbox system. *Arch-Form* **3**, 6–7.

Clark, P. (1988b) *Guidelines for the Stratigraphic Analysis of Excavation Archives*. Internal Publication, Scottish Urban Archaeological Trust.

Clark, P. (1992) Contrasts in the recording and interpretation of 'rural' and 'urban' stratification. In (K. Steane, ed.) *Interpretation of stratigraphy: A review of the art*, 17–19. City of Lincoln Archaeology Unit Report **31**.

Clark, P. (1993) Sites without Principles: post-excavation analysis of 'pre-matrix' sites. In (E. Harris, M. Brown III and G. Brown, eds.) *Practices of Archaeological Stratigraphy*, 276–292. New York: Academic Press.

Clark, P. (2000) Negative features and interfaces. In (S. Roskams, ed.) *Interpreting Stratigraphy: Site evaluation, recording procedures and stratigraphic analysis*, 103–106. British Archaeological Reports (International Series) **910**.

Clarke, A.F. (1982) The Neolithic of Kent: A review. In *Archaeology of Kent to AD 1500*, 25–30. Council for British Archaeology Research Report **48**.

Clarke, A. and Fulford, M. (2002) The excavation of Insula IX, Silchester: the first five years of the 'Town Life' project, 1997–2001. *Britannia* **32**, 129–166.

Coles, J. and Coles, B. (1986) *Sweet track to Glastonbury*. London: Thames and Hudson.

Collard, R. (1988) *The Physical Geography of Landscape*. London: HarperCollins.

Cotton, H. and Geiger, J. (1989) *Masada II, The Yigael Yadin excavations 1963–1965, Final reports, The Latin and Greek documents*. Jerusalem.

Cronin, S.J., Hedley, M.J., Neall, V.E. and Smith, R.G. (1998) Agronomic impact of tephra fallout from the 1995 and 1996 Ruapehu volcano eruptions, New Zealand. *Environmental Geology* **34**, 21–30.

Crowther, J. (1997) Soil phosphate surveys: critical approaches to sampling, analysis and interpretation. *Archaeological Prospection* **4**, 93–102.

Curtis, R. (1991) *Garum and Salsamenta: Commerce and Production in Materia Medica*. Studies in Ancient Medicine **3**, Leiden.

Davies, B.E. (1974) Loss on ignition as an estimate of soil organic matter. *Soil Science Society of America Proceedings* **38**, 150–151.

Davis, M.B. (1976) Erosion rates and land-use history in southern Michigan. *Environmental Conservation* **3**, 139–148.

de Boer, J.Z. and Sanders, D.T. (2002) *Volcanoes in Human History*. Princeton: Princeton University Press.

Dearing, J. (1999) Magnetic Susceptibility. In (J.Walden, F. Oldfield and J. Smith, eds.) *Environmental Magnetism: a practical guide*, 35–62. London: Quaternary Research Association Technical Guide **6**.

Dent, D.L. and Young, A. (1981) *Soil Survey and Land Evaluation*. London: George Allen and Unwin.

Department of Environment (1975) *Principles of Publication in Rescue Archaeology*. London: HMSO.

Devine, J.D., Sigordsson, M. and Davies, A.N. (1984) *Estimates of sulfur and chlorine yield to the atmosphere from volcanic eruptions and potential climatic effects*. Journal of Geophysical Research, **89**, 6309–6325.

Dragendorff, H. (1895) Terra Sigillata. Ein Beitrag zur Geschichte der griechischen und römischen Keramik. *Bonner Jahrbuch* **96**, 18–155.

Dragendorff, H. (1896) Verzeichnis der Stempel auf Terra-Sigillata-Gefässen, die sich in die Zeit von rund 70–250 n. Chr. datieren lassen. *Bonner Jahrbüch* **99**, 54–163.

Dunnell, R. (1984) The ethics of archaeological significance decisions. In (E. Green, ed.) *Ethics and values in archaeology*, 62–74. New York: The Free Press.

Emery, K.O. (1969) The continental shelves. *Scientific American* **221**, 106–122.

English Heritage (1991) *Management of Archaeological Projects*. London: HBMC.

Entwistle, J. and Abrahams, P.W. (1998) Multi-elemental analysis of soils and sediments from Scottish archaeological sites. The potential of inductively-coupled plasma-mass spectrometry for rapid site investigation. *Journal of Archaeological Science* **24**, 407–416.

French, C.A.I. and Whitelaw, T.M. (1999) Soil erosion, agricultural terracing and site formation processes at Markiani, Amorgos, Greece: the micromorphological perspective. *Geoarchaeology* **14**, 151–189.

French, H.M. (1996) *The Periglacial Environment*. Essex: Addison Wesley Longman.

Friedrich, W.L. (2000) *Fire in the Sea*. Cambridge: Cambridge University Press.

Gale, S.J. and Hoare, P.G. (1991) *Quaternary Sediments: Petrographic Methods for the Study of Unlithified Rocks*. London: Belhaven Press.

Gasse, A. (1990) *Catalogue des ostraca figurés de Deir el-Médineh, nos. 1676–1774*. Cairo: Institut Français d'Archéologie Orientale du Caire.

Gebhardt, A. and Langohr, R. (1999) Micromorphological study of construction materials and living floors in the medieval Motte at Werken (West Flanders, Belgium). *Geoarchaeology* **14**, 595–620.

Godwin, H. and Vishnu-Mittre (1975) Studies of the Post-Glacial history of the British vegetation. XVI. Flandrian deposits of the fenland margin at Holme Fen and Whittlesey Mere, Hunts. *Philosophical Transactions of the Royal Society* **B270**, 561–604.

Govers, G., Quine, T.A., Desmet, P.J.J. and Walling, D.E. (1996) The relative contribution of tillage and overland flow erosion to soil redistribution on arable land. *Earth Surface Processes and Landforms* **21**, 929–946.

Gressly, A. (1838) Observations geologiques sur la Jura Soleurois. *Neue Denkschr. Allg. Schweiz, Ges. Ges. Naturw.* **2**, 1–112.

Haarnagel, W. (1979) *Die Grabungen Feddersen Wierde: Methode, Hausbau, Siedlungs- und Wirtschaftsformen sowie sozialstruktur*. Wiesbaden.

Hammer, F. (2000) The structuring of records: different systems in use from excavation through to publication. In (S. Roskams, ed.) *Interpreting Stratigraphy: Site evaluation, recording procedures and stratigraphic analysis*, 143–151. British Archaeological Reports (International Series) **910**.

Hammer, F. (2002) *Archaeological post-excavation manual for the stratigraphic record: Using the single context planning method*. London: Museum of London.

Hammond, N. (1993) Matrices and Maya archaeology. In (E. Harris, M. Brown III and G. Brown, eds.) *Practices of archaeological stratigraphy*. Cambridge: Academic Press, 139–152.

Harris, E. (1975) The stratigraphic sequence: a question of time. *World Archaeology* **7**, 109–121.

Harris, E. (1979a) *Principles of Archaeological Stratigraphy*. London: Academic Press.

Harris, E. (1979b) The laws of archaeological stratigraphy. *World Archaeology* **11**, 111–117.

Harris, E. (1983) *Principi di Stratigrafia Archaeologica* (translated by A. Gabucci). Rome: La Nuova Italia Scientifica.

Harris, E. (1989a) *Zasady Stratygrafii Archaeologicznej* (translated by Z. Kobylinski). Warsaw: Osrodek Dokumentacji Zabytkov.

Harris, E. (1989b) *Nacela Arheoloske Stratigrafije* (translated by P. Novakovic and P. Turk). Ljubljana: Slovensko Archeolosko Drustvo.

Harris, E. (1991) *Principios de Estratigrafia Arqueológica* (translated by I. Trócoli). Barcelona: Editorial Critica.

Harris, E. (1992) The Central Role of Stratigraphy in Archaeological Excavation. In (Centro de Patrimonio Cultural Vasco, ed.) *Intervention Archaeology: International Sessions (Proceedings of San Sebastian Conference, December 1991)*, 113–135. Bergara-Guouzkoa: Departamento de Cultura.

Harris, E. (1995) *Principles of Archaeological Stratigraphy* (translated by K. Osawa). Toyko: Yuzankaku Publishing Company Ltd.

Harris, E. (1998) 25 years of the Harris matrix. *Newsletter of the Society of Historical Archaeology* **31**, 6–7.

Harris, E., Brown III, M. and Brown, G. (eds.) (1993) *Practices of Archaeological Stratigraphy*. New York: Academic Press.

Heathcote, J. (2000) Soil micromorphology. In Rippon, S. (2000) The Romano-British Exploitation of Coastal Wetlands: Survey and Excavation on the North Somerset Levels 1993–7. *Britannia* **31**, 107–12.

Hey, G. (2001) Yarnton. *Current Archaeology* **173**, 216–225.

Hill, D. (1986) *Canterbury Cathedral*. Harper Collins.

Hirst, S. (1976) *Recording on Excavations I: The Written Record*. Rescue.

Hosfield, R. (2001) The lower Palaeolithic of the Solent: 'site' formation and interpretive frameworks. In (R.T Hosfield and F.F. Wenban-Smith, eds.) *Palaeolithic archaeology of the Solent river*, 15–25, Lithic Studies Society Occasional Papers, **7**.

Huang, C.C., Pang, J., Ping, H., Hou, C. and Han, Y. (2002) High-resolution studies of the oldest cultivated soils in the southern Loess Plateau of China. *Catena* **47**, 29–42.

Hummler, M. and Roe, A. (1996) Sutton Hoo burials: reconstructing the sequence. In (S. Roskams, ed.) *Interpreting Stratigraphy 8: Proceedings of a conference held at the Department of Archaeology, University of York on 15th February, 1996*, 39–53. York: University of York.

Jacoby, G.C., Workman, K.W. and D'Arrigo, R.D.D. (1999) Laki eruption of 1783, tree rings, and disaster for northwest Alaska Inuit. *Quaternary Science Reviews* **18**, 1365–1371.

Jefferies, J.S. (1977) *Excavation Records Techniques in use by the Central Excavation Unit*. Department of Environment.

Karkanas, P., Rigaud, J-P., Simek, J.F., Albert, R.M. and Weiner, S. (2002) Ash, bones and guano: a study of the minerals and phytoliths in the sediments of Grotte XVI, Dordogne, France. *Journal of Archaeological Science* **29**, 721–732.

Keeling, P.S. (1962) Some experiments in the low temperature removal of carbonaceous material from clay. *Clay Mineral Bulletin* **28**, 155–158.

Keevill, G. (1992) Life on the edge: archaeology and alluvium at Redlands Farm, Stanwick, Northants. In (S. Needham and M.G. Macklin, eds.) *Alluvial Archaeology in Britain*, 27–35. Oxbow Monograph 27. Oxford: Oxbow Books.

Kenyon, K. (1952) *Beginning in Archaeology*. London: Phoenix House.

Klimek, K. (2002) Human-induced overbank sedimentation in the foreland of the eastern Sudety Mountains. *Earth Surface Processes and Landforms* **27**, 391–402.

Klimek, K. (2003) Sediment transfer and storage linked to Neolithic and Early Medieval soil erosion in the Upper Odra Basin, southern Poland. In (A.J. Howard, M.G. Macklin and D.G. Passmore, eds.) *Alluvial Archaeology in Europe*, 251–259. Lisse: A.A. Balkema.

Kopytoff, I. (1986) The cultural biography of things: Commoditization as process. In (A. Appadurai, ed.) *The social life of things*, 64–91. Cambridge: Cambridge University Press.

Leeder, M.R. (1982) *Sedimentology: process and product*. London: George Allen and Unwin.

Legrand, M. and Delmas, R.J. (1987) A 220-year continuous record of volcanic H_2SO_4 in the Antarctic ice sheet. *Nature* **327**, 671–676.

LeQuesne, C. (1999) *Excavations at Chester: The Roman and Later Defences Part 1: Investigations 1978–1990.* Chester City Council/Gifford and Partners: Chester Archaeology and Survey Report 11.

Leyton, L., Reynolds, E.R.C. and Thompson, F.B. (1967) Rainfall interception in forest and moorland. In (W.E. Sopper and H.W. Lull, eds.) *International Symposium on Forest Hydrology*, 163–177. Oxford: Pergamon Press.

Limbrey, S. (1978) Changes in the quality and distribution of the soils of lowland Britain. In (S. Limbrey and J. Evans, eds.) *The effect of man on the landscape: the lowland zone*, 21–27. Council for British Archaeology Research Report 21.

Lipe, W. (1974) A conservation model for American archaeology. *The Kiva* **39**, 213–245.

Longworth, I. and Kinnes, I. (1980) *Sutton Hoo excavations 1966, 1968–70.* British Museum Occasional Paper **23.**

Loveluck, C. (forthcoming) *Excavations at Flixborough. Vols 1 and 4.* London: English Heritage and Oxford: Oxbow Books.

Lowe, C. (1998) *Coastal erosion and the archaeological assessment of an eroding shoreline at St Boniface Church, Papa Westray, Orkney.* Edinburgh: Historic Scotland.

Macleod, D., Monk, M. and Williams, T. (1988) The use of the single context recording system on a seasonally excavated site in Ireland: A learning experience. *The Journal of Irish Archaeology* **IV**, 55–63.

Marinatos, S. (1939) The volcanic destruction of Minoan Crete. *Antiquity* **13**, 425–439.

Martin-Kilcher, S. (1994) *Die römischen Amphoren aus Augst und Kaiseraugst. Ein Beitrag zur römischen Handels- und Kulturgeschichte. 2, Die Amphoren für Wein, Fischsauce, Südfrüchte (Gruppen 2–24) und Gesamtauswertung.* Forschungen in August **7**, Römermuseum.

McCobb, L.E., Briggs, D.E.G., Carruthers, W.J. and Evershed, R.P. (2003) Phosphatisation of seeds and roots in a late Bronze Age deposit at Potterne, Wiltshire, UK. *Journal of Archaeological Science* **30**, 1269–1281.

Middleton, W.D. and Price, T.D. (1996) Identification of activity areas by multi-element characterization of sediments from modern and archaeological house floors using inductively coupled plasma-atomic emission spectroscopy. *Journal of Archaeological Science* **23**, 673–677.

Minoura, K., Imamura, F., Kuran, U., Nakamura, T., Papadopolous, G.A., Takahashi, T. and Yalciner, A.C. (2000) Discovery of Minoan tsunami deposits. *Geology* **28**, 59–62.

Needham, S. (1992) Holocene alluviation and interstratified settlement evidence in the Thames Valley at Runnymede Bridge. In (S. Needham and M.G. Macklin, eds.) *Alluvial Archaeology in Britain*, 249–260. Oxford: Oxbow Monograph 27.

ODPM (2000) *Planning Policy Guidance Note 16: Archaeology and Planning.* Office of the Deputy Prime Minister: http://www.odpm.gov.uk/stellent/groups/odpm_planning/documents/page/odpm_plan_606901.hcsp

O'Kelly, M. (1995) *Newgrange: Archaeology, Art and Legend.* London: Thames and Hudson.

Parker, A.G. and Robinson, M.A. (2003) Palaeoenvironmental investigations on the middle Thames at Dorney, UK. In (A.J. Howard, M.G. Macklin and D.G. Passmore, eds.) *Alluvial Archaeology in Europe*, 43–60. Lisse: A.A. Balkema.

Parnell, J.J., Terry, R.E. and Nelson, Z. (2002) Soil chemical analysis applied as an interpretive tool for ancient human activities in Piedras Negras, Guatemala. *Journal of Archaeological Science* **23**, 379–404.

Pearson, N. and Williams, T. (1993) Single-context planning: its role in on-site recording procedures and in post-excavation analysis at York. In (E. Harris, R. Brown and G. Brown, eds.) *Practices of archaeological stratigraphy*, 89–103. London: Academic Press.

Pennington, W. (1970) Vegetational history in North-West England: a regional synthesis. In (D.Walker and R.G. West, eds.) *Studies of the Vegetational History of the British Isles*, 41–79. Cambridge: Cambridge University Press.

Piggott, S. (1959) *Approach to archaeology.* Cambridge: Harvard University Press.

Pryor, F. (1998) *Etton, Excavations at a Neolithic causewayed enclosure near Maxey, Cambridgeshire, 1982–7.* Archaeological Report 18. London: English Heritage.

Quine, T., Govers, G., Walling, D.E., Zhang, X., Desmet, P.J.J. and Zhang, Y. (1997) Erosion processes and landform evolution on agricultural land – new perspectives from caesium-137 data and topographic-based erosion modelling. *Earth Surface Processes and Landforms* **22**, 790–816.

Raab, L.M. (1984a) Achieving professionalism through ethical fragmentation: Warnings from client-oriented archaeology. In (E. Green, ed.) *Ethics and values in archaeology*, 51–61. New York: The Free Press.

Raab, L.M. (1984b) Toward an understanding of the ethics and values of research design in archaeology. In (E. Green, ed.) *Ethics and values in archaeology*, 75–88. New York: The Free Press.

Raftery, B. (1992) Recent developments in Irish wetland research. In (B. Coles, ed.) *The Wetland Revolution in Prehistory*, 29–36. Exeter: WARP & The Prehistoric Society.

Rahtz, P. (1989) *Little Ouseburn Barrow.* York: York University Archaeology Publications 7.

Rahtz, P. (1996) Reconstructing stratigraphy within burials: the use of the planum method. In (S. Roskams, ed.) *Interpreting Stratigraphy 8: Proceedings of a conference held at the Department of Archaeology, University of York on 15th February, 1996*, 36–38. York: University of York.

Rains, M. (1985) Home Computers in Archaeology. In (E. Webb, ed.) *Computer Applications in Archaeology, 1985*, 15–26. London: University of London Institute of Archaeology.

Ramos, A. and Sopeña, A. (1983) Gravel bars in low sinuosity streams (Permian and Triassic, central Spain). In (J.D. Collinson and J. Lewin, eds.) *Modern and Ancient Fluvial Systems*, 301–312. Special Publication of the International Association of Sedimentologists. Oxford: Blackwell Scientific

Rampino, M.R. and Self, S. (1982) Historic eruptions of Tambora (1815), Krakatau (1883), and Agung (1963), their stratospheric aerosols, and climatic impact. *Quaternary Research* **18**, 127–143.

Rampino, M.R. and Self, S. (1992) Volcanic winter and accelerated glaciation following the Toba super-glaciation. *Nature* **359**, 50–52.

Rampino, M.R. and Self, S. (1993) Bottleneck in human evolution and the Toba eruption. *Science* **262**, 1955.

Randsborg, K. (1993) Kivik. Archaeology and Iconography. *Acta Archaeologica* **64**, 1. København: Munksgård.

Rankov, N., Hassall, M. and Tomlin, R. (1982) Roman Britain in 1981. *Britannia* **13**, 328–422 (Rougier Street 349).

Reading, H.G. and Levell, B.K. (1996) Controls on the sedimentary rock record. In (H.G. Reading, ed.) *Sedimentary Environments: Processes, Facies and Stratigraphy*, 5–36. Oxford: Blackwell Science.

Robinson, M. (1981) The effects of pre-afforestation drainage upon the streamflow and water quality of a small upland catchment. *Institute of Hydrology Report* 73.

Robinson, M. and Lambrick, G.H. (1984) Holocene alluviation and hydrology in the upper Thames basin. *Nature* **308**, 809–814.

Sadler, J.P. and Grattan, J.P. (1999) Volcanoes as agents of past environmental change. *Global and Planetary Change* **21**, 181–196.

Scarth, A. (1999) *Vulcan's Fury*. New Haven: Yale University Press.

Scottish Urban Archaeological Trust (1988) *Context Recording Manual*. Perth: SUAT.

Shay, T. (1989) Israeli archaeology – ideology and practice. *Antiquity* **63**, 268–772.

Shepherd, L. (1993) Interpreting landscapes – analysis of excavations in and around the southern bailey of Norwich Castle. In (J. Barber, ed.) *Interpreting Stratigraphy*, 3–10. Dundee: AOC (Scotland) Ltd.

Shotton, F.W. (1978) Archaeological inferences from the study of alluvium in the Lower Severn-Avon valleys. In (S. Limbrey and J.G. Evans, eds.) *Man's Effect on the landscape: The Lowland Zone*, 27–32. CBA Research Report 21. London: Council for British Archaeology.

Sigurðsson, H., Standford, C. and Sparks, S.R.J. (1982) The eruption of Vesuvius in AD 79: reconstruction from historical and volcanological evidence. *American Journal of Archaeology* **86**, 39–51.

Simpson, I.A. (1997) Relict properties of anthropogenic deep top soils as indicators of infield management in Marwick, West Mainland, Orkney. *Journal of Archaeological Science* **24**, 365–380.

Simpson, I.A., Vésteinsson, O., Adderley, W.P. and McGovern, T.H. (2003) Fuel resource utilisation in landscapes of settlement. *Journal of Archaeological Science* **30**, 1401–1420.

Smith, H., Marshall, P. and Parker Pearson, M. (2001) Reconstructing house activity areas. In (U. Albarella, ed.) *Environmental Archaeology: Meaning and Purpose*, 249–270. Dordrecht: Kluwer Academic Publishers.

Spence, C. (1990) *Archaeological site manual*, 2nd edition. London: Museum of London.

Spence, C. (1993) Recording the archaeology of London: the development and implementation of the DUA recording system. In (E. Harris, M. Brown III and G. Brown, eds.) *Practices of Archaeological Stratigraphy*, 23–46. New York: Academic Press.

Spriggs, J. (1992) Conservation: approach and methods. In (D. Tweddle, ed.) *The Anglian Helmet from Coppergate*. York Archaeological Trust Fascicule, **17/8**. York: Council for British Archaeology.

Stanley, D.J. and Sheng, H. (1986) Volcanic shards from Santorini (Upper Minoan ash) in the Nile delta, Egypt. *Nature* **320**, 733–735.

Stead, I. and Rigby, V. (1989) *Verulamium: The King Harry Lane Site*. Dorchester: English Heritage and British Museum.

Stead, S. (1988) The integrated archaeological database. In (C. Ruggles and S. Rahtz, eds.) *Computer and quantitative methods in archaeology 1987*, 279–284. British Archaeological Reports (International Series), **393**.

Stuck, B. (1993) Three-dimensional assessment of activity areas in a shell midden: an example from the Hoko River Rockshelter, State of Washington. In (E. Harris, M. Brown III and G. Brown, eds.) *Practices of Archaeological Stratigraphy*, 122–138. New York: Academic Press.

Terry, R.E., Fernández, F.A., Parnell, J.J. and Inomata, T. (2004) The story in the floors: chemical signatures of ancient and modern Maya activities at Aguateca, Guatemala. *Journal of Archaeological Science* **31**, 1289–1308.

Thompson, R. and Oldfield, F. (1986) *Environmental Magnetism*. London: Allen and Unwin.

Thordarson, Th. and Self, S. (1993) The Laki (Skaftár

Fires) and Grímsvötn eruptions in 1783–1785. *Bulletin of Volcanology* **55**, 233–263.

Thorpe, R. (1998) Which way is up? Context formation and transformation: The life and deaths of a hot bath in Beirut. *Assemblage* **4**, http://www.shef.ac.uk/assem/4/4rxt.html

Tucker, M. (1996) *Sedimentary rocks in the field.* Chichester: John Wiley and Sons.

van der Schriek, T., Passmore, D.G., Stevenson, A.C., Boomer, I., Mugica, F.F. and Rolão, J. (2003) The geoarchaeology of Mesolithic settlement and subsistence in the Muge Valley, Lower Tagus Basin, Portugal. In (A. Howard, M.G. Macklin and D.G. Passmore, eds.) *Alluvial Archaeology in Europe*, 15–24. Lisse: A.A. Balkema.

van Es, W. (1967) Wijster, a native village beyond the Imperial frontier. *Palaeohistoria* **11** (whole volume).

van Es, W. and Verwers, W. (1980) *Excavations at Dorestad 1. The Harbour, Hoogstraat I.* Nederlandse Oudheden 9. Amersfoort: ROB.

Wastegård, S., Wohlfarth, B., Subetto, D.A. and Sapelko, T.V. (2000) Extending the known distribution of the Younger Dryas Vedde ash into northwestern Russia. *Journal of Quaternary Science* **15**, 581–586.

Webster, G. (1963) *Practical Archaeology.* London: A. and C. Black.

Weiner, S., Goldberg, P. and Bar-Yosef, O. (2002) Three-dimensional distribution of minerals in sediments of the Hayonim Cave, Israel: diagenetic processes and archaeological implications. *Journal of Archaeological Science* **29**, 1289–1308.

Westman, A. (1990) Archive Report Manual. London: Museum of London Department of Archaeology.

Wheeler, R.E.M. (1947) The Recording of Archaeological Strata. *Ancient India* 3, 143–150.

Wheeler, R.E.M. (1954) *Archaeology from the earth.* London: Oxford University Press.

Wilkinson, K.N. (2003) Colluvial deposits in dry valleys of southern England as proxy indicators of paleoenvironmental and land-use change. *Geoarchaeology* **18**, 725–755.

Wilkinson, K.N. and Pope, R.J.J. (2003) Quaternary alluviation and archaeology in the Evrotas valley, southern Greece. In (A. Howard, M.G. Macklin and D.G. Passmore, eds.) *Alluvial Archaeology in Europe*, 187–201. Lisse: A.A. Balkema.

Williams, T. (1987) *Archive Report Manual.* London: Museum of London Department of Archaeology.

Wymer, J.J. (1976) The interpretation of Palaeolithic cultural and faunal material found in Pleistocene sediments. In (D.A. Davidson and M.L. Shackley, eds.) *Geoarchaeology.* London: Duckworth.

Zielinski, G.A., Germani, M.S., Larsen, G., Baillie, M.G.L., Whitlow, S., Twickler, M.S. and Taylor, K. (1995) Evidence of the Eldgjá (Iceland) eruption in the GISP2 Greenland ice core: relationship to eruption processes and climatic conditions in the tenth century. *The Holocene* **5**, 129–140.

3

Bioarchaeology: analysing plant and animal remains

3.0 Chapter summary

A wide range of microfossils and macrofossils play a part in enabling us to reconstruct palaeoenvironments and palaeoeconomies. This chapter addresses the concepts and methods used for all of the major groups, as well as examining some of the problems that arise from identification, analysis and taphonomy. More than one interpretation is often possible for many of the sequences in the archaeological and geological archives. Different views are discussed, along with advantages and disadvantages of the various interpretative methods. A multi-proxy approach is recommended to minimize the effects of variability, and the examples at the end of the chapter highlight the value of these techniques at a range of scales.

3.1 Definition of bioarchaeology

The field of bioarchaeology involves the analysis of fossil (and sub-fossil) plant and animal remains from both archaeological (e.g. pits, ditches, hearths, wells, cesspits, human bodies) and geological archives (e.g. lakes, peat bogs, soils, caves, rock-shelters). The goal is to improve our knowledge and understanding of the links between the environment and humans (the palaeoenvironment), as well as providing details of their diet, economy and daily life (the palaeoeconomy). To achieve this goal requires the use of a range of analytical methods (a multi-proxy approach), and the application of experimental and ethnographic data, to obtain independent lines of complementary scientific evidence (preferably from each archaeological context or geological unit), with the ultimate aim of producing precise interpretations (Bell and Walker, 1992; Coil *et al.*, 2003) and clarity when reporting the results (Gobalet, 2001). This conceptual approach is not new to modern archaeological research (Butzer, 1982) although its practical application has not been universally adopted, especially in professional as opposed to academic archaeology.

The interpretation of bioarchaeological remains follows uniformitarian principles ('the present is the key to the past'), although it is necessary to make certain assumptions about the fossil remains, and about our understanding of modern plant and animal ecology, before reconstructions and explanations can be made. These assumptions were articulated by Lowe and Walker (1984: 155) in the context of Quaternary science, but are equally applicable to bioarchaeology:

1 That we fully understand and are able to isolate the environmental parameters governing present day distributions of plants and animals.

2 That present day plant and animal distributions are in equilibrium with those controlling variables.
3 That former plant and animal distributions were in equilibrium with their environmental controls.
4 That former plant and animal distributions have analogues in the modern flora and fauna.
5 That the ecological affinities of plants and animals have not changed through time.
6 That a fossil assemblage is representative of the death assemblage and has not been biased by differential destruction of its original component parts or by contamination by older or younger material.
7 That the origin of the fossil assemblage (taphonomy) can be established.
8 That the fossil remains can be identified to a sufficiently low taxonomic level to enable uniformitarian principles to be applied.

Bioarchaeology, like Quaternary science, cannot fulfil all of these requirements, and the following sections will elucidate many of the problems while outlining the main theoretical and practical developments within the field.

3.2 Classes of bioarchaeological remains

A simple subdivision of bioarchaeological remains into two categories has been adopted in this chapter. *Microfossils* include those classes of remains that are invisible to the naked eye, except where they occur in high concentrations (clusters) or where there are exceptionally large specimens. *Macrofossils* include those classes of remains that are visible, although their identification may require microscopic analysis. Both microfossils and macrofossils occur in a wide range of geological and archaeological situations, and are preserved in both dry and wet environments.

3.2.1 Microfossils

3.2.1.1 Pollen grains and spores

The analysis of pollen grains and spores (palynology) is widely used in environmental archaeology since they frequently provide valuable information on vegetation composition, structure and succession (palaeoecology), plant migration (biogeography), climate change, human modification of the natural vegetation cover, land use (anthropogenic activity), and also diet. Pollen grains are liberated by flowering plants (angiosperms) and conifers (gymnosperms), while spores are produced by primitive and advanced vascular plants such as ferns, mosses and fungi (Fig. 3.1). They are small (5–200 μm) and dispersed by a variety of mechanisms, principally wind, insects and water, and in exceptional circumstances by humans, with c. 99 per cent being deposited within 1 km of the source (Brasier, 1980). Pollen grains and spores are highly resistant to decay, having an outer layer (exine) made of sporopollenin. Preservation, under optimal conditions, occurs in acidic (low levels of microbial activity) or anaerobic (oxygen-free) environments. These conditions may be found in geological and archaeological archives, such as soils formed in acidic substrates, peat and the sedimentary fills of lakes, river flood plains (alluvium), pits, ditches, wells, cesspits, moats and graves. In addition, pollen grains and spores may be well preserved in whole coprolites and cave sediments because of the relatively dry conditions. To illustrate the range of information obtainable from pollen analysis, four of these archives are discussed below: peat, caves, soils and graves.

Once extracted from the sub-sample of soil, sediment, peat or faecal material (see Section 3.3), identification is achieved with modern reference collections, keys and photographs noting characteristics such as the orientation (equatorial or polar views), size, shape, apertures, furrows, wall structure, sculpture and size using a transmitted light microscope (Moore *et al.*, 1991) or occasionally scanning electron microscopy (see Lee *et al.*, 2004). Useful websites concerning pollen and its analysis

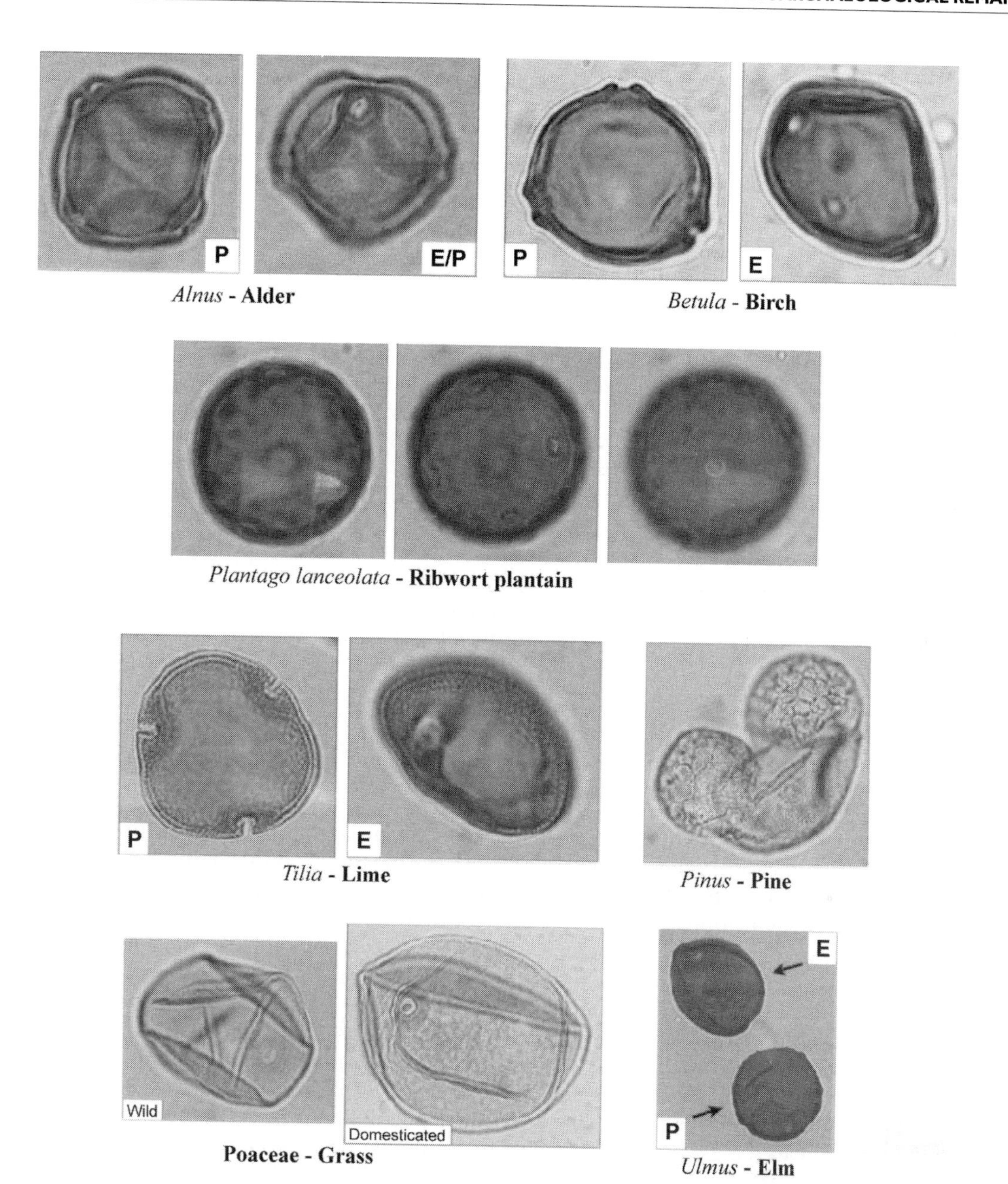

FIGURE 3.1 A selection of pollen grains found in European archaeological and geological archives (E = equatorial view; P = polar view). The photographs were all originally taken at ×1000 magnification but have been resized by varying amounts to illustrate the different shapes, apertures and sculpturing.

can be found at http://www.geo.arizona.edu/palynology/polonweb.html and http://www.ngdc.noaa.gov/paleo/epd/epd_main.html. A pollen count consists of recording identifiable pollen grains and spores within each sub-sample by systematically traversing the microscope slide until a predetermined total has been achieved. The total, known as the pollen sum, may vary according to: (a) the type of sub-sample being analysed; (b) the specific research question being addressed; and (c) the geographical origin of the sample. For many European and North American studies the pollen sum consists of 300–500 pollen grains representing tree, shrub and herb taxa (total land pollen sum). Aquatic pollen grains and spores are also counted, but are excluded from the main pollen sum because they often represent the local (microscale) vegetation cover and hence tend to be over-represented.

Traditionally the pollen and spore values from each sub-sample are tabulated or presented diagrammatically as percentages (Fig. 3.2). The main problem with percentage pollen diagrams is that an increase in values of one taxon (e.g. *Quercus*, oak) will lead to the decline in another taxon because of the mutually dependent nature of the individual pollen graphs. Since these changes in pollen percentages may not correspond to actual changes in plant ecology, pollen concentration (grains/ml) or total pollen influx (grains/ml/yr) diagrams are also employed (so-called 'absolute pollen diagrams'; Fig. 3.3). Producing them depends on adding an exotic pollen to the sub-sample during the laboratory extraction, enabling an estimate of the pollen concentration for each taxon by calculating the ratio of fossil to exotic pollen. These values can be converted into pollen influx values by calculating the rate of sediment accumulation using an appropriate number of radiocarbon dates or other suitable techniques (see Chapter 4). Used in conjunction, these various methods for statistical representation of the pollen data provide a valuable means of assessing changes in past vegetation.

The interpretation of pollen data from both geological and archaeological situations is highly problematical and requires an understanding of:

1 Variations in pollen production between plant taxa
2 The source area of the pollen
3 Mechanisms of pollen transportation
4 Modes of pollen deposition
5 Preferential survival of certain pollen grains and spores
6 The relationship between the pollen record and plant communities.

Records of the amount of pollen produced by different plants are often vague and provide only a relative guide. For example, *Pinus* (pine), *Betula* (birch), *Alnus* (alder) and *Corylus* (hazel) produce high amounts of pollen; *Picea* (spruce), *Quercus* (oak), *Fagus* (beech) and *Tilia* (lime) produce moderate amounts; and *Ilex* (holly) and *Viscum* (mistletoe) produce very little. Although there seems to be little differentiation between wind and insect pollinated plants in terms of pollen production, in theory higher pollen producers should be better represented in pollen diagrams, whereas self-pollinating plants (e.g. domesticated grasses) should be under-represented. This is of course dependent on the proximity of the plant to the sampling location, the type of sample and the transportation mechanism to the site of deposition.

Models detailing the complex processes responsible for pollen transportation and deposition onto peat surfaces (e.g. Andersen, 1970, 1973, 1979, 1992) – probably the most intensively studied geological archive – suggest that site selection and the sampling strategy are critical for determining whether the pollen record is representative of a predominantly micro-, meso- or even macroscale vegetation cover. It is also evident that the pollen catchment area and mechanisms of pollen transportation to a peat surface (e.g. bog or fen) will vary through time. Under optimal conditions, therefore, the surface area of the peat should be restricted to 100–200 m in diameter ensuring that pollen from mainly the trunk space component, reflecting sub-regional vegetation communities, will dominate the pollen record (Bradshaw, 1981). In addition, the sampling location should be within 30 m of the likely pollen source to detect the presence of human activity at the edge of a peat bog (Tauber, 1965, 1967; Jacobson and Bradshaw, 1981;

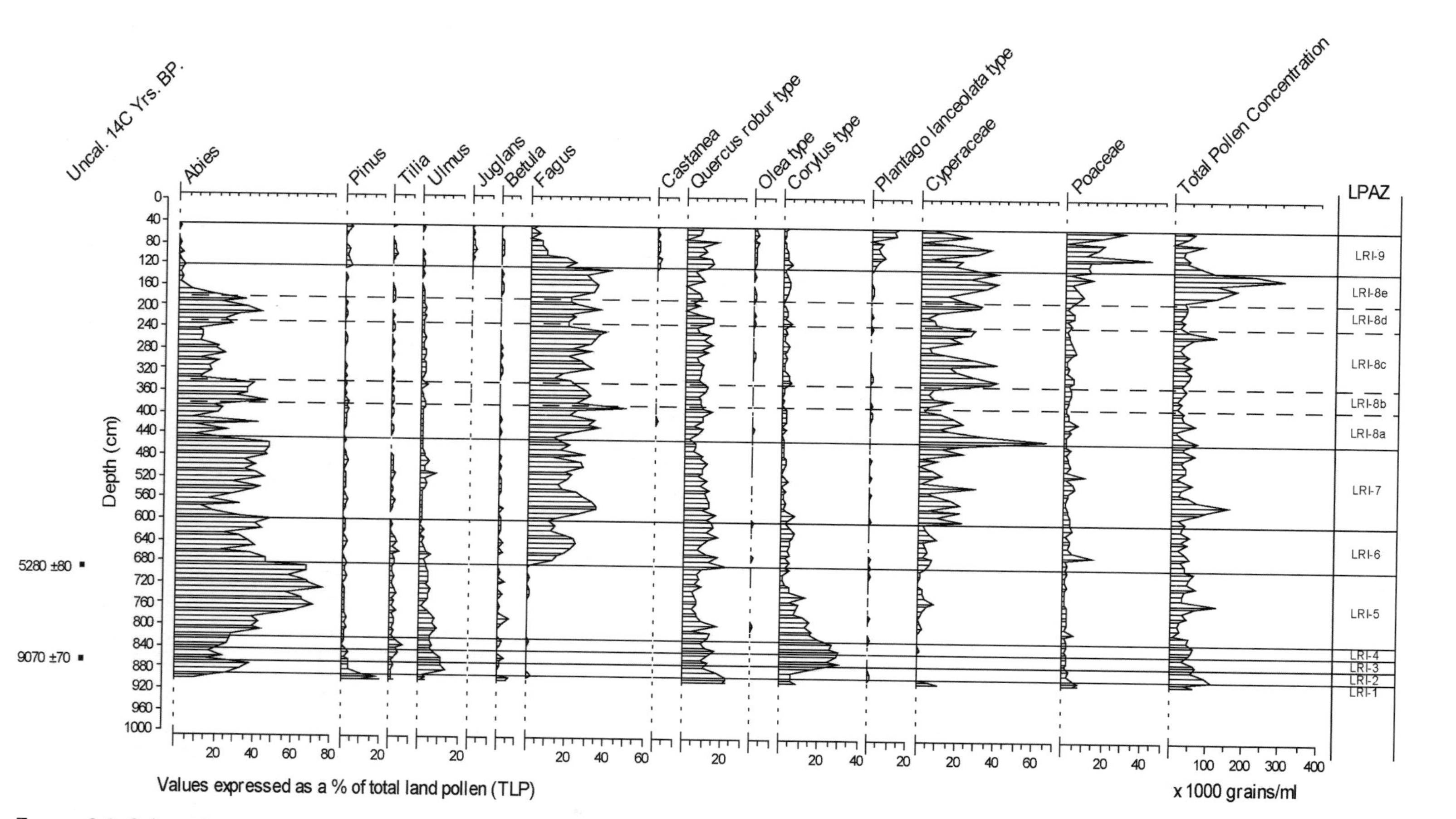

FIGURE 3.2 Selected taxa percentage pollen diagram from Lago Riane, Liguria, Italy. (Branch, 1999)

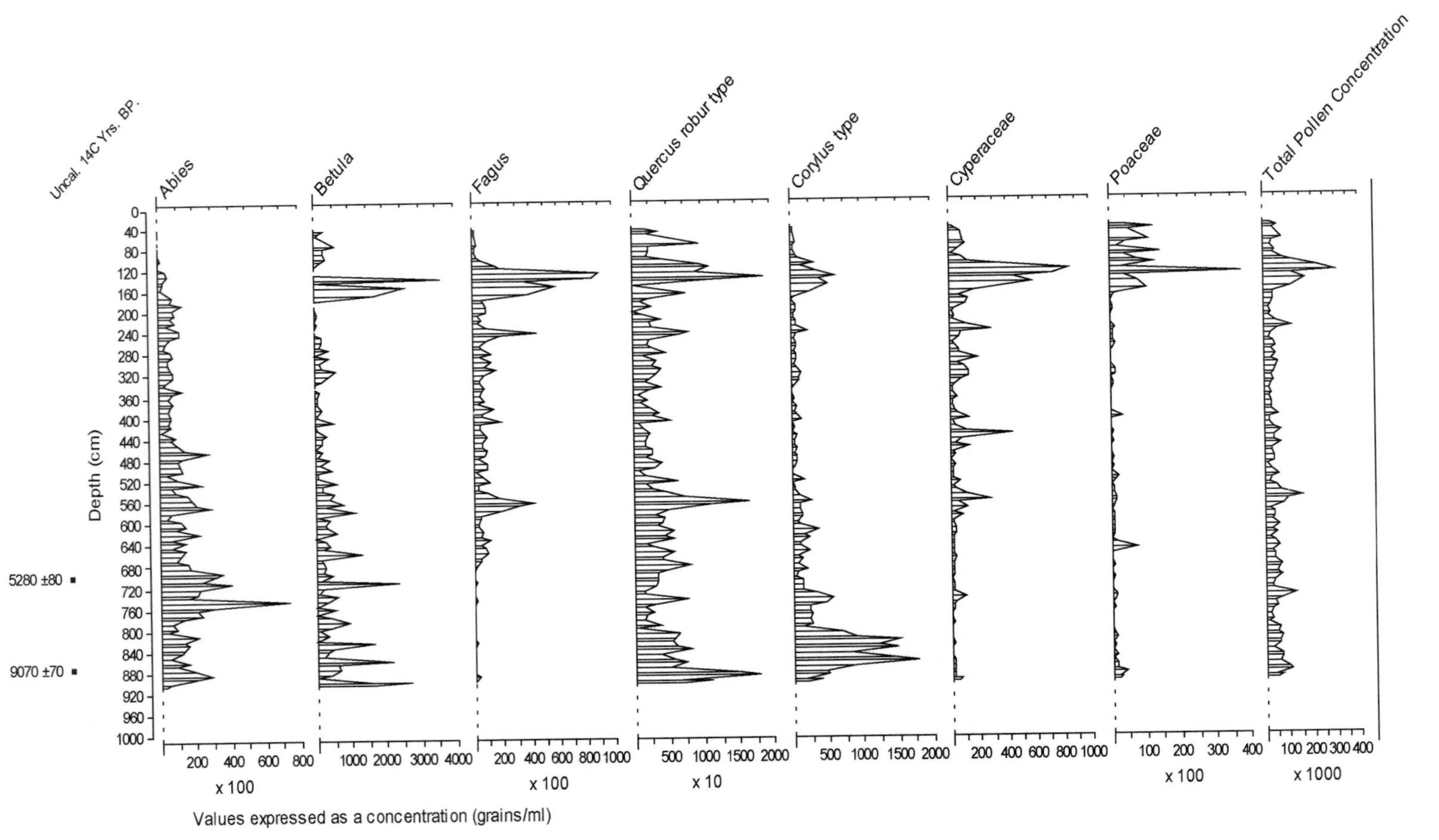

FIGURE 3.3 Selected taxa concentration pollen diagram from Lago Riane, Liguria, Italy. (Branch, 1999)

Edwards, 1983; Turner, 1986; Bradshaw, 1991). Thus mire dimension and sampling strategy provides the greatest potential for detecting changes in vegetation cover that may result from human activities (Edwards, 1979, 1982, 1989, 1991; Janssen, 1986; Berglund, 1994; Waller, 1998).

Identifying the extent and nature of human interference in natural vegetation succession requires the recognition of anthropogenic pollen indicators, often subdivided into indicator species or groups (Moore, 1980, 1983, 1985b, 1986a, b; Behre, 1986, 1990; Turner, 1986). The 'indicator species' approach designates certain pollen taxa as indicative of plants associated with particular forms of land use, such as cultivation, pasture and meadow. However, few of the taxa are diagnostic of a single type of land use and the taxonomic resolution required (to species level) is frequently problematic. Despite these difficulties the concept of 'indicator species' can be extended to the notion of 'indicator groups', which can be defined on the basis of the ratio of arboreal (AP) to non-arboreal pollen (NAP) or using various multivariate statistical tests (Birks, 1990). In addition, qualitative indices of the importance of 'indicator species' have included:

1 Records showing a decline in arboreal pollen values or individual tree taxa (e.g. the Neolithic *Ulmus* decline in the UK), and a corresponding increase in heliophilous taxa (e.g. *Fraxinus*).
2 The presence of certain ruderal weed species, considered as taxa strongly associated with cultivation or pastoralism (e.g. *Plantago lanceolata*).
3 The presence of cultivated species, including crops (e.g. *Hordeum*, *Triticum*, *Zea*) and taxa indicated managed woodland (e.g. *Castanea*, *Juglans* and *Olea*).
4 The increase in taxa strongly associated with human disturbance and pioneer re-colonization (e.g. *Corylus*).

The application of pollen analysis to soils buried beneath archaeological remains, such as ancient monuments, artefact scatters, field boundaries or hearths, has received a great deal of attention in the literature, but requires a very different analytical approach to that used on mires and lakes (Dimbleby, 1957, 1961a, b, 1962, 1977). Soil pollen analysis provides information on the local vegetation composition and structure, and may provide a more detailed picture of the dryland vegetation than is obtainable from bogs and lakes (Dimbleby, 1985; Bradshaw and Millar, 1988). The constraints on the study of soil pollen are, however, immense. The pollen-stratigraphic analysis of deposits associated with mires and lakes relies on the basic assumption that pollen grains and spores found at any single stratigraphic level are coeval, and that no upward or downward movement of pollen occurs. This does not hold true for pollen analysis in soils, where only a crude stratification can be assumed. The controlling factors governing the distribution and diversity of pollen in a soil profile are: (a) the properties of the soil, such as clast particle size and organic matter content; (b) faunal activity; (c) the rate of pollen stratification; and (d) the morphological characteristics of pollen grains and spores. Unfortunately, the relationship between these variables is difficult to quantify and only qualitative statements about the apparent stratification of pollen in soils can be made, which undoubtedly limits any interpretations of the data (Branch and Canti, 1994; Tipping *et al.*, 1994).

Preferential preservation of pollen grains and spores due to differences in their complex biochemical structure may lead to distortions in the pollen spectra obtained from some soils where optimal conditions are not found, e.g. low pH (Havinga, 1964, 1967; Kerney *et al.*, 1980). These problems are compounded by the properties of the soil structure, e.g. particle size, and artificial stratification due to faunal mixing. Tipping *et al.* (1994) demonstrated the clear need for soil micromorphologists and pollen analysts to work in unison, with field sampling strategies coordinated and laboratory results integrated to check the validity of the pollen-stratigraphic data. They found that knowledge of pedological processes had significant implications for interpretation of the pollen record, and also that the pollen record may assist in the interpretation of pedological features (e.g. ground disturbance by cultivation or turf stripping). The methodology should also include a systematic procedure to assess the reliability of pollen assemblages, building upon several studies conducted in the 1980s. This involves assessing the condition of preservation of each identifiable pollen grain and spore (Fig. 3.4), and recording whether it is: (a) amorphous (i.e. loss of exine features and often distortion of the grain/spore shape); (b) corroded (i.e. differential preservation of exine features); (c) broken (i.e. fragmentation of the pollen grain or spore); or (d) folded (i.e. creases in the pollen grain or spore) (Cushing, 1967; Lowe, 1982). Although categories (c) and (d) may be due to laboratory extraction procedures, categories (a) and (b) provide a valuable means of assessing preferential preservation.

Pollen analysis has been widely used in archaeological cave and rock-shelter studies; although both the theoretical and practical approaches differ radically to those employed in the routine investigation of mires and lakes (Butzer and Freeman, 1968; Triat-Laval, 1978; Gale and Hunt, 1985; Carrión, 1992; Carrión and Dupré, 1994; Carrión *et al.*, 1995, 1998; Branch, 1999; Navarro Camacho *et al.*, 2000). In addition, further experimentally based taphonomic studies are required (Schmid, 1969; Legge, 1972; Laville, 1976; Turner, 1985; Coles *et al.*, 1989; Hale and Noel, 1991; Piperno and Giacobini, 1992; Hunt, 1993; Coles and Gilbertson, 1994). Figure 3.7 provides a hypothetical model of the mode of transportation of pollen into cave sediments, although both components (natural and anthropogenic) could equally include other classes of biological remains. The model, while highlighting the various sources of pollen, nevertheless gives the false impression that both components have an equal chance of being present. This is highly unlikely in cave sites that have evidence for long-term human occupation. Preservation of pollen is also difficult to predict in caves due to the highly variable physical and geochemical properties of the sediments, which may be a consequence of natural variations in pH and particle size, or alteration of these natural properties by human activities including stabling of animals and burning. Nevertheless, important information may be obtained on the nature of the local environment and human activities.

Another example of the successful application of pollen analysis to environmental archaeology is that of graves. Recovery of bioarchaeological

Box 3.1 Soil pollen analysis at Lagorara archaeological site, Liguria, Italy

Located near La Spezia, the Lagorara Valley archaeological site indicates intensive human activity during the Copper and Bronze Ages (4700–2800 BP), including the quarrying of jasper and preparation of various tools (Fig. 3.5). Geoarchaeological records indicate that during the early Holocene, extensive colluviation resulted in the deposition of thick mineral sediments on the valley sites. In one location, these deposits contained interbedded organic matter (topsoil and humus) formed in a small localized depression, overlain by a thick occupation surface. The organic deposits, representing locally eroded topsoil and woodland detritus, provided a unique opportunity to reconstruct the local dryland vegetation cover prior to the main period of occupation. Analysis of pollen grains and spores, together with an assessment of their reliability (Fig. 3.6), soil micromorphology and a range of physical tests (magnetic susceptibility, particle size and organic matter content), was able to establish that the woodland comprised *Fagus* with *Abies*, *Ulmus*, *Corylus* and *Tilia* from c. 6700 BP (Early Neolithic); this was the earliest record of *Fagus* colonization in the northern Apennines. Immediately prior to the Copper and Bronze occupation, there is unequivocal evidence for the decline of the woodland cover (mainly *Abies*, *Tilia* and *Ulmus*) and an increase in non-arboreal taxa (*Plantago*, *Rumex*, *Potentilla*, *Centaurea* and *Serratula*), possibly representing the earliest signs of human interference in the natural vegetation succession. (Branch, 1999, 2002)

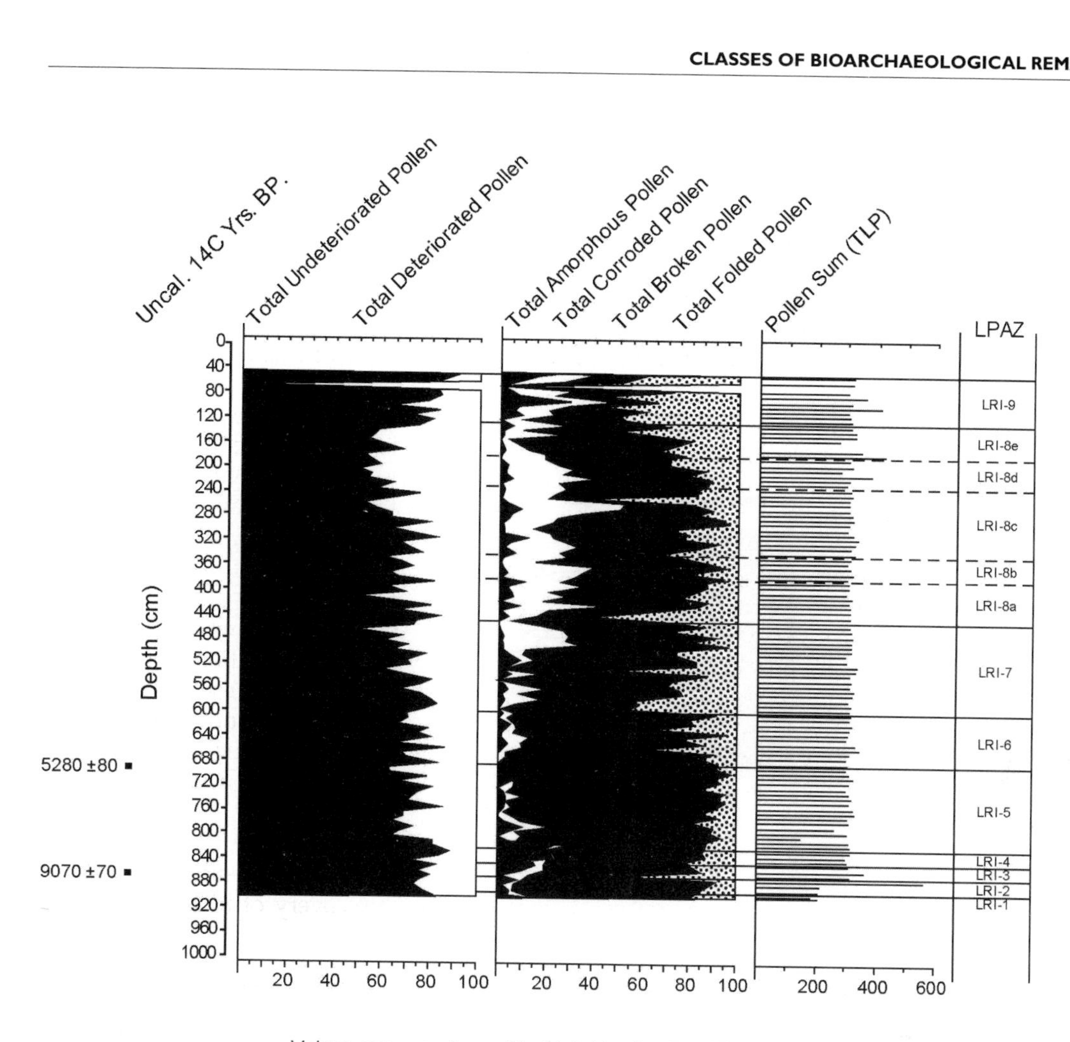

FIGURE 3.4 Pollen preservation diagram from Lago Riane, Liguria, Italy. (Branch, 1999)

FIGURE 3.5 The archaeological site of Lagorara, Lagorara Valley, Liguria, Italy. (Branch, 1999)

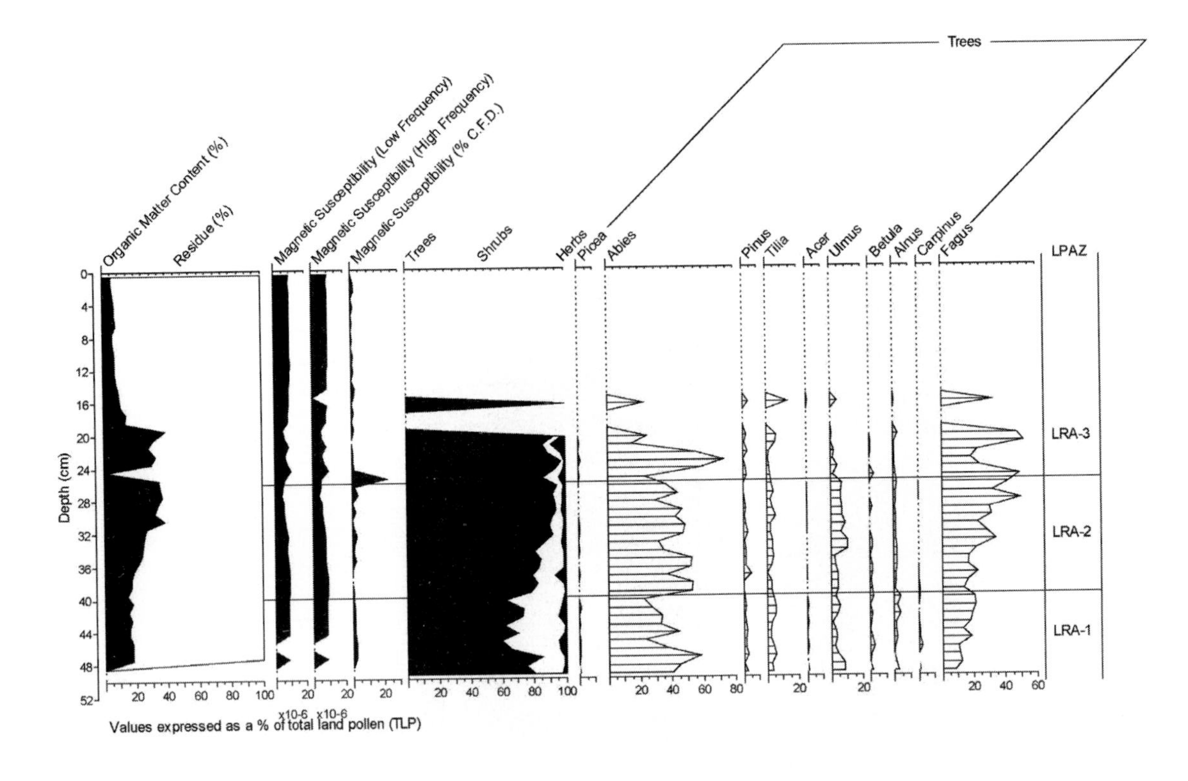

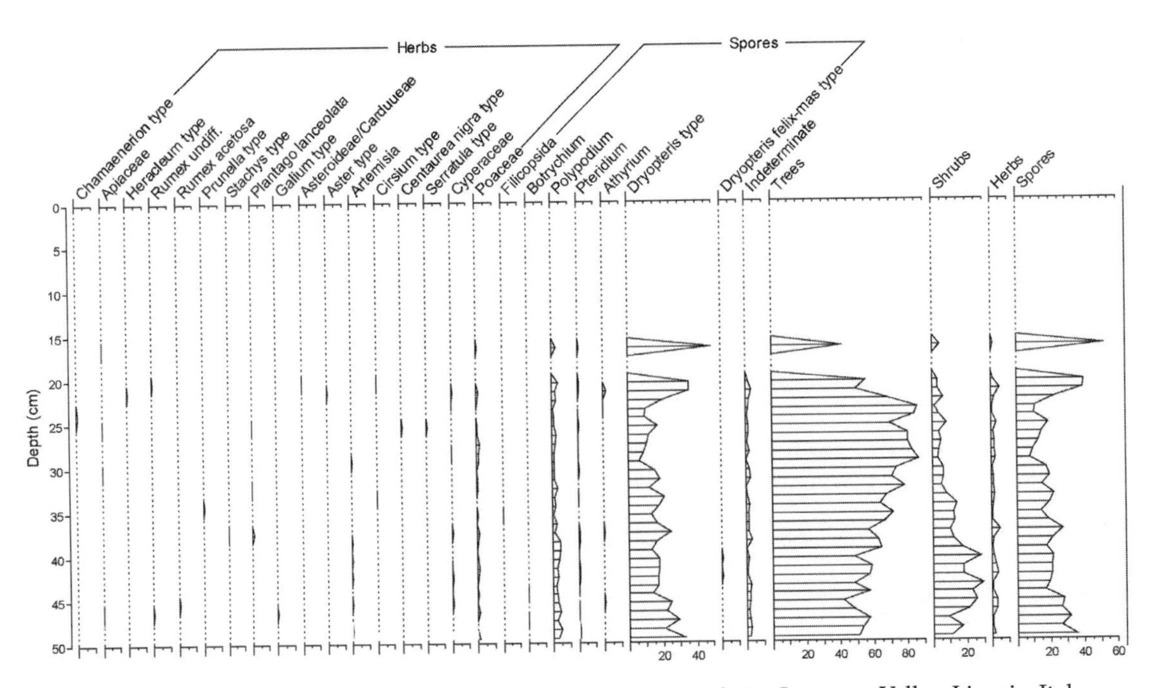

FIGURE 3.6 Percentage pollen diagram from Lagorara archaeological site, Lagorara Valley, Liguria, Italy. (Branch, 1999)

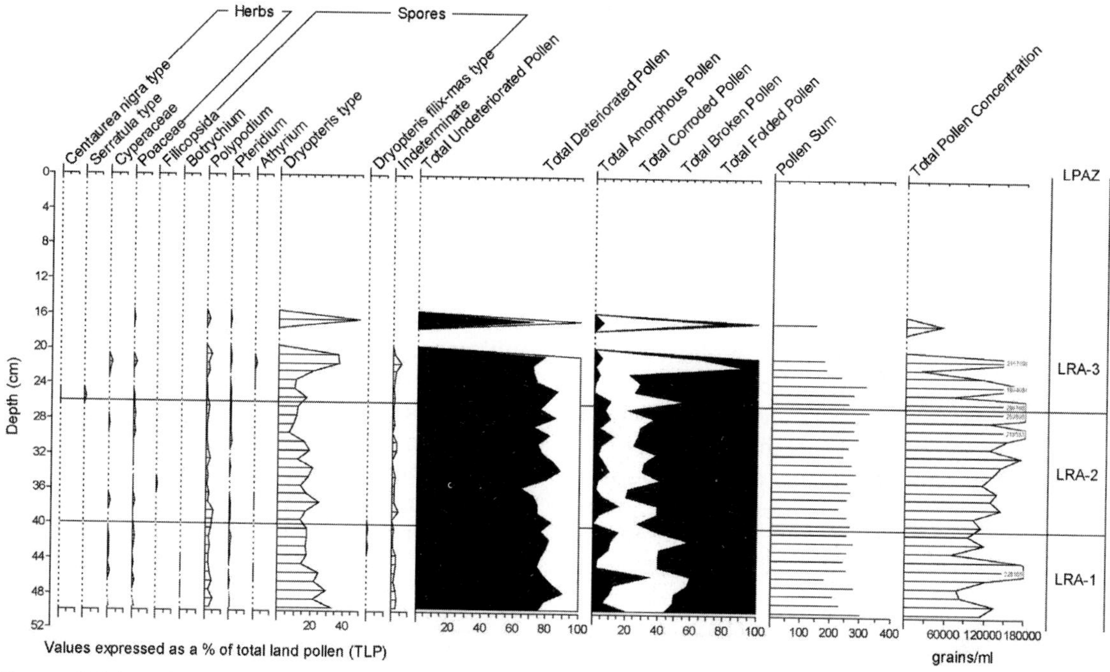

FIGURE 3.6 *(cont.)*

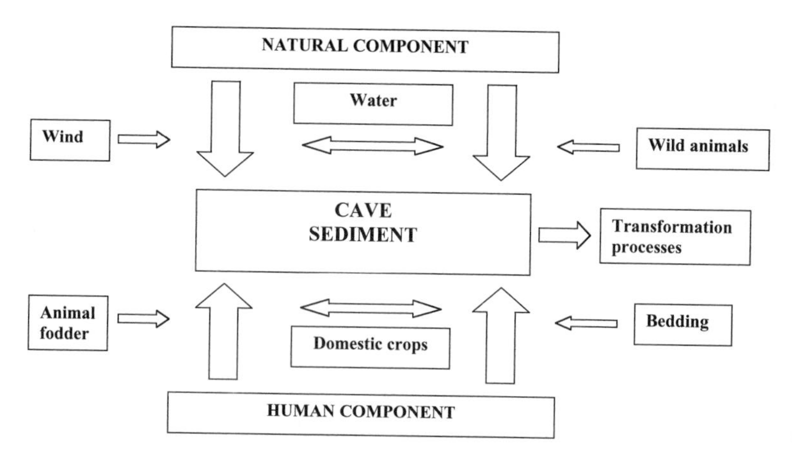

FIGURE 3.7 Hypothetical model of pollen transportation into cave sediments.

remains, other than human bone, from graves is relatively rare. When present, they can provide important information on ritual/religion practices and diet (Dickson, 1978; Whittington, 1993; Clarke, 1999). This is exemplified by the pollen analyses of Bronze Age (c. 3800–3300 uncal BP) cist burials in Scotland (Tipping, 1994; Bunting *et al.*, 2001). At Linga Fiold, Orkney (UK), soils beneath the burials recorded an open landscape with meadow or pasture, whereas pollen and plant macrofossil analysis of the cist fills indicated the presence of plant taxa probably associated with fuel used in the pyre (e.g. turves or peat blocks). Study of the cist 'floors' was complicated by problems associated with understanding the pollen taphonomy. Bunting *et al.* identified four possible sources: (a) wind transporting pollen into the open grave; (b) mixing of pollen from the subsoil and fill; (c) pollen from ecofacts and artefacts placed on the grave floor; and (d)

Box 3.2 The cave of Arene Candide, Liguria, Italy

The Jurassic limestone cave of Arene Candide, excavated since 1864, has revealed evidence for human occupation from the Upper Palaeolithic to the Roman period, although it is the Neolithic deposits that have received the greatest attention (Fig. 3.8). The bioarchaeological study of these deposits, involving the analysis of pollen, phytoliths, molluscs, small mammal, charcoal, charred plant remains and animal bone, in collaboration with soil micromorphology, indicates that they consist almost entirely of anthropogenic deposits derived from the gathering of fodder, bedding, burning and trampling (Fig. 3.9). During the Early Neolithic (6900–6150 BP), evidence for herbivore stabling and domestic occupation involved the exploitation of *Quercus ilex*, deciduous *Quercus*, *Pistacia terebinthus*, *Phillyrea*, *Olea*, *Taccus baccata* and *Ulmus*, among others, for branch fodder (Nisbet, 1997) as well as cereal cultivation (Branch, 1997). During the Middle Neolithic (6000–5400 BP), the overwhelming evidence for stabling includes the exploitation of *Erica arborea* and *Genista*, and the utilization of *Triticum dicoccum* and *Hordeum vulgare* var *hexasticum*. The multi-proxy analysis has also concluded that outside the cave, the Early Neolithic vegetation cover consisted of diverse undisturbed mixed coniferous-deciduous woodland, although there is clear evidence for a reduction in woodland cover during the late Middle Neolithic and an increase in 'macchia' type vegetation, typical of the Mediterranean coast. (Branch, 1997, 1999)

FIGURE 3.8 Stratified sequence of sediment reflecting mainly anthropogenic activities inside the cave of Arene Candide, Liguria, Italy. (Branch, 1999)

operator errors, such as contamination or sampling problems. The study highlighted the importance of establishing the possible presence of each component, but its main contribution was the identification of high pollen values of *Plantago lanceolata* in three cists thought to be associated with deliberate placement of the flowering plant into the open grave. Tipping recorded similar evidence involving the deposition of several taxa, especially *Filipendula* (*F. ulmaria* (meadowsweet) and/or *F. vulgaris* (dropwort)), and concluded that the high concentrations of well preserved *Filipendula* in clusters and spatially discreet areas of the grave indicated deliberate placement of the flowering plant rather than the presence of honey, mead (i.e. *Filipendula ulmaria* being used as flavouring) or food.

3.2.1.2 Diatoms

Diatoms are unicellular algae and comprise a silicified (opaline silica) cell wall (frustule) with two overlapping valves (epivalve and hypovalve) (Fig. 3.10). The cell ranges in size between 5 and 2000 μm, although most species are 20 to 200 μm (Brasier, 1980; Battarbee, 1986). Valve sculpture, and the area of overlap between the

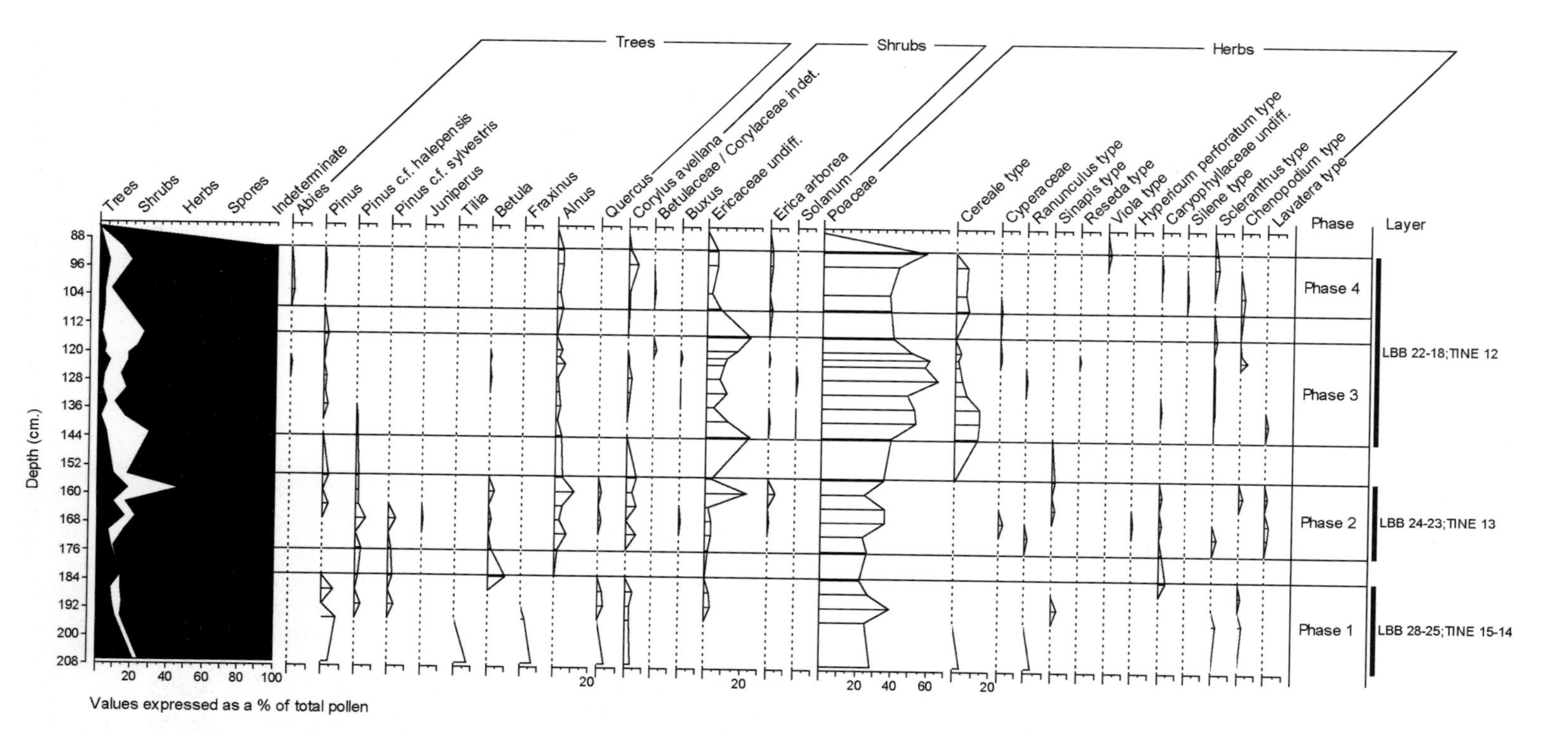

FIGURE 3.9 Selected taxa percentage pollen diagram from Arene Candide archaeological site, near Finale Ligure, Liguria, Italy. (Branch, 1999)

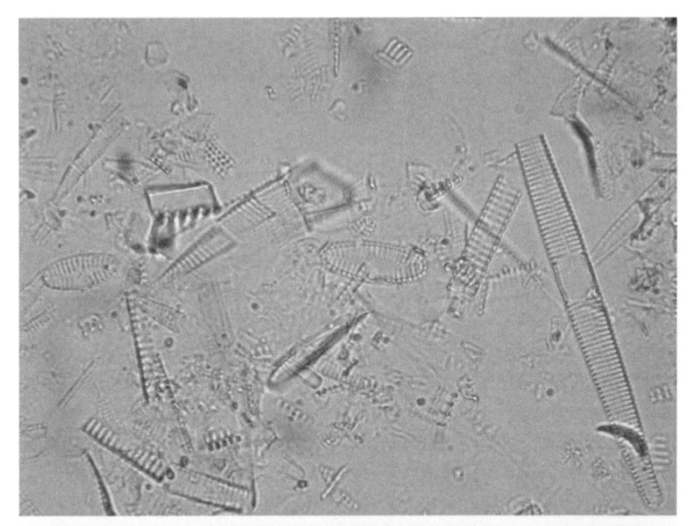

FIGURE 3.10 Selected diatoms commonly found in European archaeological and geological archives (magnification ×1000).

Achnanthes lanceolata and *Synedra pulchella*

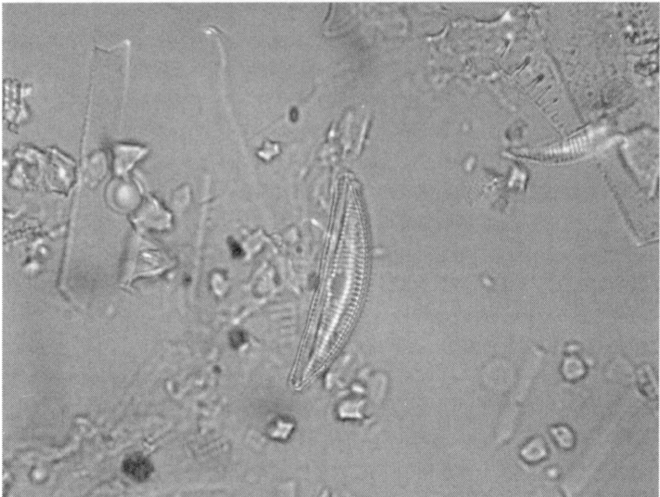

Amphora libyca

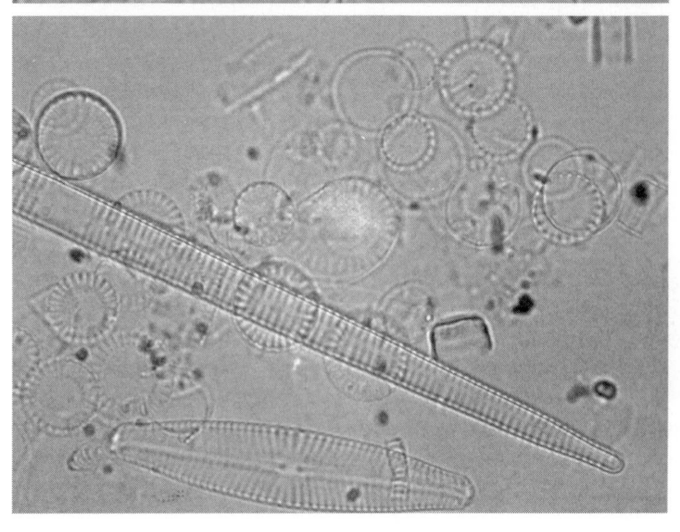

Synedra ulna

FIGURE 3.10 (*cont.*)

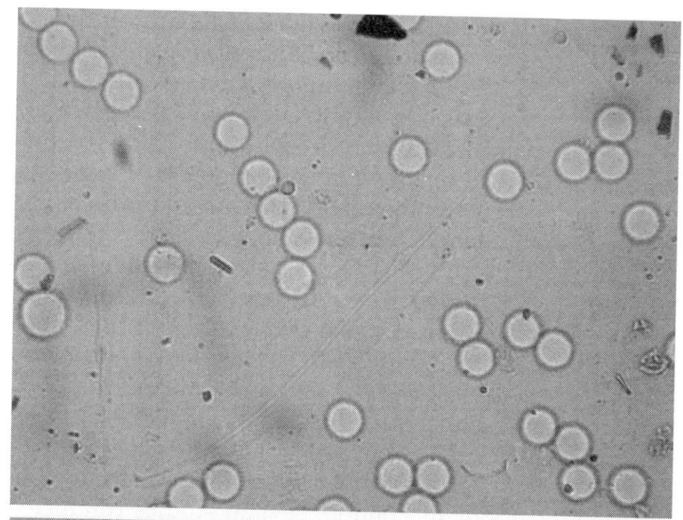

Gyrosigma acuminatum (with microspheres for calculating diatom concentration)

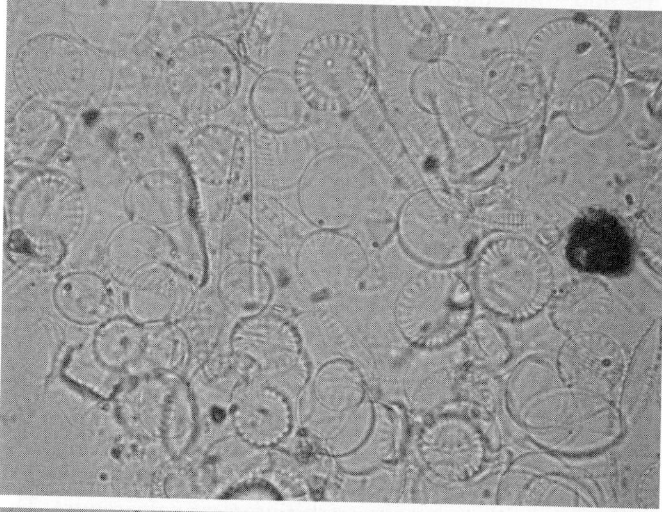

Cyclotella meneghiniana

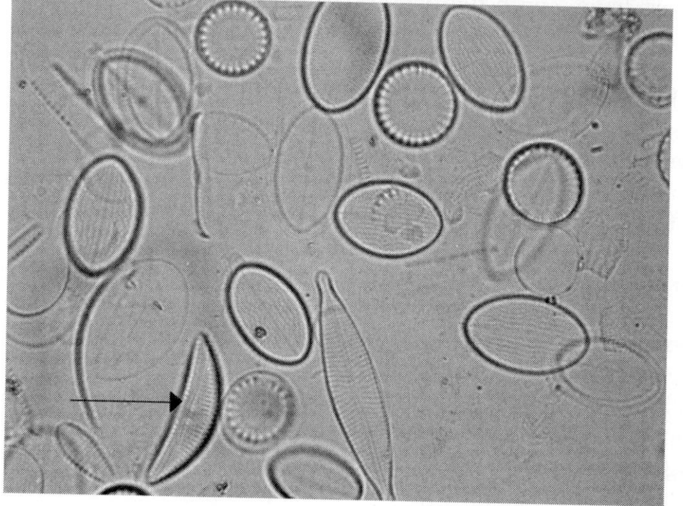

Amphora libyca (indicated) and *Cocconeis placentula* (all modern specimens)

valves (girdle), are especially important for diatom identification, as are the punctae (perforations in the valve surface) and the valve shape (either pennate (elliptical) or centric (circular)). Diatom taxonomy and identification is based upon several guides, including keys and photographs. The following are a few examples and are referenced in Battarbee (1986): Schmidt (1874–1959); Hustedt (1930, 1930–66); Huber-Pestalozzi (1942); Cleve-Euler (1951–55); Patrick and Reimer (1966, 1975). Websites can be found at http://www.soton.ac.uk/~ibg/, http://www.calacademy.org/research/diatoms/, http://www.bgsu.edu/departments/biology/facilities/algae_link.html and http://www.ucmp.berkeley.edu/chromista/bacillariophyta.html. Diatom taxonomy (two orders: Pennales and Centrales; Hendey, 1964) and ecology are well known, with different species occupying the bottom of (benthic), or floating within (planktonic), water bodies (e.g. oceans, lakes, ponds, rivers, salt marshes, ditches), and living in soil and on trees (epiphytic). They are valuable in archaeology because species are indicative of a wide variety of environmental conditions (e.g. marine, brackish or freshwater), reflecting temperature, salinity (level of common salt in solution), pH (acidity or alkalinity), oxygen and mineral content (e.g. silica, phosphate, nitrate and iron). However, depositional processes, and strongly acidic or alkaline conditions, may lead to the dissolution of diatoms, especially the weakly silicified forms, resulting in a biased assemblage. In addition, several studies have also highlighted the problems associated with poor palaeoenvironmental resolution brought about by the mixing of marine, brackish and freshwater taxa. Once extracted from a sedimentary sequence, counted diatoms are presented diagrammatically as percentages, although diatom concentration and influx diagrams have also been used (Fig. 3.11).

Diatoms have been used routinely in environmental archaeological research to reconstruct changes in the natural environment, possibly as a consequence of human activities. The geological archives most frequently studied include lake sediments and alluvium. Diatoms found in these sediments are often indicative of changes in the lake or river catchment, such as accelerated erosion consequent upon deforestation, which may produce fluctuations in trophic (nutrient) status. Deep alluvial sequences in lowland river valleys (e.g. lower Thames, UK) have also recorded changes in diatom assemblages due to fluctuations in relative sea level during the Holocene. Marine transgressive phases are indicated by the dominance of marine diatoms, whereas the transition to marine regressive phases (reduction or stabilization in relative sea level) is shown by a progressive increase in freshwater and brackish water taxa. Studies in the London Thames have also elucidated the timing and rate of the westward movement (upstream) of the tidal head, which has important consequences for our understanding of the development of the Roman harbour in London. Several studies conducted on ditch fills have provided useful information on water depth, presence of running water and pollution from faecal material. In conclusion, diatoms have the potential to record local and regional environmental changes as a consequence of natural as well as human induced processes. Once integrated with other proxy data and archaeological records, they provide the means of assessing the response of humans to changes in coastal, estuarine, riverine and lacustrine environments (Battarbee, 1988).

3.2.1.3 Ostracods

Ostracods (Ostracoda) are Crustacea (aquatic invertebrates) and comprise a shell (carapace) with two valves (chitinous or calcareous). The shell is 'ovate, kidney-shaped or bean-shaped and from 0.3 mm to 30 mm long' (Brasier, 1980: 123), with one valve slightly larger and overlapping the other, hinged along the dorsal margin (Fig. 3.12). Shell shape and features along the posterior, anterior, dorsal and ventral views, as well as on the inside, enable identification of adult specimens. Species are divided into two groups, those occupying the benthic zone (the bottom of a water body) and those within the pelagic zone (the open water body). The benthic ostracods are found in freshwater and seawater, while the pelagic species are almost exclusively marine. They are highly sensitive to changes in salinity with three main assemblages identifiable: freshwater (<0.5‰;

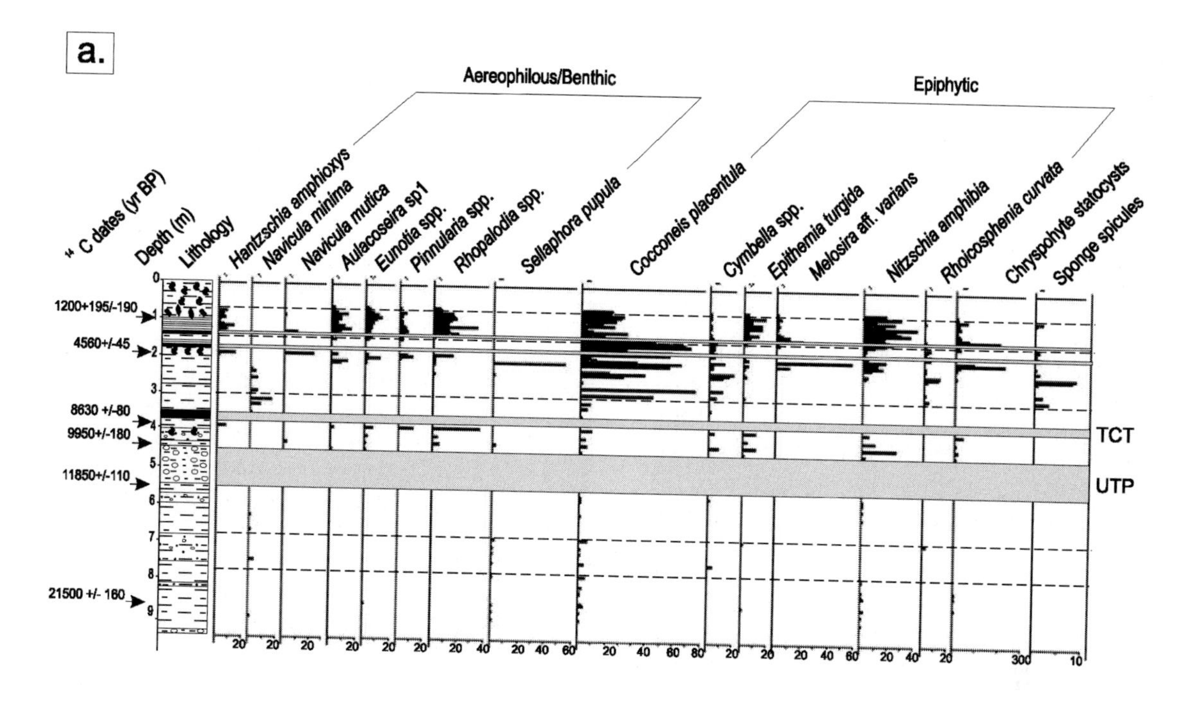

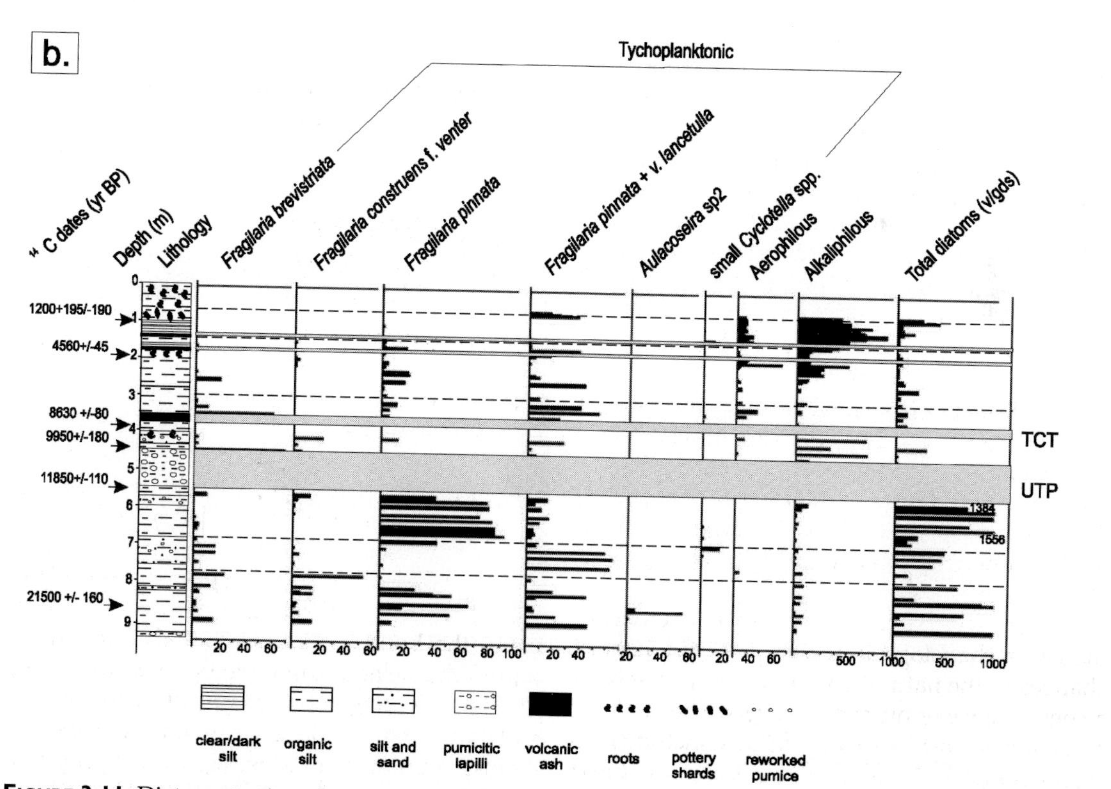

FIGURE 3.11 Diatom stratigraphic sequence from the Upper Lerma Basin, Central Mexico. (Caballero *et al.*, 2002)

FIGURE 3.12 Selected Ostracoda found in European archaeological and geological archives (scanning electron micrographs).

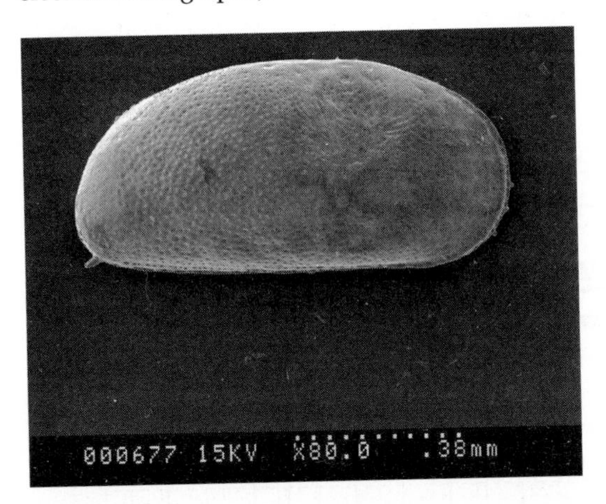

Cyprideis torosa

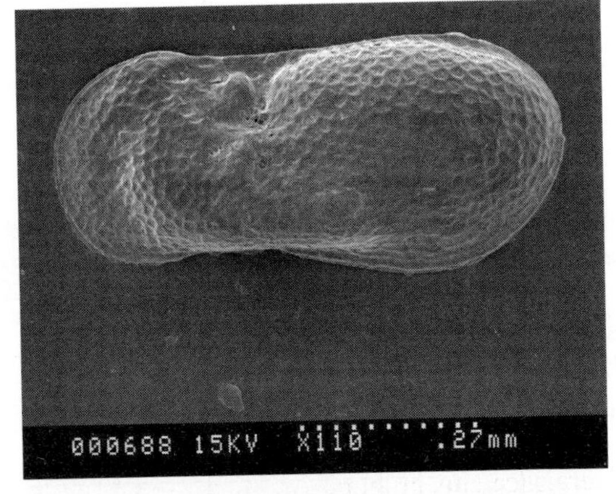

Limnocytherina sanctipatricii

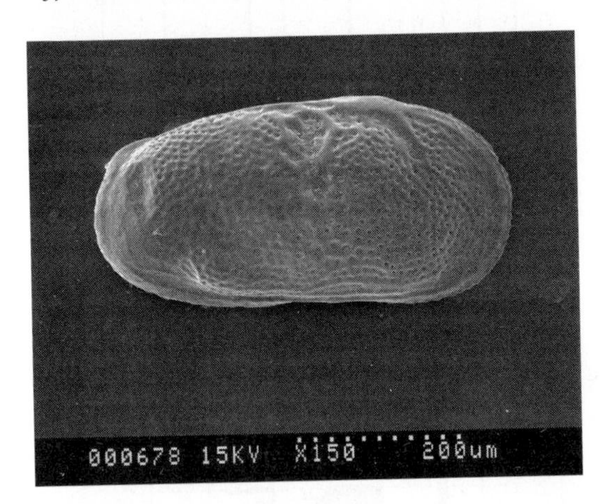

Leptocythere castanea

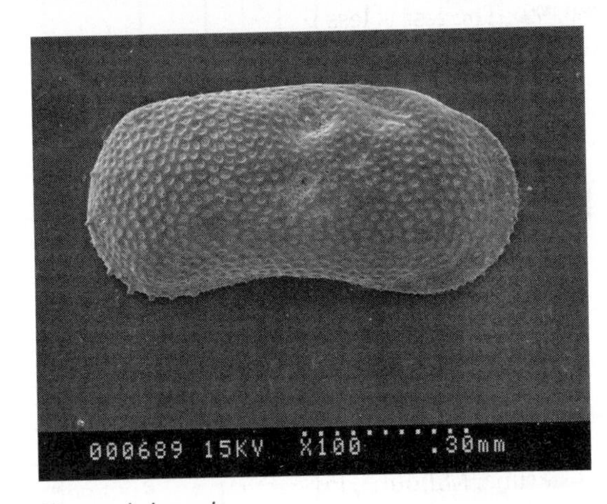

Ilyocypris inermis

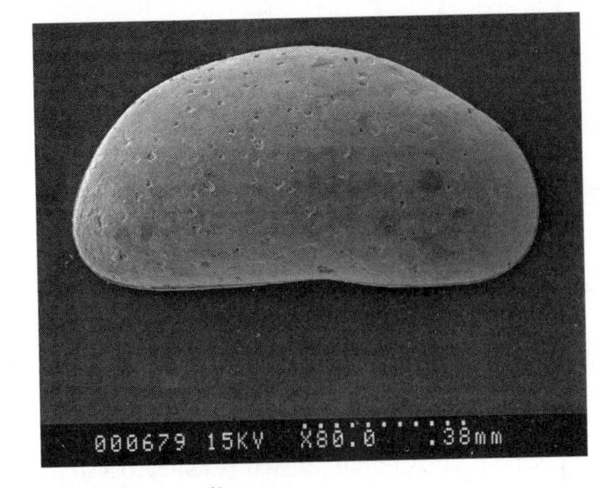

Psychodromus olivaceus

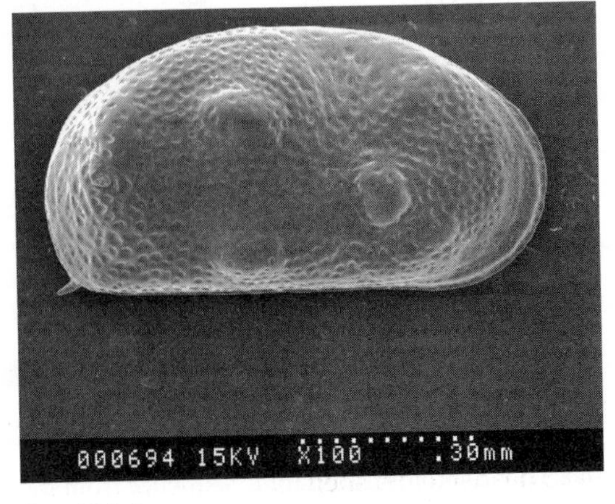

Cyprideis torosa

lakes, rivers and ponds), brackish water (0.5–30‰; lagoons and salt marshes) and marine (30–40‰; oceans and seas), as well as rainfall, temperature and alkalinity. They are valuable in archaeology as indicators of changing coastline morphology and the height of relative sea level (Fig. 3.13). Websites for ostracod studies are http://www.ucl.ac.uk/GeolSci/micropal/ostracod.html and http://www.gre.ac.uk/schools/nri/earth/ostracod/index.htm.

3.2.1.4 Foraminifera

Forams (Foraminiferida) are unicellular organisms and comprise a shell (test) composed either of secreted organic matter (tectin) and secreted minerals (calcite, aragonite or silica), or of agglutinated particles from the environment (Brasier, 1980: 90). The test is less than 1 mm in size, with single (unilocular) or multiple (multilocular) chambers, openings (foramen/foramina) and apertures (Fig. 3.14). Test sculpture, wall structure and composition, chamber development and growth plan, aperture and foramen shape and position, are all very important for foraminifera identification (see http:/www.emidas.ethz.ch/, http://rin.hiroba.org/foraminifera/ and http://www.ucmp.berkeley.edu/foram/types/types.html). For individual species, careful differentiation between the gamont and schizont generations of the life cycle may also be required. Their taxonomy (five suborders: Allogromiina, Textulariina, Miliolina, Fusulinina and Rotaliina) and ecology are well known, with different species occupying niches in marine environments according to food availability, levels of predation, type of substrate, light penetration, water temperature and salinity. Thus, foraminifera are useful indicators of changes in water depth, salinity and climate (Fig. 3.13).

3.2.1.5 Cladocera

Cladocera are microscopic animals (arthropods), comprising a hard exoskeleton that is periodically discarded (exuviae) to enable the organism to grow in size (see www.cladocera.fsnet.co.uk). These discarded elements (skeletal fragments) include headshields, shell, post-abdomen, post-abdominal claws, antennules, antennal segments, mandibles and trunk limbs. Although it should be possible to identify individual species from these fragments, in reality only headshields and shells enable reliable morphological differentiation. The species are divided into two broad groups based upon their ecological preferences: offshore planktonic species; and inshore littoral species. The second group can be subdivided into those associated with macrophytes (algae and detritus), and those with sediments. Cladocera are therefore useful for reconstructing the 'responses of aquatic ecosystems to external changes in climate and watershed processes' (Frey, 1986: 667). These external factors, e.g. changes in temperature and precipitation (e.g. Hofmann, 2000), erosion and human activity (Jeppesen *et al.*, 2001), may cause water level fluctuations and changes in sediment accumulation rates, which in turn will result in changes to the Cladocera species composition. Detecting palaeoecological changes may be limited by taphonomic factors, especially those associated with natural variations over time in Cladocera concentration between the littoral and offshore zones.

Quantification of the species concentration for each sample (typically 1 cm^3) normally involves counting a minimum of c. 200 exuviae per sample. Data analysis includes abundance diagrams of the individual taxa, a summary of percentage planktonic and littoral species, and a diversity index and equitability ratio. These approaches allow an assessment of changing diversity by comparing palaeoecological data with a predicted model. Changes in species diversity may indicate a response to reduced macrophyte abundance and light penetration because of increased mineral sediment input to the lake. Significant changes in assemblages may be used to compile Cladocera zones, established by visual examination of the various data analyses, or statistical tests including cluster (dendrograms) and discriminant analysis. In addition, indicator species may be used to establish specific environmental conditions, e.g. eutrophication.

The application of Cladocera analysis to environmental archaeology is still in its infancy, but the technique clearly has enormous potential (Jeppesen *et al.*, 2001). This is exemplified by the studies of Szeroczyńska

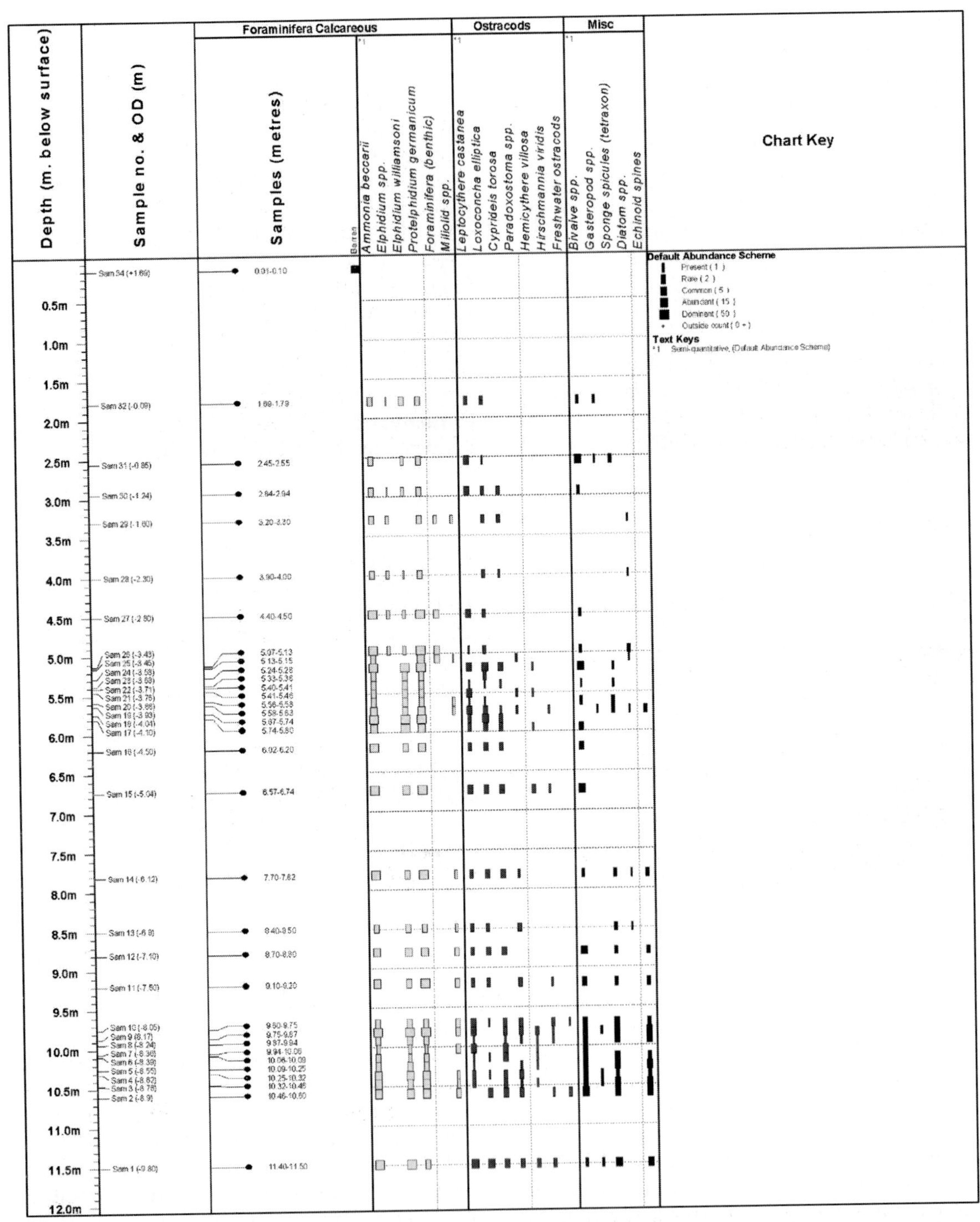

FIGURE 3.13 Ostracoda and Foraminifera abundance for samples obtained from a borehole located on the margins of the Isle of Sheppey, Kent (UK) as part of an archaeological investigation aimed at reconstructing the environmental history of the region. (Branch *et al.*, 2003)

FIGURE 3.14 Selected Foraminifera found in European archaeological and geological archives (scanning electron micrographs).

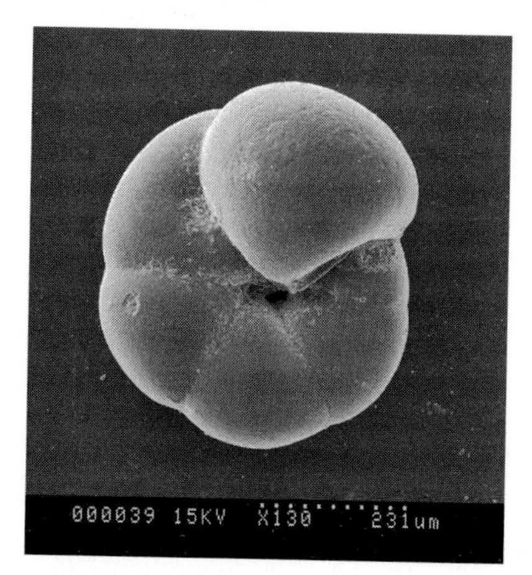

Trochammina inflata

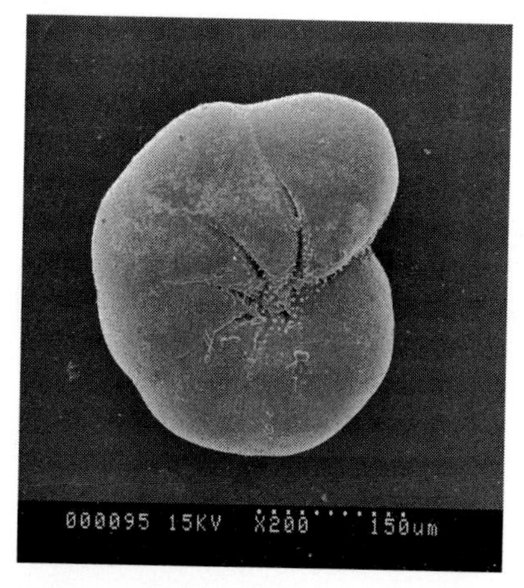

Haynesina germanica

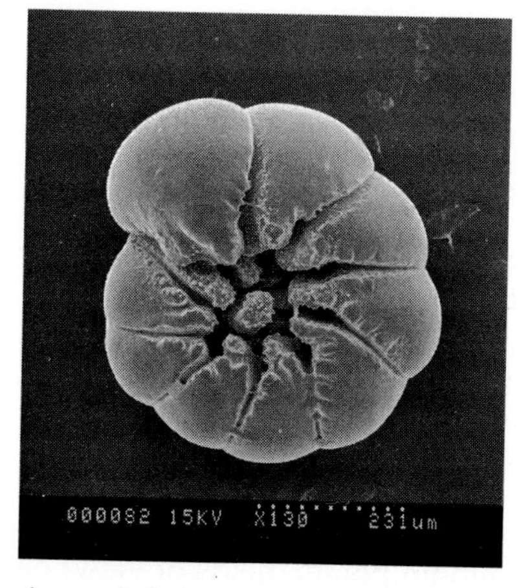

Ammonia batavus

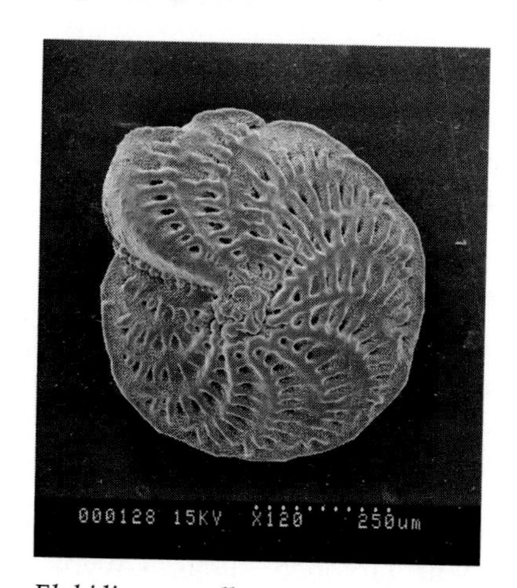

Elphidium macellum

(1998, 2002), who analysed Cladocera from 26 lakes in Poland, and 1 in Croatia. The work demonstrated the value of Cladocera in reconstructing climate change, lake level oscillations, trophic status and chemistry. It was concluded that increased eutrophication accompanied the expansion of the Cladocera *Bosmina longirostris* during phases of human settlement due to the deposition of refuse into the lake. To detect the activities of humans, therefore, requires a similar approach to that used in pollen analysis, with the location of the lake sediment sample proximal to dryland. The success of these studies are mirrored by those of Guilizzoni *et al.* (2002) who conducted diatom and Cladocera analyses on varved crater lake sediments in Italy, and these have provided valuable information on climate change and human activity. The investigations of lakes Albano and Nemi in central Italy reveal a diatom flora dominated by centric, planktonic taxa (>80 per cent), the main change being an increase in *Stephanodiscus* sp during the last 2000 years, which has been linked to increased lake productivity and human impact from the Roman period

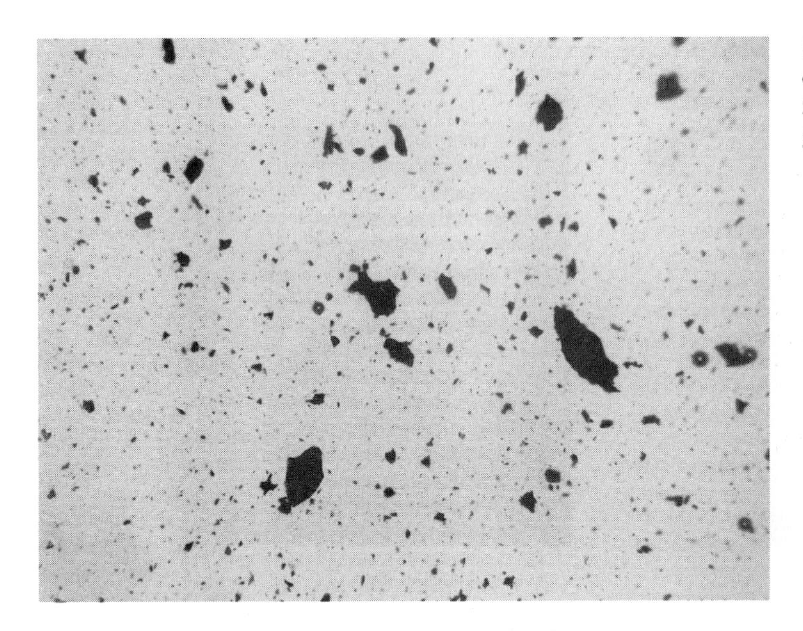

FIGURE 3.15 Microscopic charred particles commonly found in archaeological and geological archives.

onwards. Both lakes also show a shift from *Daphnia* dominated to *Bosmina* dominated Cladocera communities that may be correlated with forest clearance during the last 3000 years.

3.2.1.6 Microscopic charred particles

The presence of microscopic charred particles (MCP <100 μm; Fig. 3.15) in archaeological and geological archives often provides useful information on natural wild fires (including climate history), human-induced woodland clearance, agricultural practices and woodland management (Tolonen, 1986; Cwynar, 1987). Distinguishing between human and natural fires is problematical, and often relies on supporting information from macroscopic charcoal, pollen and sedimentological studies. The interpretation of MCP is complicated further by difficulties involved in establishing the source area (including multiple simultaneous sources), the dispersal mechanisms (although this is thought to be mainly by wind suspension, saltation and traction), the mode of deposition (e.g. wind, water), and identifying diagenetic effects (e.g. fragmentation). These variables are also known to be affected by the type of fuel, the size of the fire and the prevailing weather conditions (Clark, 1988; Morrison, 1994). For these reasons, the usefulness of quantitative MCP data has been questioned. Nevertheless, two procedures are conventionally used, while the third is highly innovative. Extraction of MCP follows procedures conventionally used in pollen analysis, and necessitates counting particles to provide an estimate of their total concentration.

1 Counting particles to establish the total 'charcoal area', 'charcoal concentration' (particles/ml) and 'charcoal influx' (particles/ml/year).
2 The point-count method (Clark, 1982).
3 Select 500 random points (fields of view) on the microscope slide, and count particles measuring a minimum of 15 μm in diameter with a minimum area of 156 μm^2 using an image analysis system (magnification ×160) (Morrison, 1994; Fig. 3.16); the system is programmed to select grey shade values between 0 and 63 (0 = black, 63 = white), and by analysing the pixels in each particle image, mean grey shade values and standard deviations can be calculated.

In this case, only particles having a value of <3.5 and a standard deviation of <5 were recorded, and the concentration was calculated using *Lycopodium* tablets (after Stockmarr, 1971).

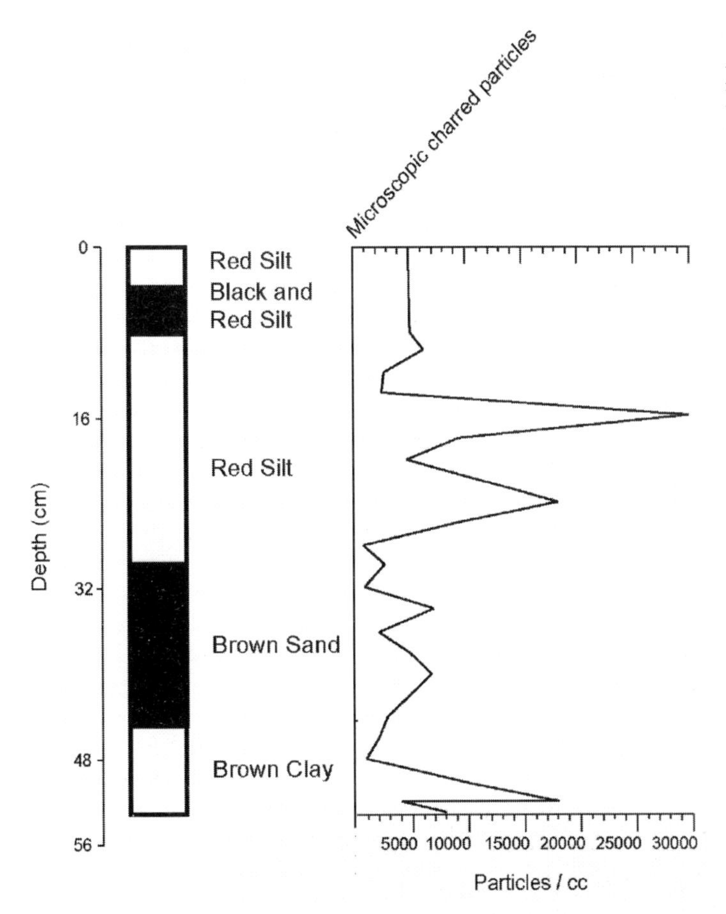

FIGURE 3.16 Microscopic charred particle analysis of sediments from Kamalapuram reservoir, South India. (Modified from Morrison, 1994)

Because of these various methods, and inconsistencies between operators, results are rarely comparable between sites, and differential fragmentation of charred particles makes estimates of concentration meaningless. For these reasons, the presence or absence of charred particles (ubiquity) is an equally valid approach for recording fire history. Site selection is also critical for producing reliable results, and studies have demonstrated that small bogs (<100 m diameter) are ideal due to the proximity of dryland and absence of bioturbation (Smart and Hoffman, 1988). Nevertheless, multi-proxy investigations of blanket peat have frequently provided a rich source of information on human activity, especially woodland disturbance caused by burning (Simmons, 1969a, b, 1975a, b, 1993; Simmons and Innes, 1987, 1996a, b; Innes and Simmons, 1988). For example, pollen and charcoal data from several sequences of early Holocene age in the UK and North America have recorded small-scale clearance of short duration resulting in pronounced changes in woodland ecology (Delcourt *et al.*, 1998; Moore, 2000). These changes have involved removal of woodland by burning, colonization of the woodland gap by heliophilous (light-loving) plant taxa (e.g. *Melampyrum*) followed by the regeneration of shrubs and finally trees (Edwards, 1998).

At North Gill, Yorkshire (UK), radiocarbon-dated fine (high) resolution pollen analysis (1 mm intervals) combined with plant macrofossil and charred particle analysis has provided detailed information on woodland clearance (*Quercus* and *Tilia*) over a period of c. 131 years, cereal cultivation (*Triticum*) over 50 years, and secondary woodland colonization over c. 69 years (Simmons and Innes, 1996a). Although it is widely accepted that localized disturbances in natural vegetation succession would have increased the availability of edible plant and animal resources (e.g. deer, elk, aurochs and pig) for human groups (Legge and Rowley-Conwy, 1988), these inferences are largely based on theoretical modelling and modern analogue studies rather than unambiguous palaeoecological data (Innes and Blackford, 2003). However, pioneering studies conducted at North Gill, involving the use of fungal spore analysis, especially the identification of spore taxa associated with herbivore dung and biomass burning, combined with pollen and charcoal data have provided unequivocal evidence for linkages between woodland disturbance and forest grazing. The evidence indicates multiple phases of fire disturbance of oak-alder woodland, the succession of light-demanding plant taxa and variable amounts of grazing pressure. The latter was determined by fluctuations in the grass pollen record and the presence of several spore taxa, such as *Sporormiella*, *Cercophora*, *Coniochaeta* and *Chaetomium*, associated with herbivores and their dung (Innes and Blackford, 2003).

3.2.1.7 Faecal spherulites

Faecal spherulites are microscopic mineral features found in animal dung (Brochier, 1983, 1996; Brochier *et al.*, 1992). They typically occur in a size range of 5 to 15 μm, and are formed of microcrystalline calcium carbonate (Canti, 1997) radially organized to produce a fixed cross of extinction in cross-polarized light (Fig. 3.17) and other distinctive optical characteristics (Canti 1998a). Faecal spherulites are produced in the gut of many different animals, but appear to be more abundant in herbivores such as cow, sheep, goat (Canti, 1999), llama and alpaca (Coil *et al.*, 2003). The reasons for this production are not known and many modern dung samples do not contain any. However, among samples that do contain spherulites, there is a strong tendency for

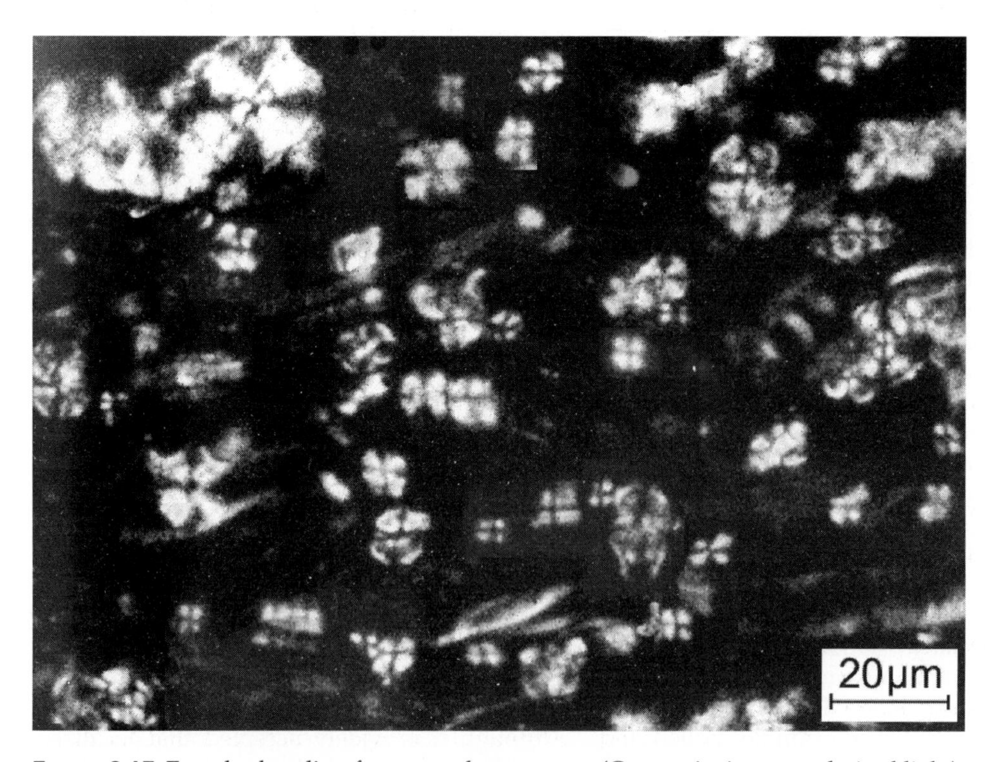

FIGURE 3.17 Faecal spherulites from a modern cow pat (Composite image, polarised light).

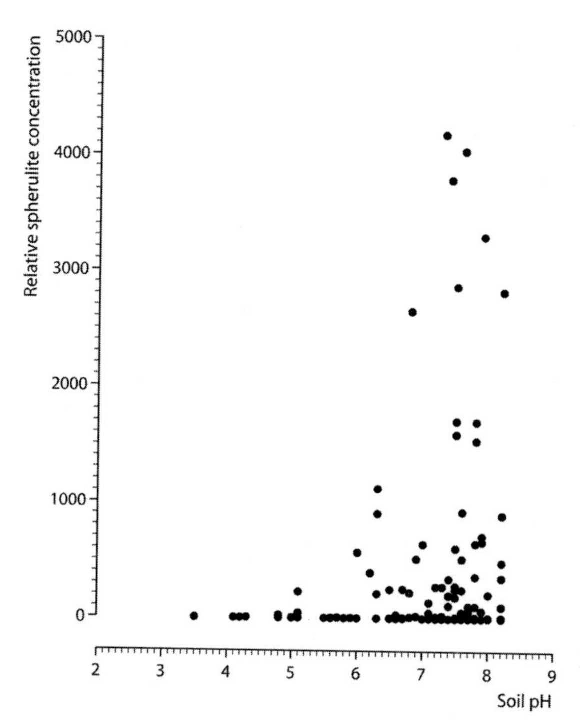

FIGURE 3.18 Relative faecal spherulite concentration in 134 UK dung samples plotted against pH of the soil that the animals were grazing on. (From Canti, 1999)

higher numbers to be produced by animals grazing plants from soils higher than pH 6 (Fig. 3.18). Preservation is partly pH dependent with matrix values above pH 7.5 offering the best opportunity for discovery. Spherulites are valuable indicators of dung in studies of sediments and are part of a growing suite of microscopic materials used for elucidating context processes and functions (Brochier, 1990, 1996; Coil *et al.*, 2003).

3.2.1.8 Earthworm granules

Calcium carbonate is produced by many earthworms in special glands close to the mouth (Darwin, 1881; Robertson, 1936) and some species release granules of calcite aggregates into the soil (Canti, 1998b). The granules vary in size from individual crystals of less than 100 μm to clusters up to 2.5 mm in size (Fig. 3.19). They show distinctive morphological characteristics, but are recognizable to species level only in some cases (Canti and Piearce, 2003). The bigger granules (especially those of the large worms such as *Lumbricus terrestris*) are cast at the surface and accumulate distinctive concentration patterns in the soil profile (Canti, 2003; Preece *et al.*, 1995, 1998). They can, therefore, yield valuable information about stillstand and soil development phases in stratigraphic accretion (Fig. 3.20).

3.2.1.9 Phytoliths and starch grains

Phytoliths are small (5–50 μm) opaline silica bodies produced by plant cells from silica and water (Figure 3.21). They have characteristic shapes enabling identification to various levels, depending on plant taxa. Phytoliths are preserved where other microfossils are commonly absent including dry, alkaline and aerobic conditions. This makes them valuable for archaeological contexts, such as hearths, caves, pits and coprolites. In addition, phytoliths have been found in meal residues, tooth calculus and flint tools. Samples for phytolith analysis are recovered from these contexts by spot sampling (20 gm/sample), separated from the matrix by laboratory extraction, then identified and counted. The data are presented graphically either as a percentage (e.g. % of total count) or estimate of the total phytolith concentration (e.g. phytoliths/gm). Phytoliths may provide valuable information on grassland (e.g. the presence of meadow and pasture), cultivation of crops (e.g. maize) and the presence of herbivores (e.g. within dung), the identification of sub-species (e.g. rice), and in combination with soil micromorphology may be used to assess the extent of bioturbation of soils and sediments infilling archaeological features (Grave and Kealhofer, 1999; Zheng *et al.*, 2003). Useful websites concerning phytoliths can be found at http://www.missouri.edu/~phyto/index.shtml, http://webput.byu.net/tbb/ and http:/www.geo.arizona.edu/palynology/pphyolth.html.

One of the most detailed recent studies of phytoliths has been concerned with the history of maize (*Zea mays*) domestication and cultivation in the New World. The dietary importance

FIGURE 3.19 SEM photograph of typical earthworm granules found in many soils and sediments.

of maize has been conclusively demonstrated by the analysis of phytoliths from residues in prehistoric pottery and dental calculus. These studies have provided valuable information on maize consumption and preparation, as well as tracing the history of maize domestication. Recent work has attempted to widen the scope of this research by the analysis of soil and sediment samples from archaeological and geological (lakes and mires) archives, and establishing specific criteria for the three-dimensional morphological recognition of maize cob phytoliths (Pearsall *et al.*, 2003a). These criteria have been successfully applied to the analysis of stone tool residues from the Real Alto archaeological site in Ecuador, confirming the presence of maize in Ecuador from 2800 to 2400 cal BC, and possibly before (Pearsall *et al.*, 2003b). Iriarte (2003) has extended this study outside neotropical South America and has conclusively identified maize cultivation in south-eastern Uruguay. However, the procedures used in the identification of maize phytoliths, especially cob phytoliths, has not received universal acceptance and several authors have seriously questioned the criteria proposed (Rovner, 2004; reply to Rovner by Pearsall *et al.*, 2004). These concerns cast some doubt on the 'accepted' model for the dispersal of maize into Central America and northern South America, possibly as early as c. 5000–3000 uncal BC, and has important implications for our understanding of the history of this important global crop.

Starch grains recovered from stone tools, soils and sediments are derived from the breakdown of plant resources commonly utilized by human groups, such as seeds, roots and tubers. In combination with phytoliths, and other macroscopic remains of plants, they provide a valuable means of reconstruction of past

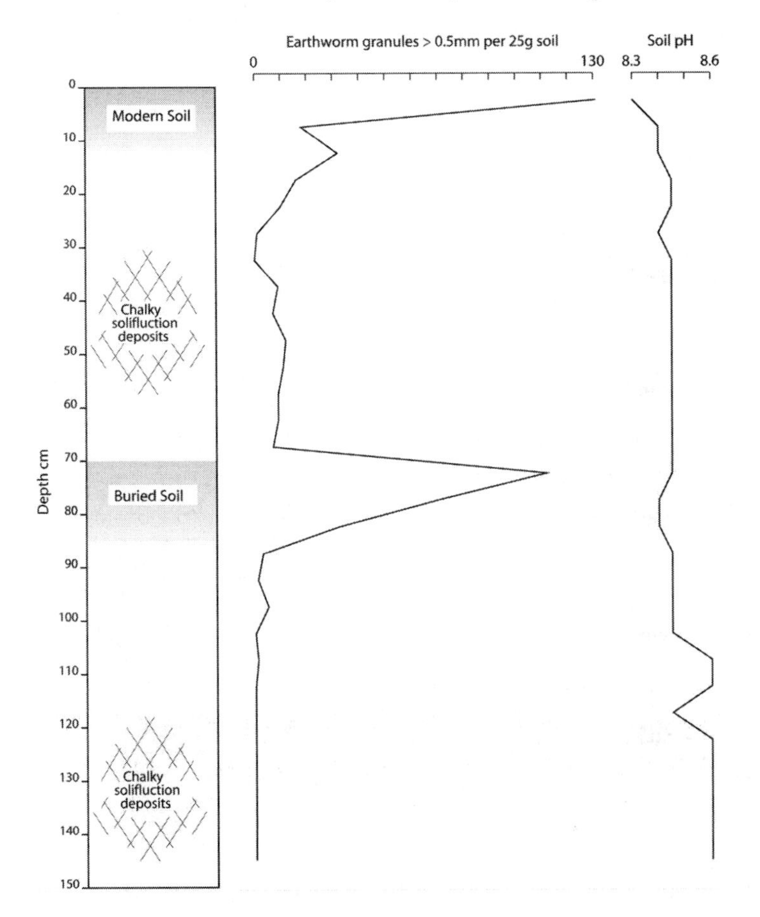

FIGURE 3.20 Concentration of earthworm granules and pH through a modern and buried soil profile at Ventnor, Isle of Wight, UK.

human economy and diet (Loy, 1994; Barton *et al.*, 1998; Piperno and Holst, 1998; Piperno *et al.*, 2000) (Fig. 3.22).

3.2.2 Macrofossils

3.2.2.1 Seeds and associated plant components (charred, waterlogged and mineralized)

The most common plant remains found on archaeological sites are charred (exposure to heat, e.g. hearth) and waterlogged (exposure to excess moisture, e.g. ditch, moat, well) seeds, and to a lesser extent mineralized (replacement by salts, e.g. cesspits) and desiccated (exposure to extreme aridity) seeds, and those preserved as pottery impressions (see http:/www.ucl.ac.uk/archaeology/research/profiles/butler/annbute.htm and http://www.dur.ac.uk/~drk0www5/ena.html). In geological archives, seeds are almost exclusively preserved in a waterlogged state (e.g. lakes, peat). Preservation by charring (and/or carbonization) is related to several factors, such as temperature of the fire, length of exposure, moisture content (seeds and fuel) and the type of fuel, and may involve several plant components associated with the seeds (Fig. 3.23). For example, Old World cereal grains (e.g. *Triticum dicoccum* (emmer), *T. spelta* (spelt) and *T. monococcum* (einkorn); Fig. 3.24) might be found with straw and chaff, such as spikelets and spikelet forks (part of the cereal that encloses the grain), glumes (part of the spikelet; Fig. 3.24) and rachis (top of the cereal plant stem). The process of charring undoubtedly leads to differential

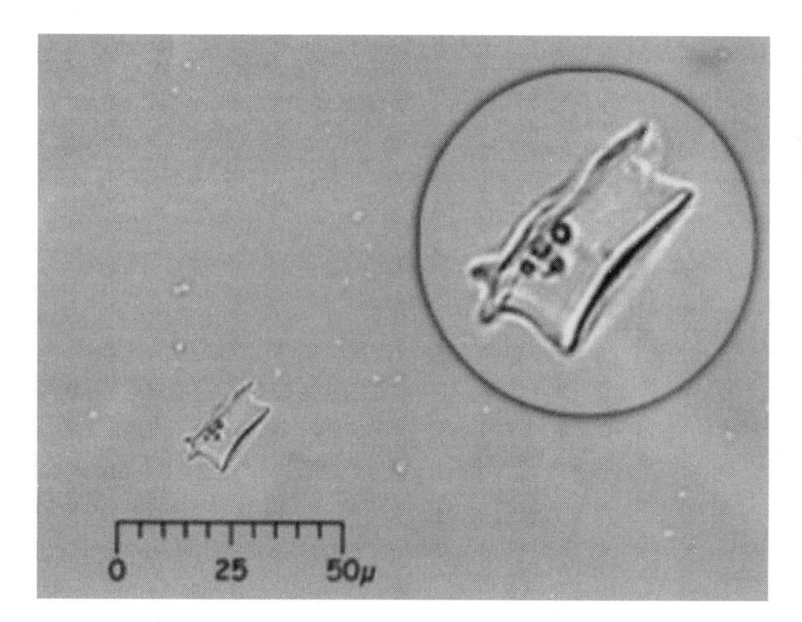

FIGURE 3.21 *Zea mays* (maize) phytolith. (Reproduced by kind permission of the University of Missouri, Palaeoethnobotany Laboratory)

FIGURE 3.22 Modern starch grains of *Solanum tuberosum* (potato) photographed using polarized light. (Reproduced by kind permission of Chris Thomas)

FIGURE 3.23 Selection of charred seeds commonly found in archaeological and geological archives.

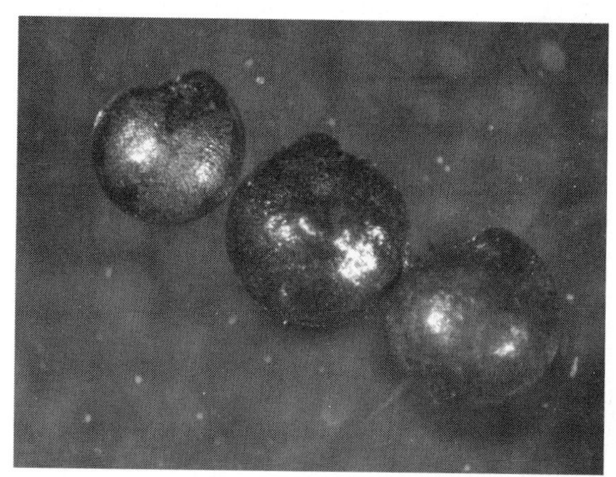

Chenopodium album

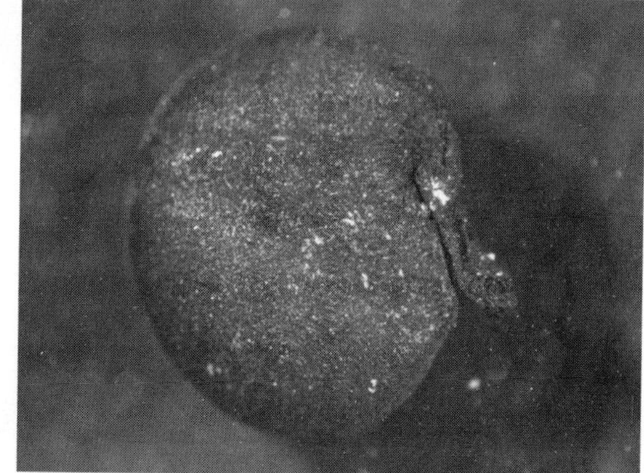

Lens sp.

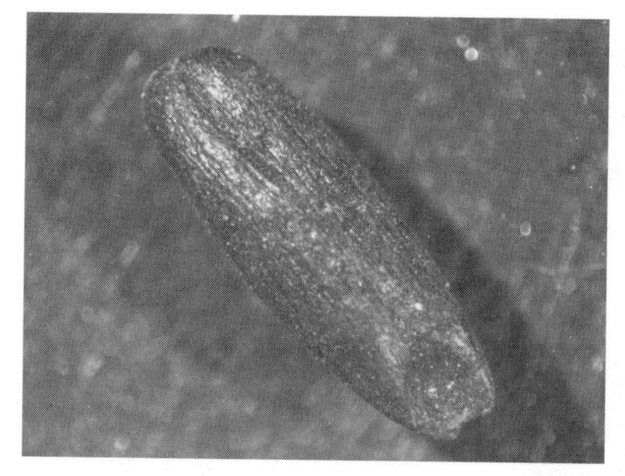

Poaceae

Vitis vinifera

FIGURE 3.24 Small selection of modern and charred cereal grains and glumes of *Hordeum* and *Triticum*.

Modern *Hordeum* (magnification ×10)

Charred *Triticum* (magnification ×5)

Charred glumes of *Triticum spelta* (magnification ×5)

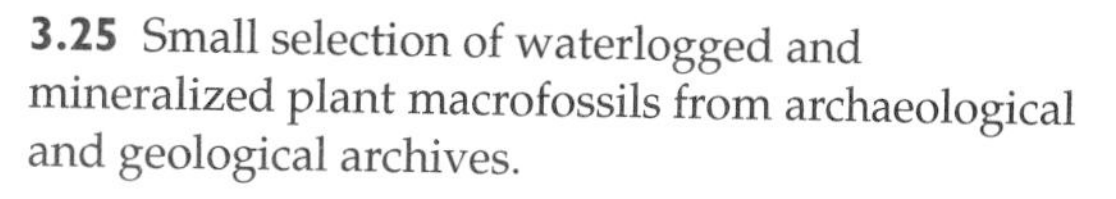
3.25 Small selection of waterlogged and mineralized plant macrofossils from archaeological and geological archives.

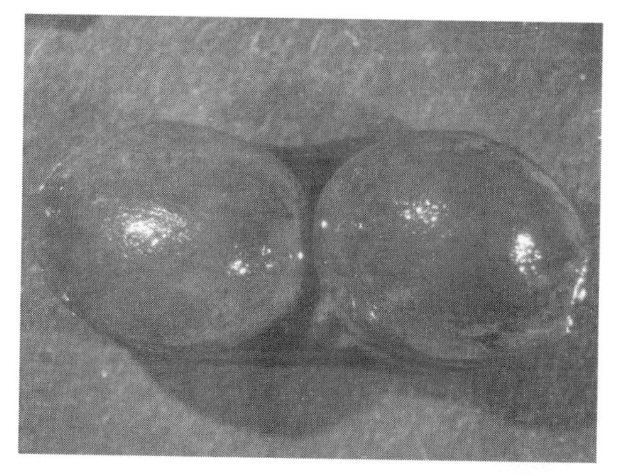

Prunus avium (waterlogged, magnification ×6)

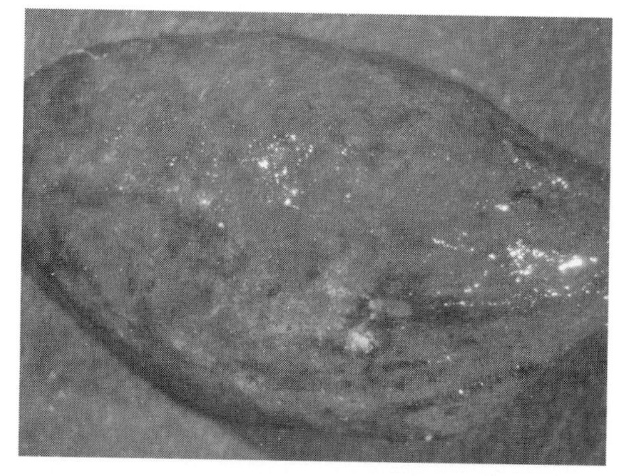

Prunus domestica (waterlogged, magnification ×4)

Rubus fruticosus (waterlogged, magnification ×20)

Rubus sp. (mineralized, magnification ×20)

preservation, distortion (in length, width and thickness) and different levels of charring depending upon the properties of the seed (Wilson, 1984; Smith and Jones, 1990; Gustafsson, 2000; Wright, 2003). Preservation by waterlogging occurs in anoxic conditions, which retards the decay process and results in the loss of internal anatomical structures (Fig. 3.25), while mineral replacement (mineralization) is a complex process and is mainly dependent on having waterlogged conditions and an abundance of phosphate (e.g. cesspits; Carruthers, 2000; see also McCobb *et al.*, 2001, 2003).

Depending upon the archive from which the seeds and their components (e.g. stems, leaves, buds) are recovered, they will represent either plants growing locally (autochthonous) or plants growing at an uncertain distance from the point of deposition (allochthonous). For example, remains found in peat indicate plants growing on the surface of the mire, while those found in lakes represent plants found within the catchment and the lake margins (Table 3.1). These remains are of limited interpretational value for environmental archaeology, but used in conjunction with other proxies (e.g. pollen, insects) may provide valuable information on climate change or vegetation history. In contrast, remains found in archaeological contexts provide essential information on the economy, diet and daily life of the site's occupants, irrespective of whether they are hunters and gatherers or from agricultural communities. In the case of the latter, palaeoeconomic, experimental and ethnographic studies of Old World cereals and their associated components have made significant contributions to our understanding of the origin, spread, distribution, methods of cultivation and processing techniques of these important food resources (Zohary and Hopf, 1994; Nesbitt and Samuel, 1996). Seeds and associated components found in these archives are therefore primarily allochthonous, with each assemblage representing single or multiple depositional events according to the nature of the human activities. For example, the contents of a hearth may be cleaned on a regular basis and re-deposited into a rubbish pit, ditch or simply spread on the ground surface. Therefore, most remains found in archaeological contexts do not represent the local vegetation cover (with the exception of plants growing on the margins or bottom of a feature, e.g. ditch), and have been moved from their primary depositional position. Only in special circumstances are remains found in their primary context (e.g. catastrophic fire or 'cleansing' of a storage pit by fire).

TABLE 3.1 Taphonomy of seeds and their associated plant components

Archive	Autochthonous	Allochthonous
Peat	++	
Lake	+	++
Palaeosol	++	+
Hearth		++
Well	+	++
Moat	+	++
Cesspit		++
Crannog	+	++
Cave and rock-shelter	+	++
Lake dwelling	+	++
Ditch	+	++
Rubbish pit		++
Storage pit		++
Grave	+	++
Pottery vessel		++
Floor/Street surface		++

Key:
+ = Secondary source
++ = Primary source

The identification of seeds and their associated plant components is achieved using a range of modern and fossil reference collections, keys and photographs (e.g. Berggren, 1981), and requires good preservation of morphological and anatomical criteria. One particular problem that has proved impossible to resolve using grain morphology is distinguishing between species of 'modern' free-threshing ('naked') wheat grains: tetraploids *Triticum durum* and *T. turgidum*, and the hexaploid *T. aestivum* (Hillman *et al.*, 1996). Although well-established criteria exist for the accurate identification of whole rachis or spikelets from free-threshing wheat, these parts are often less well preserved (Maier, 1996). In order to overcome this problem, polymerase chain reaction (PCR) has been successfully used to amplify and analyse ancient DNA (Deoxyribonucleic acid) from charred free-threshing wheat grains, and has led to the confident identification of hexaploid *T. aestivum* (Allaby *et al.*, 1994; Schlumbaum *et al.*, 1998).

The results from the standard analyses of the remains are normally tabulated (Table 3.2), providing a list of identified species and either a qualitative assessment of their ubiquity (presence or absence) or semi-quantitative assessment of their abundance (e.g. rare, occasional, frequent or abundant). Alternatively, a count may be provided of the total concentration of seeds and components per species within a sample of known volume (e.g. concentration of seeds/kg or species/kg). All of these methods permit diagrammatic representation of the data. Several studies have also employed statistical tests to demonstrate similarities/dissimilarities within and between datasets. For example, Mangafa and Kotsakis (1996) presented a new method using discriminant analysis that enabled the identification of wild and cultivated charred grape seeds (*Vitis vinifera* ssp *sylvestris* and *V. vinifera* ssp *vinifera* respectively) based upon the analysis of modern and sub-fossil specimens. The potential implications of this research for our understanding of viticulture in the Mediterranean region are immense.

The interpretation of plant remains requires the use of modern botanical data derived from either autecological or phytosociological surveys. In addition, where there is good preservation of charred remains, on many Old World archaeological sites complex ethnographic models detailing traditional agricultural practices have been used to interpret ancient agronomic practices, in particular the various stages and sub-stages involved in the cultivation and processing of cereals (e.g. wheat, barley):

1 Status of the arable field (e.g. soil type, moisture content)
2 System of cultivation (e.g. intensive, extensive)
3 Manuring regime (present or absent)
4 Tillage practices (e.g. ard or mouldboard ploughing)
5 Sowing methods (e.g. broadcasting)
6 Harvesting methods (e.g. uprooting, plucking or cutting with a sickle or scythe)
7 Processing methods, in particular free-threshing cereals (FTC: 'modern' bread and durum wheat, and naked barley), hulled barley (HB) and hulled or glume wheat (GW: 'ancient' emmer, spelt and einkorn).

The identification of one or more of these stages and sub-stages is based upon the overall composition of the charred plant assemblage, i.e. the prime grain, weed seeds, straw and chaff. For example, the species of plants represented by the weed seeds will provide detailed information on the quality of the arable field (e.g. those species that are common to damp, clay-rich soils), the intensity of cultivation and tillage methods (e.g. the ratio of perennial to annual plants), sowing methods (e.g. the diversity of the flora) and the harvesting methods (e.g. the presence or absence of climbing and low growing plants) (Hillman, 1991). However, although the interpretation of ancient weed floras associated with cultivated fields has great potential, there are also concerns with the validity of the standard autecological and phytosociological approaches: 'if we are to use modern weed studies to interpret archaeological data, where different species and different husbandry practices may be represented, there is a pressing need to identify the ecological principles underlying species distribution' (Charles *et al.*, 1997: 1152). The use of 'Functional Interpretation of Botanical Surveys' ('FIBS') in archaeobotanical research has therefore been advocated, and involves recording basic modern ecological information ('functional attributes' such as leaf area, canopy height, epidermal cell size, seed bank and longevity) associated with particular species growing within specific agricultural regimes (e.g. dry-farmed, irrigated fields, crop rotation), and applying the database of information to different groups of species that have

TABLE 3.2 List of plant remains from Southwark Cathedral, London, UK (Vaughan-Williams, 2004)

Charred assemblage

Family	Genus	Species	English name	Habitat	Sample	24	31	43	20	21	22	23	25	26
					Context	630	843	953	402	562	523	533	618	619
					Feature	BE cut	Dump	Ditch	Pit	Pit	Pit	Pit	Pit	Pit
					Period	R	R	R	SN	SN	SN	SN	SN	SN
					Sample Vol (L)	1	30	?	10	10	20	20	19	20
Papaveraceae	*Rumex*	*acetosella*	Sheep's Sorrel	A, G		1								
	Sorbus	sp.	Whitebeam	Wo		1								
Fabaceae	*Vicia*	sp.	Vetch	A							1			
	Indet		Pea family	A		2								
Poaceae (Cereals)	*Hordeum*	sp.	Barley grain	A				1			2			
	cf. *Hordeum*	sp.	Barley grain	A							5			
	Triticum	sp.	Wheat grain	A						2	1		1	
	Hordeum/Triticum	sp.	Barley/ wheat grain	A							73		3	
Poaceae	Indet		Grasses	A, G							3			

Mineralized assemblage

Family	Genus	Species	English name	Habitat	Sample	24	31	43	20	21	22	23	25	26
					Context	630	843	953	402	562	523	533	618	619
					Feature	BE cut	Dump	Ditch	Pit	Fill	Pit	Pit	Pit	Pit
					Period	R	R	R	SN	SN	SN	SN	SN	SN
					Sample Vol (L)	1	30	?	10	10	20	20	19	20
Papaveraceae	*Papaver*	sp.	Poppy	A, Wa		2								
Polygonaceae	*Rumex*	*acetosella*	Sheep's Sorrel	A, G		14								
Rosaceae	cf. *Rubus*	sp.	Bramble	Wa						1				
	Alchemila	sp.	Agrimony			8								
	Malus	sp.	Crab apple	Wo									3	
	cf. *Malus*	sp.	Crab apple	Wo										
	Sorbus	sp.	Whitebeam	Wo						10			1	
Fabaceae	*Vicia*	sp.	Vetch	A							1			1
	Vicia/ Lathyrus	sp.	Vetch/pea	A								2		
	cf. *Lens*	sp.	Lentil	A								1		
	Pisum	*sativum*	Pea	A								1		

	Indet		Pea family sp.1	A	6								
	Indet		Pea family sp.2	A	11								
Linaceae	*Linum*	*usitatassimum*	Flax	A									1
Geraniaceae	cf. *Geranium*	sp.	Crane's bills	G, Wa							1		
	Indet		Carrot family								1		
Lamiaceae	*Prunella*	*vulgaris*	Selfheal	G, Wa	3								
	Indet		Dead nettle family						1				
Plantaginaceae	*Plantago*	*lanceolata*	Ribwort plantain	G	4								
Rubiaceae	*Galium*	*cf. serum*	Bedstraw	W	3								
Asteraceae	*Artemisia*	*sp.*	Mugwort	G									
Poaceae	*Triticum*	*dicoccum*	Emmer wheat	A							1		
	Hordeum/Triticum	sp.	Barley/wheat grain	A							3		
Poaceae	Indet		Grasses	A, G							2		
			Grass caryopses	A, G								2	1
			Bran, mashed seeds								A	O	F
			Bracken pinicle										1
			Root nodule									1	
			Indet		10								6

Waterlogged assemblage

Family	Genus	Species	English name	Habitat	Sample	24	31	43	20	21	22	23	25	26
					Context	630	843	953	402	562	523	533	618	619
					Feature	BE cut	Dump	Ditch	Pit	Fill	Pit	Pit	Pit	Pit
					Period	R	R	R	SN	SN	SN	SN	SN	SN
					Sample Vol (L)	1	30	?	10	10	20	20	19	20
Ranunculaceae	*Ranunculus*	*repens*	Creeping buttercup	W, A				48						
	Ranunculus	*sardous*	Hairy buttercup	G, A				4						
	Ranunculus	*sceleratus*	Celery-leaved buttercup	W				2						
	Ranunculus	*cf. ficaria*	Lesser celandine	W, Wo				1						
	Ranunculus	sp.	Buttercup	G			66	36						
Moraceae	*Ficus*	*carica*	Fig	Ex			2	114						
Urticaceae	*Urtica*	*dioica*	Common nettle	Wo, Wa, A			3							

Table 3.2 (*cont.*)

Waterlogged assemblage (*cont.*)									
Betulaceae	*Corylus*	*avellana*	Hazlenut	Wo, Wa		5			1
Chenopodiaceae	*Chenopodium*	*album*	Fat hen	Wa	15				
	Chenopodium	sp.	Goosefoot	Wa					2
Caryophyllaceae	*Stellaria*	*media*	Common chickweed	A, Wa	300				
	Stellaria	*holostea*	Greater stitchwort	Wo		5			
	Stellaria	*gramineae*	Lesser stitchwort	G	21	16			
	Cerastium	sp.	Mouse-ears			10			
	Agrostemma	*githago*	Corncockle	A, Wa		4			
Polygonaceae	*Polygonum*	*lapathifolia*	Pale persicaria	A, Wa, W			1		
	Polygonum	*hydropiper*	Waterpepper	W	3	8			
	Polygonum	*lapathifolia/ minor*	Pale persicaria/ Small waterpepper	W		5			
	Polygonum	*Sect. Avicularia*	Knotgrasses	G		12			
	Rumex	*crispus*	Curled dock	A, W, Wa		26			
	Rumex	*acetosa*	Common sorrel	G		11			
	Rumex	sp.	Docks	W	6	8			
Malvaceae	*Malva*	*sylvestris*	Common mallow	Wa	2				
Violaceae	*Viola*	sp.	Violet	Wo		1			
Brassicaceae	*Thlaspi*	*arvense*	Field pennycress	A, Wa		2			
	Brassica/Sinapsis	sp.	Cabbage/mustard	A		2	1		
Resedaceae	*Reseda*	*alba*	White mignonettes	Wa		7			
Rosaceae	*Rubus*	sp.	Bramble	Wa		9			
	Potentilla	sp.	Cinquefoil	W, Wa		2			
	Fragaria	*vesca*	Wild strawberry	Wo		6			
	Agrimonia	*eupatoria*	Agrimony	G		1			
	Prunus	*spinosa*	Sloe (Blackthorn)	Wo		1			
	Prunus	*avium/spinosa*	Wild cherry/sloe	Wo		6			
	Prunus	*cerasus/avium*	Dwarf/wild cherry	Wo					
	Malus	cf. *sylvestris*	Crab apple	Wo				3	
	Sorbus	sp.	Whitebeam	Wo	3				
Fabaceae	*Medicago*	sp.	Medicks	G, A	46				
Vitaceae	*Vitis*	*vinifera*	Grape	A, Ex		1			

Apiaceae	*Coriandrum*	*sativum*	Coriander	Wa, A			8			
	cf. *Aethusa*	sp.	Fool's parsley	A, Wa			2			
	Bupleurum	sp.	Hare's ears				4			
	cf. *Ammi*	sp.	Bullwort	Wa			2			
Solanaceae	*Atropa*	*belladonna*	Deadly nightshade	Wo			2			
	Solanum	sp.	Nightshade			5	3			
Lamiaceae	*Stachys*	*sylvatica*	Hedge woundwort	Wo, Wa			3			
	Prunella	*vulgaris*	Selfheal	G, Wa			52			
Caprifoliaceae	*Sambucus*	*nigra*	Elder	Wo, Wa		2				
	Bidens	sp.	Bur-marigolds	W			1			
Potamogetonaceae	*Potamogeton*	sp.	Pondweed	W			1			
Cyperaceae	*Eleocharis*	*palustris*	Common spike-rush	W		3	122			
	Carex	sp.	Sedge	W		60	11			
			Awns	A, G				A		
			Straw frag	A, G	O			A	O	F
			Bud				1			

Key:
A – arable (incl weeds) Ex – exotics BE cut – brickearth lined cut O – occasional
W – wet G – grassland R – Roman
Wa – wasteground Wo – woodland SN – Saxo-Norman F – frequent

TABLE 3.3 Processing of cereals (After Hillman, 1984; Jones, 1984)

Sub-stage	Aim	Result	Product	By-product
Threshing and raking	Detach ears from straw	1. FTC: separates grain from rachis 2. GW: ear split into spikelets	Misc products and by-products	Coarser weed seeds, straw and unbroken ears
Winnowing	Removal of fine fraction	1. FTC: removal of paleas, lemmas, awns, rachises and some weed seeds 2. GW and HB: paleas and lemmas still attached	Misc products and by-products	Light chaff
Coarse-sieving	Removal of coarse fraction	1. FTC: unbroken ears, weed seeds, straw, rachis 2. GW: unbroken ears, weed seeds, straw	Grain, spikelets, weed seeds, rachis (FTC only)	Coarse by-products
Fine-sieving	Removal of fine fraction	1. FTC and HB: retention of semi-clean grain 2. GW: retention of semi-clean spikelets	Semi-clean grain and spikelets	Fine by-products
Pounding	Release GW from spikelet, and HB from palea and lemma	1. GW and HB: semi-clean grain	Semi-clean grain	Spikelets, paleas and lemmas
Hand-sorting	Remove small pieces of chaff, small stones and occasional weed seeds	Clean grain	Clean grain	

similar attributes (Bogaard *et al.*, 1999, 2001; Jones *et al.*, 2000). This approach not only improves understanding of why certain species are associated together within particular agricultural regimes but also allows the investigator to make informed decisions about the reliability of modern analogues in archaeobotany (Charles *et al.*, 1997).

The most detailed application of the ethnographic records has been to the interpretation of processing methods (Table 3.3). Each sub-stage produces a distinctive signature in the fossil record that may be used to understand similarities/differences in processing methods throughout the Old World, identify (arguably) whether the site was a producer (growing the crop), consumer (utilizing the crop) or both, and elucidate the nature of storage practices. However, at the beginning of this section, it was highlighted that plant remains in archaeological archives are allochthonous in origin and in many cases assemblages of charred remains have been mixed following re-deposition (secondary contexts). Therefore, many of the sub-stages outlined in Table 3.3 may not be clearly identifiable unless the various products and by-products have been: (a) transported to the point of primary deposition and used as fuel; (b) stored on site as clean grain and charred in a catastrophic fire, deliberate fire or accidentally during cooking; (c) stored as semi-clean grain or spikelets and processed on site, and charred in a catastrophic or deliberate fire.

There are numerous examples of the use of charred plant remains in bioarchaeological research. Although they are found in many situations, some of the most exciting information has come from the investigation of a range of archives on a single archaeological site. These studies reveal a wealth of information about diet and health of the occupants through the preservation of crop plants, associated weed seeds (cereals, pulses and fruits) and possible medicinal plants. As an example, the reader is directed to the study of the Philistine port city of Ashkelon (Israel) by Weiss and Kislev (2004), which has provided valuable information about the economic activities of the city and its hinterland through the study of several archaeological archives, including the administrative centre, counting house, shops, streets, silos and warehouses.

Waterlogged seeds can provide much of the same information outlined above, as well as detailed records of the general environment, from a range of situations including cesspits, ditches, wells, moats and crannogs. In contexts where excellent preservation occurs, the seeds, combined with other remains such as insects, animal bone and pollen, may provide useful information on local flora, presence of standing water, animal husbandry and the presence of cultivation (Ponel *et al.*, 2000). This is exemplified by the multi-proxy investigations at several lake-village sites in France and Switzerland, which have provided evidence for transportation to the site of plants exploited by humans for animal fodder, human food and building materials (Richard, 1993). For example, At Egolzwil 3, a Neolithic (c. 5600–5200 uncal BP) site in Switzerland, there is bioarchaeological evidence in the form of charred grain and animal bone for arable agriculture and animal husbandry, combined with hunting, fishing and gathering. The analyses show that domesticated animals accounted for 43 per cent of the assemblage and included goats/sheep, pigs and cattle, while roe deer, red deer and wild boar were present in greater numbers, accounting for 57 per cent of the total assemblage. Pollen and plant macrofossil (e.g. wood, bark, budscales, anthers) analysis of animal faecal remains provided direct evidence for the exploitation of woodland for leaf fodder (e.g. hazel (*Corylus*), alder (*Alnus*) and birch (*Betula*)) between February and May, and stabling of animals within the settlement (Rasmussen, 1993).

Faecal remains (including stomach and intestinal contents as well as coprolites) can survive whole or partially intact in desiccated (dry) environments, highly fragmented in wet (waterlogged or mineralized) environments, as amorphous lumps by charring, and in the stomach, small intestine or large intestine of dead humans, such as desiccated bodies (Holden and Núnez, 1993), ice bodies and bog bodies (Stead *et al.*, 1986). According to Hillman (1986: 99), human palaeofaecal remains are 'pearls beyond price' because of the wealth of information they can provide on past behaviour, such as diet, cooking methods and medicine, as well as disease.

The study of macrofossils preserved within human palaeofaeces offers two major benefits for bioarchaeology:

- they provide evidence of plant (wild and domesticated) and possibly animal (e.g. fish and large mammals) species actually consumed by humans;
- they provide information on those plant and possibly animal parts often missing from the normal bioarchaeological assemblages, such as leafy food, edible bark, stems and roots (Hillman, 1986).

These studies are complemented by microfossil analyses, such as pollen and phytoliths, elucidating the season of occupation, cultivation methods, basic health (e.g. presence of intestinal parasite ova) and quality of diet (Jones, 1986; Rhode, 2003). Scaife (1986) identified three routes by which pollen may enter the human body and become incorporated within faecal material:

- direct ingestion of floral parts (e.g. drinks and food);
- secondary ingestion of pollen in food (e.g. cereal processing); and
- indirect ingestion with food or drinks (e.g. background airborne component).

The only remaining route is breathing-in of pollen followed by accidental swallowing as a component of mucous, a process that may add significant amounts of pollen to faecal material (Branch, 2003). Animal palaeofaeces also provide a rich source of information, from insects, diatoms, phytoliths, pollen and plant macrofossils (e.g. twigs and leaves), on the composition of animal fodder, season of occupation and animal husbandry practices (Rasmussen, 1993; Panagiotakopulu, 1999; Horrocks *et al.*, 2003).

3.2.2.2 Waterlogged wood and charcoal

Wood preserved by anaerobic, waterlogged conditions or burning is often found in both archaeological (e.g. trackways, platforms, hurdles, ditches, pits) and geological archives (e.g. blanket peat, mire basins). It provides primary data on woodland composition, and hence vegetation history, woodland management, agricultural practices (e.g. fodder and bedding for animals), woodland exploitation for domestic fires (fuel), human impact on the natural environment, catastrophic, natural wild fires, material culture (wooden artefacts), time of woodland exploitation, local environmental conditions, preservation and bias in wood assemblages and technological sophistication. Identification of both waterlogged wood and charcoal is based on micro- and macroscopic characteristics, and comparison with reference collections, including keys and photographs (Schweingruber, 1978; and see Fig. 3.26). Quantification of waterlogged wood and charcoal assemblages is problematical due to taphonomy (preservation or fragmentation), hence the relationships between taxa and woodland composition and structure, and human activities, are often difficult to ascertain. Normal practice is simply to count the total number of fragments for each species and convert these to percentages to permit graphical representation.

In order to achieve a statistical analysis of the taphonomic characteristics of charcoal assemblages in the rock-shelters at Pinarbaşi (Turkey), Asouti (2003) used density (total weight of charcoal per litre of floated sediment) and diversity (Shannon-Weaver) indices. The former assesses variations in the 'deposition, preservation and rates of recovery of charcoal remains' (Asouti, 2003: 1192), while the latter may be able to distinguish between 'generalized assemblages' (e.g. domestic rubbish pits) and 'specialized assemblages' (e.g. hearths). In addition, Asouti constructed a 'fragmentation/preservation index' that provides a general assessment of preservation and context type by comparing the number of identifiable charcoal fragments with the number of indeterminate fragments. Hence, index values of <0.5 indicated good preservation, while 1–5 indicates many indeterminate fragments. Spearman's Rank Correlation Coefficient and Correspondence Analysis was used to compare the three indices, and highlight context related variation. Field-based studies of ancient woodland composition and structure preserved within peat stratigraphy have been attempted using archaeological box and quadrant excavation methods with the aim of providing a three-dimensional reconstruction of the former woodland cover and more refined interpretations of vegetation succession (Lageard *et al.*, 1995).

The complex of rock-shelters at Pinarbaşi (Turkey) investigated by Asouti (2003) demonstrate the detailed information it is possible to obtain from robust microscale analysis of charcoal assemblages. Asouti was able to reconstruct the taphonomic history of the assemblages, determine the range of species used throughout the period of site occupation from the Early Neolithic to the Late Neolithic/Chalcolithic (*Pistacia*, *Amygdalus*, *Rosa*, *Celtis*, *Tamarix* and *Fraxinus*), the nature of the woodland exploitation, i.e. opportunistic gathering of firewood, and the general environmental history of the site. The charcoal analysis concluded that the vegetation cover consisted of woodland-steppe made up of drought-resistant trees, marshes, pools and saline lakes over a range of biophysical environments.

3.2.2.3 Mollusca

Mollusca are preserved on land (e.g. soil and mires), and in freshwater (e.g. lakes), brackish water (e.g. high salt marsh) and marine (e.g. estuaries) sediments where there is an adequate amount of calcium carbonate, e.g. areas having limestone or chalk geology. In these situations they can be found in high concentrations, and identification is achieved by recording shell morphology (Fig. 3.27) and anatomical characteristics (e.g. univalves or bivalves) using modern reference collections and a range of guides including Kerney (1963), Evans (1972), Sparks and West (1972) and Preece (1981). Websites concerning molluscs and their analysis can be found at http://wwwinfochembio.ethz.ch/links/en/zool_weicht_schnecken.html

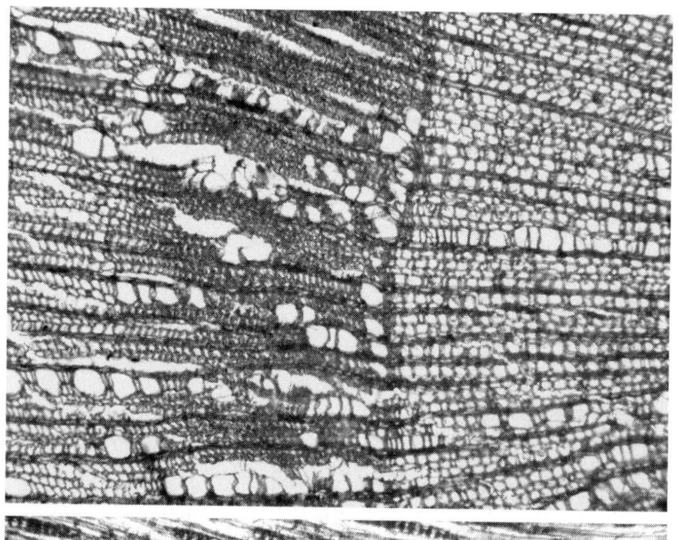

FIGURE 3.26 Wood anatomy of selected European tree taxa.

Alnus glutinosa (transversal section, magnification ×40)

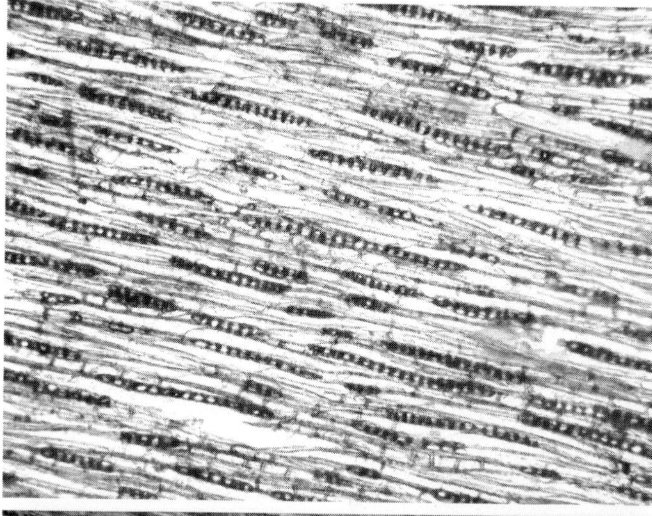

Alnus glutinosa (tangential section, magnification ×40)

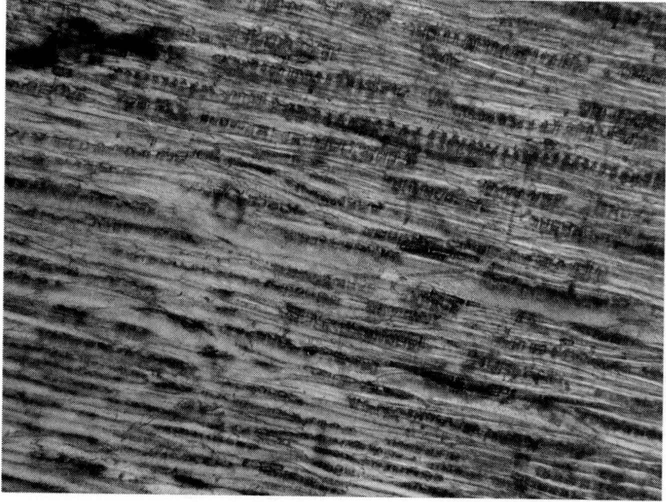

Alnus glutinosa (radial section, magnification ×40)

FIGURE 3.26 *(cont.)*

Picea abies (transversal section, magnification ×40)

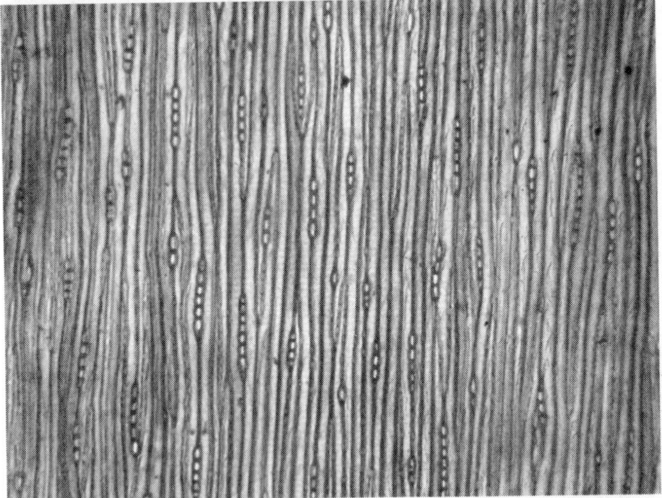

Picea abies (tangential section, magnification ×40)

Picea abies (radial section, magnification ×40)

FIGURE 3.27 Selection of Mollusca found in European archaeological and geological archives.

Trichia hispida (magnification ×6)

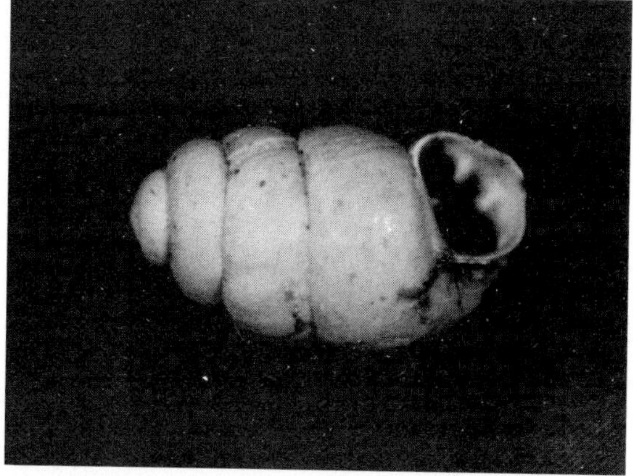

Vertigo pygmaea (magnification ×30)

Vallonia costata (magnification ×6)

Succinea oblonga (magnification ×5)

and http://files.dnr.state.mn.us/ecological_services/nongame/projects/consgrant_reports/200_barthel_sign.pdf. Individual species, or groups of species based upon their ecological affinities (e.g. land, freshwater), once identified and counted are tabulated, presented as histograms either as relative abundance (percentage) or total (absolute) concentration (number of individuals/kg, species/kg) or converted into estimates of the Minimum Number of Individuals (MNI). The main problem with the quantification of molluscs is the fragmentary nature of the remains, requiring that each identifiable fragment (based upon the apical or aperture fragment) is counted as an individual.

Molluscs have the potential to provide three categories of information useful to environmental archaeology:

1 Broad palaeoenvironmental reconstruction, which is dependent on recording species with particular climatic or habitat ranges.
2 Human impact on the natural environment, which has the same requirements as (1) and may provide useful information on woodland clearance and land use.
3 Human economy and diet, which is mainly (but not exclusively) confined to shellfish.

For both (1) and (2), assigning species to precise ecological groups is especially important, and species will vary in their significance between geographical areas. For example, Table 3.4 presents the so-called 'index species' and their ecological groups for central Europe according to Ložek (1986). These species have been invaluable in reconstructing environmental change during the last cold stage and the present interglacial. Palaeoecological studies of this type have been significantly improved by investigations into the response of mollusc communities to recent cultural landscape changes (e.g. land snails within field systems) and taphonomic processes (e.g. braided river systems) (Briggs *et al.*, 1990; Magnin *et al.*, 1995).

TABLE 3.4 Ecological groups for central Europe (Modified from Ložek, 1986)

Main ecological groups	Ecological groups	Index species
A: Woodland species	1. Closed forest	*Acicula polita, Ruthenica, Isognomostoma, Bulgarica cana, Aegopis verticillus*
	2. Predominantly woodland	*Alinda biplicata, Helix pomatia, Cepaea hortensis, Bradybaena fruticum*
	3. Moist woodland	*Perforatella bidentata, Monachoides vicina, Vestia turgida, V. gulo*
B: Open ground species	4. Steppe	*Chondrula*, Helicellinae, *Granaria, Cecilioides, Chondrina, Pyramidula*
	5. Open ground (general)	*Pupilla muscorum, Vallonia pulchella, Vertigo pygmaea*
C: Indifferent species	6. Dry	*Cochlicopa lubricella, Euomphalia strigella*
	7. Medium (catholic)	*Cochlicopa lubrica, Punctum, Euconulus fulvus, Vitrina pellucida*
	8. Damp	*Carychium tridentatum, Vertigo substriata, Nesovitrea petronella*
D: Aquatic marshland species	9. Marshlands, banks	*Carychium minimum, Oxyloma, Vertigo antivertigo, Zonitoides nitidus*
	10. Aquatic habitats	Lymnaeidae, Planorbidae, *Viviparus, Bithynia, Valvata, Bivalvia*

In Britain, colluvium (Bell, 1982, 1983, 1986) and calcareous tufa (spring-chalk) have provided rich sources of information on past human environments. Although pollen and plant macrofossils can occasionally be well preserved, Mollusca and Ostracoda are found in abundance. Preece and Robinson (1984) investigated tufas in Lincolnshire (UK), and provided records for the migration of taxa following the end of the last cold stage, succession, local environmental conditions and the presence of molluscs that are extinct in the British Isles today. At Castlethorpe site 3 (Fig. 3.28), the following succession is suggested by the molluscs: (a) 190–180 cm, open woodland (e.g. *Vallonia costata*); (b) 180–85 cm, increase in shade-demanding species (e.g. *Discus rotundatus*); (c) 85–65 cm, decline in shade-demanding species and increase in dry, open grassland species (e.g. *Pupilla muscorum*); (d) above 65 cm, increase in shade-demanding species (e.g. *D. rotundatus*). The change in the molluscan assemblage between 85–65 cm was interpreted as evidence for prehistoric woodland clearance (3410 ± 80 BP).

More unusual examples of the application of mollusc analysis come from shipwrecks. Bioarchaeological investigations of these archives are normally confined to insects and plant macrofossils (Haldane, 1991, 1992), although the identification of land molluscs from the 3300 year old shipwreck at Uluburun (Turkey), provided information on its route as well as taphonomic issues regarding the differentiation between taxa originally on-board and those accumulating within the wreck (e.g. near shore species) after it had sunk (Welter-Schultes, 2001).

The exploitation of marine molluscs, and their utilization within coastal and inland locations, has provided valuable information on subsistence practices, including diet, seasonality, technological developments and trade (e.g. Rick and Erlandson, 2000; Cabral and da Silva, 2003). For example, at Ysterfontein (South Africa), a coastal cliff shelter, evidence has been provided for the exploitation of shellfish, including *Choromytilus meridionalis* (black mussel) and *Patella granatina* (granite limpet), as well as *Chersina angulata* (tortoise), *Bathyergus suillus* (cape dune molerat) and *Spheniscus demersus* (jackass penguin) together with many other animals, enabling the reconstruction of human exploitation strategies and population levels during the Middle Stone Age (c. 60,000–40,000 years ago) (Halkett *et al.*, 2003).

In Britain, some of the most influential studies have been conducted on Oronsay (southern Hebrides), where molluscan, fish bone and isotope evidence from human bone suggests Late Mesolithic (c. 5000–4000 cal BC) sedentary occupation and a dependence on marine food (Mellars and Wilkinson, 1980; Richards and Mellars, 1998). At Culverwell (Isle of Portland), Mannino and Thomas (2001) conducted a detailed investigation of an Early Mesolithic (c. 6000–5200 cal BC) shell midden and concluded that despite the complexities of studying midden formation and transformation through time, frequent and intensive human exploitation had a significant impact on shellfish resources, especially edible periwinkle, limpets and thick top shell (Table 3.5). Indeed they have suggested from the age classifications of these taxa, that predation may have occurred annually, possibly from late autumn to winter (Mannino *et al.*, 2003).

The complexities of shell midden interpretation alluded to above have been investigated in detail by Bird *et al.* (2002) and others. They suggest that 'the inferences about paleo-diets that we might draw from these analyses hinge on a key question . . . is the archaeological record as represented in shellfish midden assemblages an accurate reflection of past prey choice and selectivity?' (Bird *et al.*, 2002: 458). There are three reasons for these justified concerns:

1 The composition of the midden is likely to be a biased representation of the relative importance of particular resources because of the pre-depositional processing and transportation strategies of the human groups.
2 Shell middens have low visibility in the archaeological record due to changes in the height of relative sea level causing inundation and erosion of coastal areas previously habitable by human groups; hence shell midden records to date may be unrepresentative of the full range of resources exploited and their relative importance in the diet.
3 The excavation and sampling strategy for the midden may bias the bioarchaeological

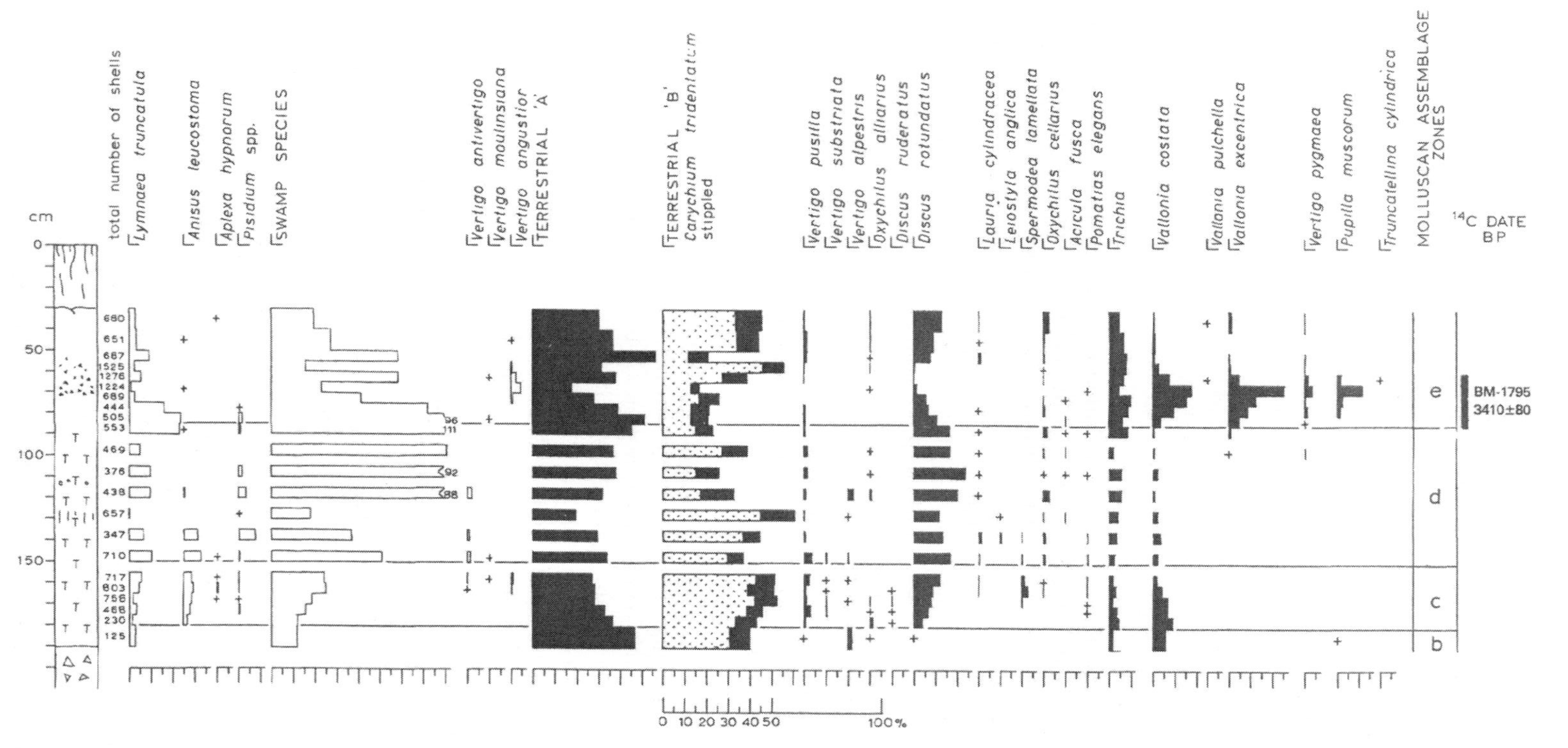

FIGURE 3.28 Mollusc diagram from Castlethorpe site 3, Lincolnshire, UK.

TABLE 3.5 Shellfish from Culverwell shell midden (after Mannino and Thomas, 2001)

Mollusca	Gastropoda	Common name	Assessment
Family Patellidae	*Patella* species	Limpets	A;T
Family Trochidae	*Monodonta lineata* (da Costa)	Thick top shell	A;T
	Gibbula umbilicalis (da Costa)	Purple top shell	F;T
	Calliostoma zizyphinum (Linnaeus)	Painted top shell	R
Family Littorinidae	*Littorina littorea* (Linnaeus)	Edible periwinkle	A;T
	Littorina obtusata (Linnaeus)	Flat periwinkle	F;T
	Littorina saxatilis (Olivi)	Rough periwinkle	R
Family Eratoidae	*Trivia monacha* (da Costa)	European cowrie	R
Family Muricidae	*Nucella lapillus* (Linnaeus)	Dog whelk	F;T
	Ocenebra erinacea (Linnaeus)	Sting whelk	I
Family Buccinidae	*Buccinum undatum* (Linnaeus)	Whelk	I
Family Rissoidae	*Rissoa* sp.	No common name	R
Family Cerithiidae	*Bittium reticulatum* (da Costa)	Needle whelk	I
Family Nassariidae	*Hinia reticulata* (Linnaeus)	Netted dog whelk	I
	Hinia incrassata (Ström)	Thick-lipped dog whelk	I
Mollusca	**Bivalvia**	**Common name**	**Assessment**
Family Arcidae	*Arca tetragona* (Poli)	Ark shell	R
Family Mytilidae	*Mytilus edulis* (Linnaeus)	Common mussel	I
Family Ostreidae	*Ostrea edulis* (Linnaeus)	Common European oyster	R
Family Pectinidae	*Pecten maximus* (Linnaeus)	Great scallop	I
Family Cardiidae	*Cerastoderma edule* (Linnaeus)	Common edible cockle	F;T
Family Veneridae	*Tapes decussata* (Linnaeus)	Carpet-shell	F;T
Crustacea	**Decapoda**	**Common name**	**Assessment**
Family Cancridae	*Cancer pagurus* (Linnaeus)	Common edible crab	F;T

Key:
A = abundant
F = frequent
I = infrequent
R = rare
T = throughout

record, and the subsequent analyses may fail to fully appreciate the degree of post-depositional preferential preservation.

Employing excavation and ethnographic methods to study modern and ancient shell middens and exploitation practices in the Meriam Island (Australia), Bird *et al.* (2002) concluded that the contemporary shellfish diets of the islanders are not reflected in the modern and ancient midden records, and that the most important prey are extremely poorly represented. The biased assemblages are thought to be due to the different procurement and consumption locations, i.e. field processing resulted in only flesh being delivered to the main occupation area.

3.2.2.4 Insects

Insect remains are found in archaeological and geological archives across a range of wet and dry environments. Their robust chitinous exoskeletons are often found as well-preserved fragments, and species identification is achieved using a reference collection of modern and sub-fossil specimens, keys and photographs (Fig. 3.29). Useful

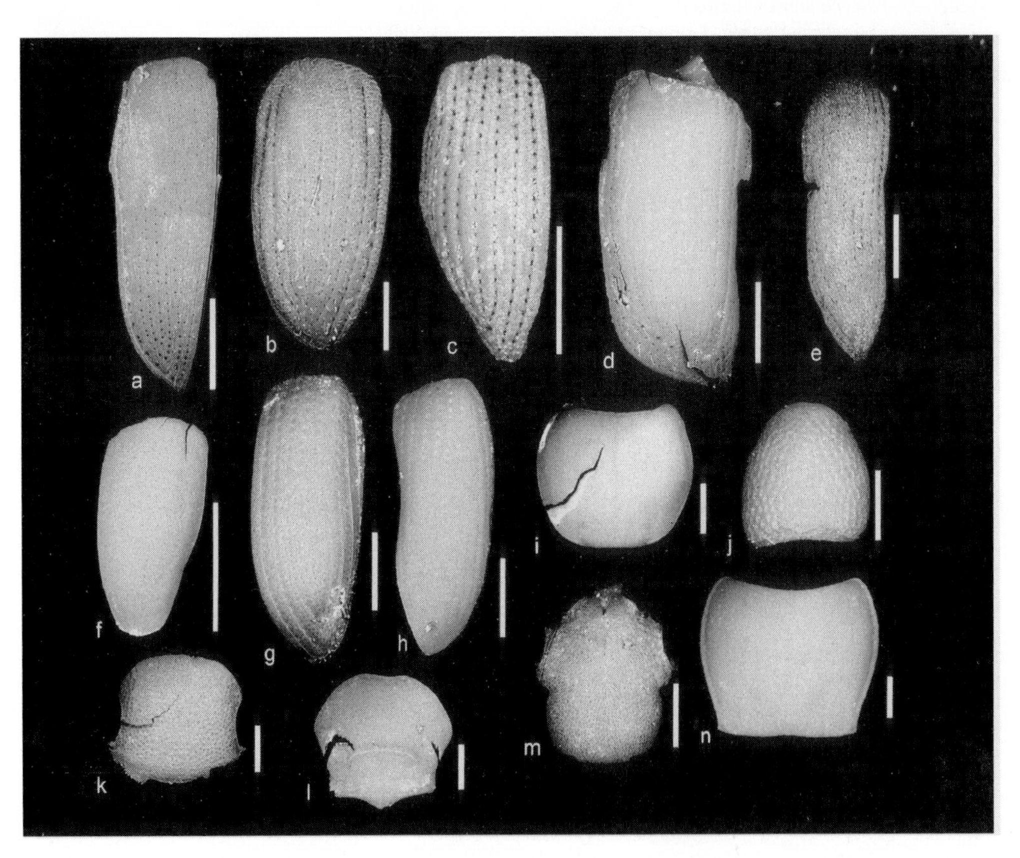

FIGURE 3.29 Selection of insect remains from Woolwich Manor Way archaeological site, London, UK. (Branch *et al.*, 2003)

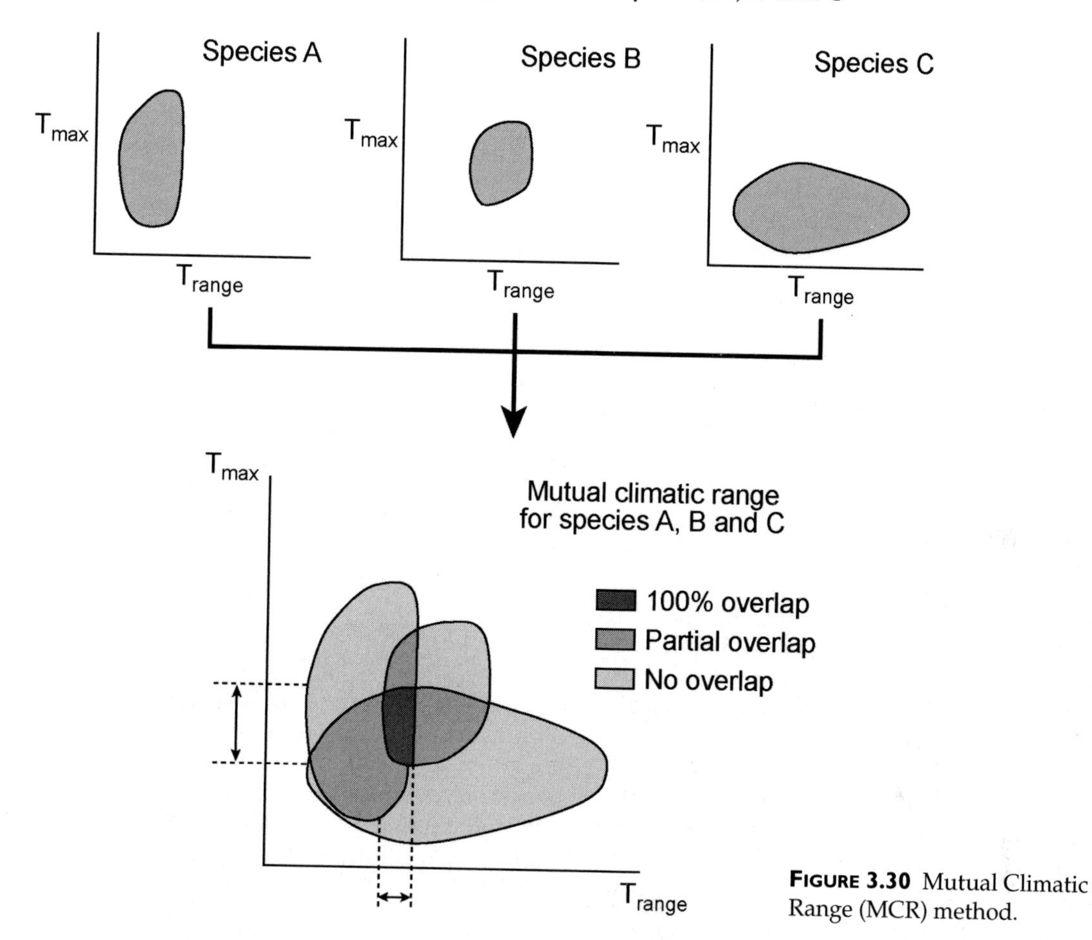

FIGURE 3.30 Mutual Climatic Range (MCR) method.

websites are http://www.ent.iastate.edu/List/Images.html, http://www.insectimages.org/ and http://entomology.unl.edu/images/images.htm. Insects are unique in providing quantitative terrestrial palaeoclimatic records using the Mutual Climatic Range (MCR) method (Atkinson *et al.*, 1986), based upon the 'mutual intersection of modern climatic ranges of selected species in the fossil record' (Hellqvist and Lemdahl, 1996: 875; and see Fig. 3.30). They also provide valuable information on regional and local environmental conditions, the local human environment, human and animal diet, the function of archaeological features, condition of human and animal mummified remains, and the contents of offerings (Kenward, 1978; Buckland *et al.*, 1994, 1996; Elias, 1994; Ashworth *et al.*, 1997; Schelvis, 1997; Panagiotakopulu, 2001). These applications require detailed records of modern groups of insect species and their ecological preferences, and the ability to differentiate between those species indicative of the general environment (allochtonous species) and local area (autochtonous species) (Kenward, 1976, 1985; Hall and Kenward, 1990).

Several authors have argued that this simple ecological approach is flawed in many archaeological settings because species that were once abundant in intensively occupied human environments, especially urban contexts, are now rare (Carrott and Kenward, 2001). Nevertheless, detailed research has demonstrated that in addition to compiling insect data into 'indicator groups' or ecological groups to aid interpretation, e.g. aquatic, fast flowing/aquatic, waterside, dung and foul terrestrial, grassland/terrestrial and woodland terrestrial (Robinson, 1981, 1983; Kenward and Hall, 1997; Smith and Howard, 2004), the identification of recurrent groups of taxa that occurred within specific contexts in the past provides a more precise level of interpretation. Carrott and Kenward (2001) analysed remains from York (UK) using statistical tests including detrended correspondence analysis (DCA). They established six main groups: (A) inside buildings and dry ('house fauna'), such as dry plant debris, wood borers, the human flea and stored products; (B) foul decaying organic matter, rich in ammonia, commonly found in stable manure; (C) foul decaying organic matter, associated with moist ground; (D) very foul conditions; (E) cesspits; (F) wool cleanings. This approach, while successfully identifying taxa commonly co-occurring in specific contexts in the past, highlighted the complex interpretational problems associated with single archaeological contexts (e.g. floors) containing insect assemblages from several habitats, reflecting a range of human activities as opposed to species occupying a diverse range of habitats ('compound taxa').

In non-urban archaeological contexts, several studies have clearly demonstrated the wide-ranging usefulness of insect analyses and the plethora of information that may be obtained. For example, Smith and Howard (2004) have shown that insect remains extracted from alluvial sediments representing high-energy (fast flowing) and low-energy (slow flowing) fluvial conditions have the potential to accurately reflect the different regimes. One of the most important and comprehensive studies of insect remains was undertaken at the Neolithic lake settlement at Weier, Switzerland (dated from c. 5600 cal BP). This study uncovered the remains of a possible byre (stable) consisting of compressed layers of manure, twigs and leaf fragments separated by wooden floors (Overgaard Nielsen *et al.*, 2000). The deposit consisted of three main categories of insect remains: dipterans (puparia or larvae only), mites and beetles. The species indicated the presence of decomposing foul organic matter, and the absence of insects associated with drier decomposing matter suggesting that the byre was cleaned periodically, probably for manuring of ploughed fields. In addition, variations in the physical properties and composition of the deposits, as indicated by the pollen and plant macrofossil evidence, suggested a mixture of fermenting excreta and compost, creating diverse microenvironments for insects, possibly rich in toxic gasses such as ammonia. The presence of liver fluke eggs (*Fasciola hepatica*) and biting louse (*Damalinia bovis*) suggests the presence of ruminant dung, an interpretation confirmed by the zooarchaeological evidence from cattle bones. The latter has also provided evidence for the seasonal occupation of the byre, since the ectoparasite is prolific during the winter months. This interpretation is confirmed by the macroscopic plant remains, which suggest leaf foddering during the winter months.

3.2.2.5 Vertebrate remains

The most common vertebrate remains found within archaeological and geological archives are bones (e.g. fish, small and large mammals). Bones are composed of organic (protein collagen) and mineral (calcium phosphate) materials, making them strong but susceptible to degradation depending on the 'levels of calcium, protein and fat as a result of factors such as animal age, sex, genetics and nutritional state' (Nicholson, 1996: 513). Identifying the genus or species, the type of bone (e.g. femur), and the sex and age of the animal is dependent on having a comparative reference collection, which may include both modern and sub-fossil specimens (Fig. 3.31). However, it is not without problems. For example, establishing the sex and age of animals from bones recovered from contexts including caves, middens, rubbish pits and alluvial sediments, involves measuring diagnostic skeletal features; in particular the: (a) length to breath ratio of particular bones (e.g. metatarsals and metacarpals of cattle); (b) age of tooth eruption; (c) age of the transition from milk (deciduous) to adult dentition; (d) rate of tooth wear; and (e)

Fish and bird bone

Fish bone

Thorn backed ray

FIGURE 3.31 Small selection of animal bone from archaeological and geological archives.

FIGURE 3.31 *(cont.)*

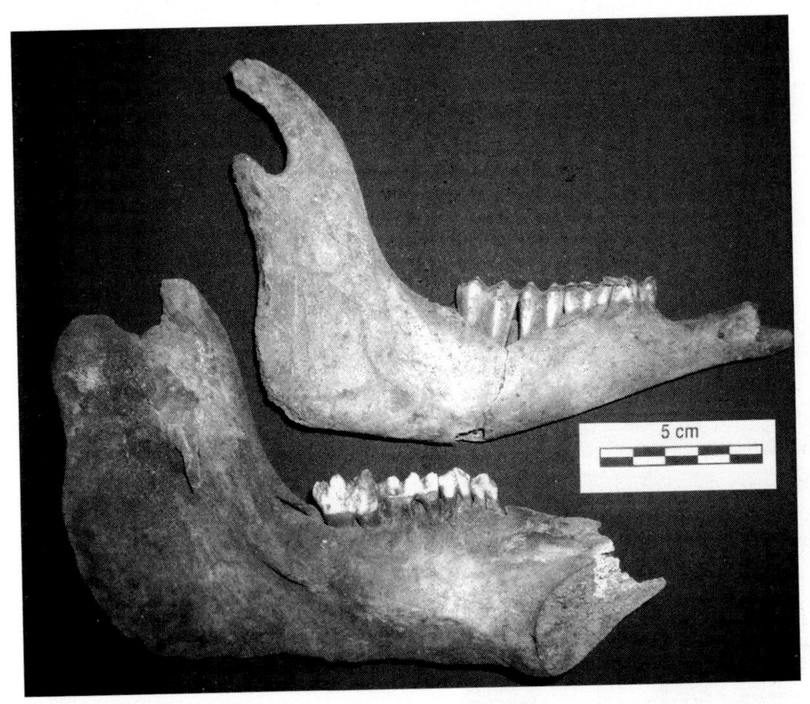

Sheep (upper) and pig (lower) jaws

Mixture of cow, sheep, pig and cod bones

presence of epiphysial fusion between the ends of bones (epiphysis), across the main shaft of the bone (diaphysis) and between cranial plates of the skull. The usefulness of these anatomical features varies significantly between animals, and can limit our ability to differentiate confidently between wild and domesticated taxa.

Similarly, the quantification of the entire animal bone assemblage is controversial, but necessary to enable comparison between contexts, periods and sites. There are three main methods (Rackham, 1986; Davis, 1987; Moreno-García *et al.*, 1996) (Table 3.6):

1 Number of Identified Specimens (NISP): records the number of identifiable bones for each species. There are problems with this method, including over-representation of animals having larger numbers of bones in their bodies, over-representation of complete skeletons, bias towards larger bones during field sampling and analytical error (e.g. identification of the most easily recognizable species).
2 Minimum Number of Individuals (MNI): records the minimum number of each species in an assemblage based upon three criteria: (a) the bone type can be identified, (b) each bone only occurs once in a skeleton, and (c) each bone can be identified to its precise skeletal position. This enables the analyst to calculate the MNI based upon the bone type that occurs most commonly in the assemblage. Weaknesses associated with this method include, over-representation of particular species and hence an over-estimation of their importance, and analytical error.
3 Bone weight: records the approximate meat weight yield from an animal species based upon the assumption that bone weight and animal body weight are a constant ratio for each species. A questionable basis of the method is the assumption that the animals were being eaten, rather than being used for dairy products or as pets.

Nevertheless, because of the ubiquitous nature of animal bone in the form of articulated and disarticulated skeletal remains in archaeological situations, there is an overwhelming amount of literature available for consultation.

TABLE 3.6 A selection of taxa commonly found as animal bone in archaeological archives in the UK

	Latin name	Common name
Mammals	*Ovis aries*	Sheep
	Capra hircus	Goat
	Ovis/capra	Sheep/goat
	Bos taurus	Cow
	Sus scrofa	Pig
	Oryctolagus cunniculus	Rabbit
	Sorex sp.	Shrew
	Rattus sp.	Rat species
	Mus sp.	Mouse
	Felis catus	Domestic cat
	Canis familiaris	Dog
Birds	*Gallus gallus*	Domestic chicken
	Anser anser	Domestic goose
	Anas crecca	Teal
	Turdus	Blackbird or thrush (genus)
	Corvid	Crow family
	Charadriidae	Wader (snipe sized)
Amphibians	*Rana tempororia*	Frog
	Rana/Bufo	Frog or toad
Fish	*Scomber scombrus*	Mackerel
	Clupea harengus	Herring
	Merlanogrammus aeglefinus	Haddock
	Anguilla anguilla	Eel
	Conger conger	Conger eel
	Gadidae	Cod family
	Salmonid	Salmon family
	Raja clavata	Thornbacked ray
	Raja	Ray (genus)

These data have permitted identification and differentiation between wild and domesticated animals, understanding of human economy and diet, in particular hunting and husbandry practices (e.g. butchery and gnawing marks; Gautier, 1993; Lupo, 1994; Greenfield, 1999), provided pathological information (e.g. disease; Brothwell, 1981), and enabled reconstruction of past environments and taphonomic processes. The following sections provide a few examples of many of these areas of animal bone analysis.

The identification of domesticated livestock in Old World archives (mainly species of *Capra*, *Ovis*, *Bos* and *Sus*) containing bone assemblages has primarily been based on three characteristics: morphological change (e.g. size reduction), demographic profiling and species abundance. However, the usefulness of morphological criteria to differentiate between wild and domesticated animals is limited by natural variations in animals due to habitat diversity and species' response to environmental factors within their distributional area (Zohary *et al.*, 1998), and variations due to age and sex (Zeder, 2001). For these reasons, size reduction may not be an effective method for identifying domesticated livestock (e.g. *Capra*). Demographic profiling, based upon the recognition of differences in age and sex between animals resulting from hunting as opposed to herding practices, is limited by our ability to establish the age and sex of animals from faunal assemblages especially those species with very similar morphologies (e.g. *Capra* and *Ovis*). Presence/absence, or abundance, of species is certainly of limited value inside the distributional area of the wild progenitor and therefore the distinction between early husbandries, in contrast to wild animal hunting, is difficult (Zeder, 2001). In cases where species identification may not be possible on critically important sites, the use of PCR/DNA (polymerase chain reaction/deoxyribonucleic acid) analysis may provide the answer, and future developments may permit differentiation between wild and domesticated animals of the same genus (Newman *et al.*, 2002).

Reconstructing human economy and diet, especially the relative proportions of animal and plant food in past diet, and processing methods, is a key objective of bioarchaeological research (e.g. Davis, 1987). In addition to the quantitative methods outlined above, measurements of stable carbon ($\delta^{13}C$ ‰) and nitrogen ($\delta^{15}N$ ‰) isotope ratios of collagen from human and animal remains (using mass spectrometry; Gilbert and Mielke, 1985; Mays, 1997; Polet and Katzenberg, 2003), lipids (using gas chromatography and gas chromatography/mass spectrometry; Evershed *et al.*, 1995) and cholesterol (using isotope ratio monitoring-gas chromatography/mass spectrometry; Stott *et al.*, 1999) have improved understanding of the importance of terrestrial and marine food, the use of exotic food, and changing dietary patterns, e.g. importance of marrow yields (Madrigal and Capaldo, 1999). For example, at Star Carr (UK), carbon isotope analysis of dog remains indicates that the Mesolithic human occupants may not have had a coastal base, as previously thought, but may have occupied the site over longer periods rather than simply during the summer (Day, 1996).

Using the bone assemblage to gain a more precise understanding of hunting and herding strategies is not always simple. For example, studies conducted on hunting strategies and, in particular, selective transportation of body parts, have largely been concerned with analysing the simple statistical relationship between skeletal parts and their economic use, although there are several problems with this approach which centre around variations in the size of the sample population and post-depositional alteration in the assemblage. One attempt to overcome these problems has been to utilize a statistical method known as abcml or 'Analysis of Bone Counts by Maximum Likelihood', which requires detailed information about the likely *agents of deposition* that have created the faunal assemblage (e.g. human hunters), the 'survivability' of the bone assemblage and the species identified. These data enable calculation of the amount of faunal material contributed by each agent and which species are represented, thereby creating a more robust approach for establishing whether the assemblage represents unselective (entire animal transported), more selective (animal parts of low utility discarded) or highly selective (only parts of highest utility retained) transportation practices by hunters (Rogers, 2000; Rogers and Broughton, 2001).

Another linked aspect of faunal analysis that has received critical review concerns the use of mortality profiles (based upon the ages of jaws and teeth) to elucidate these practices, e.g. preferences for hunting over scavenging, selective hunting and herding systems. This method has been challenged by a series of experimental taphonomic studies (e.g. Payne and Munson, 1985) which demonstrate that physical changes in mandibles (bone density, size and biomechanical strength) from infancy to adulthood, e.g. in white tailed deer (*Odocoileus virginianus*) and domestic sheep (*Ovis aries*), together with moderate animal ravaging of the mandibles results in a significant preservational bias in the mortality profiles towards older animals (Munson and Garniewicz, 2003).

Recent advancements in the use of animal bone to reconstruct broad environmental changes, in particular the past climate, have been achieved by the analysis of oxygen isotopes ($\delta^{18}O$) in animal bone phosphate. The isotope values in bone correspond to local meteoric water values (drinking water of the animals) and mean annual air temperature, and thereby provide quantitative reconstructions of past climatic regimes (Stephan, 2000). The study of faunal assemblages from cave and rock-shelter archaeological sites has, in particular, enabled the reconstruction of the climatic history of specific regions (e.g. Spain), as well as providing information on major changes in the pattern of human resource exploitation and diet (e.g. transition to the Neolithic cultural period) (Auban *et al.*, 2001). For example, in the Eastern Highlands of South Africa and Lesotho, faunal analyses conducted by Plug (1997) suggest a dramatic increase in species abundance from 21,000 BP onwards, with large bovids including wildebeest and hartebeest grazing on open grassland. During the Late Pleistocene and Holocene, independent palaeoclimatic data indicate that multiple phases of climate change between 14,000–10,000 BP and 9000–4000 BP can be correlated with faunal extinction (e.g. *Equus capensis, Megalotragus priscus, Antidorcas bondi*), although for much of the Holocene the faunal assemblages dominated by *Tragelaphus strepsiceros, Cercopithecus aethiops, Connochaetes* cf. *taurinus* and *Hippotragus* can be correlated with stable climatic conditions and a wooded environment.

In order to address, and possibly overcome, many of the problems with animal bone assemblages highlighted above, detailed taphonomic studies are required. These studies should be concerned with the transportation, processing, deposition and diagenesis of animal bone, in particular micro- and meso-faunal scavenging (e.g. insects and rodents), chemical and biological deterioration, and the effects of general ecosystem processes (Kos, 2003). These studies are fundamentally important for the future of animal bone studies, because quantification is based on the assumption that the relative abundance of species recorded in an archaeological (or geological) archive is 'representative of that originally deposited' (Nicholson, 1996: 513). For example, in a series of experiments conducted to examine the variation of susceptibility to bone decomposition between different taxonomic groups of animals, Nicholson (1996) concluded that:

1 Microbial activity, supplemented by hydrology (drainage) and pH, is the main factor determining bone preservation.
2 The depth of burial may have a significant effect on bone preservation.
3 The structure of mammal cortical bone degrades more rapidly than smaller bird and fish bone.
4 Cooking of the bone enhances the rate of degradation.

These results have important implications for the interpretation of faunal assemblages, especially in soils that have a naturally higher level of microbial activity, where differences in the state of preservation between large and small bones may be mistaken for contamination, and where cooking meat on the bone is normal practice (e.g. fish). Therefore, 'bone degradation does not follow a simple predictable pathway, and particularly if the susceptibility of different kinds of bone varies depending on burial environment, then the application of now generally routine comparative quantification methods (fragment counts, MNI, epiphyses counts, bone weights) and statistics based upon them will be inappropriate and misleading' (Nicholson, 1996: 529–530). The results suggest that fully quantitative analyses of faunal remains are unnecessary, and that semi-quantitative reporting is more appropriate. Nicholson recommends that for inter-taxon comparisons, a scale of 'rare', 'common' and 'abundant' or simply 'present' or 'absent' could be used, except where preservation conditions are optimal. To investigate the burial environment in greater detail, Nicholson

(1998) conducted an interesting experiment on the degradation of small animal bones in a compost heap, with the aim of simulating conditions occasionally recorded on archaeological sites where thick organic-rich layers have accumulated. The results of the study showed that animal remains are extremely well preserved, even though the environment was oxygenated and there was an abundance of micro-organisms. This was attributed to cross-linking of collagen with humic substances, increasing the survivability of the faunal remains. Nicholson concluded, however, that, without mineralization of the remains, complete degradation will occur over time.

Rich faunal assemblages recovered from alluvial sediments have, as outlined above, the potential to provide valuable information on environmental change and human exploitation strategies. In most depositional contexts, however, there are immense taphonomic problems with their interpretation particularly when they are also being used as a chronological marker in Pleistocene sedimentary sequences. These issues were addressed by Sept (1994) who studied land surfaces along the River Ishasha (Zaire) to consider modern bone distributions in riverine forest and on river flood plains. The study unsurprisingly concluded that there are complex relationships between assemblages and ecosystem processes that have implications for the interpretation of fossil assemblages, in particular surface weathering and fragmentation, bone density, bias towards larger mammals, and mixing of habitat-specific faunal assemblages due to rapid ecosystem changes. Preferential bone fragmentation in mainly young animals has been identified in a number of studies, and has wider implications for accurate measurement and identification, and therefore our ability to recognize important attributes of the faunal assemblage (e.g. sex and age of the animals).

Evidence for differential degradation as a consequence of taphonomic processes has led to the utilization of a range of analytical techniques to study bone density (including the measurement of bone mineral content, volume density, bone width and bone thickness), bone porosity and the effects of microbial attack. These methods have been applied to both experimental and archaeological datasets, and include photon densitometry (Ioannidou, 2003), mercury intrusion porosimetry (HgIP) (Jans *et al.*, 2004), computed tomography (Lam *et al.*, 1998) and nitrogen porosimetry (Robinson *et al.*, 2003). For example, Robinson *et al.*, (2003) have shown that pig (*Sus scrofa*) degrades faster than other ungulates (e.g. sheep (*Ovis aries*), red deer (*Cervus elaphus*), cow (*Bos taurus*), cockerel (*Gallus gallus*) and wood pigeons (*Columba palumbus*)). In addition, the development of analytical techniques to understand preferential selection of animal parts by human groups based upon the composition of skeletal assemblages (utility indices) together with quantitative assessments of the density-mediated destruction of faunal assemblages have, in combination, significantly enhanced our understanding of the relationship between human behaviour and faunal assemblages (Lam *et al.*, 1998). According to Lam *et al.*, this is illustrated by the 'reverse utility curve' at several archaeological sites, which arguably indicates the preferential destruction of faunal remains representing high utility body parts, and clearly has significant implications for the interpretation of faunal data.

3.3 Recovery of bioarchaeological remains

3.3.1 Field methods and sampling

3.3.1.1 Borehole samples

Borehole or core samples are useful for a range of bioarchaeological analyses, principally microfossils such as pollen, diatoms, foraminifera and ostracods. There are two main methods:

1 Manual devices such as peat/sediment samplers (also known as 'Russian' or 'Jowsey' corers) or piston corers (e.g. Livingstone), that obtain 5–10 cm diameter semi-circular (D section) samples that are 50–100 cm in length; these devices should be used in very unconsolidated sediment and peat (Fig. 3.32).

FIGURE 3.32 Borehole core sampling using manual equipment.

2 Motorized piston corers (e.g. Stitz) or cable percussion, that obtain 5–10 cm diameter circular samples that are 50–100 cm in length; these devices should be used in unconsolidated and consolidated sediment and peat.

For further information on coring techniques, see Section 2.2.3.2.

3.3.1.2 Column samples

Column sampling is useful for a range of bioarchaeological analyses, especially pollen, phytoliths and diatoms. It requires the use of plastic or metal sample holders (100 × 10 cm or 50 × 10 cm), and involves either hammering or carving the column into the section depending on the physical properties and composition of the matrix. The column should be clearly labelled using a permanent marker pen with the orientation ('Top' and 'Bottom') and depth ('cm' or 'm Ordnance Datum'), and recorded on the section drawing and environmental archaeological sheet. Column sampling is a central procedure enabling transportation of undisturbed material to the laboratory for analysis, where decisions about resolution and detailed sampling can be carried through away from the difficulties found in field work situations (Fig. 3.33).

3.3.1.3 Bulk samples

Bulk sampling is useful for a range of bioarchaeological analyses, especially molluscs, animal bone, insects, charred, mineralized and waterlogged plant macrofossils. It requires the use of plastic containers or plastic bags, and the sample size varies between 40 and 70 litres. The sample is added to the containers or bags using a clean trowel or spade, and clearly labelled using a permanent marker pen with the depth ('cm' or 'm Ordnance Datum') or context number. The samples should be recorded on the section drawing and environmental archaeological sheet (Fig. 3.34).

3.3.1.4 Spot samples

In situations where column samples are difficult to recover from geological and archaeological deposits due to the physical properties and composition of the matrix (e.g. dry, friable sediments in caves and rock-shelters) or accessibility of the section (e.g. small, restricted cave area), small bulk samples (spot samples) are recommended. These samples are useful for a range of bioarchaeological analyses, especially pollen, phytoliths and diatoms. The sampling technique requires the use of a trowel (4″ pointing trowel), small measuring tape (e.g. 3 m), nails (e.g. 6″), small plastic bags (21 × 14 cm [A5 size]), permanent marker pen, large spoon spatula, tissue paper and water. Clean the section from the top to the base using the trowel and attach the measuring tape. Mark the sampling positions with the nails, at regular intervals (e.g. every 10 cm) or for each archaeological context (or geological unit). Remove approximately 20 gm of the matrix from each position (from the base upwards) using a clean spatula (cleaned with water and paper towel), and place the sample in a labelled plastic bag. The samples should be recorded on the

FIGURE 3.33 Column (monolith) sampling.

FIGURE 3.34 Bulk sampling in 5 cm 'spits'.

section drawing and environmental archaeological sheet. The disadvantages of this method are twofold: the potential for contamination between samples is increased because of the difficulty of keeping equipment clean under field conditions; and the resolution of the sampling, and hence the analysis, must be determined during the fieldwork.

3.3.1.5 Flotation

The most rapid and efficient method for the recovery of charred plant remains involves the use of a flotation machine (Williams, 1973). The component parts are easily assembled but require a water supply, something that is often unavailable in rural locations. In these circumstances, water will need to be brought onto site (e.g. water tank) or pumped from a local source (e.g. petrol driven water pump). A filter can be fitted to the water pump, tank or flotation machine to prevent contamination (e.g. plant detritus) of the sample. A further advantage of the technique is that the sample residue can be dried (air dried or using a heated laboratory drying cabinet) and sorted, by eye or using a low-magnification microscope (in the laboratory), for other bioarchaeological remains: bone (human and animal), charcoal and mineralized plant remains.

3.3.1.6 Wet sieving

Wet sieving can also be conducted in the laboratory, and requires a clean water supply, graduated bucket or bowl, sieves of various mesh sizes, plastic storage containers or plastic bags. The recovery of waterlogged plant macrofossils and small pieces of wood involves dispersion of a known volume of sample in water (e.g. 1 litre), wet sieving through 4 mm, 1 mm and 300 μm mesh sizes, and wet storage in labelled (using permanent marker pen) plastic containers. The recovery of animal bone (including small fish and bird bone) involves processing approximately 40–70 litres of sample, wet sieving through a 1 mm or 500 μm mesh and drying (air dried or using a heated laboratory drying cabinet). Clay or silt-rich samples can be dispersed in 1 per cent sodium pyrophosphate or hexametaphosphate (or 'Calgon') before processing (Fig. 3.35).

3.3.1.7 Dry sieving

Dry sieving is commonly used on archaeological sites in arid climates. It involves processing bulk samples (40–70 litres) for the recovery of molluscs, animal bone, charred and mineralized plant macrofossils. The sample is placed

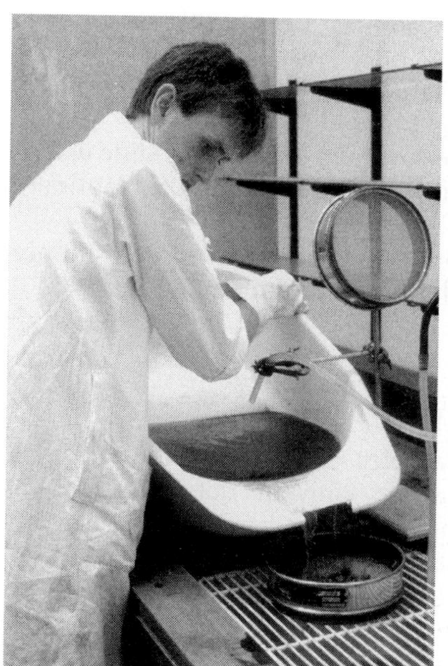

FIGURE 3.35 Wet sieving for the recovery of waterlogged plant macrofossils and insects.

on a large sieve with (usually) a 1 mm mesh size (100 × 100 cm) suspended above the ground surface. The sample is agitated by gently rocking the sieve from side to side, and the greater than 1 mm fraction transferred to labelled plastic bags.

3.3.2 Laboratory methods and sub-sampling

3.3.2.1 Pollen grains and spores

Pollen samples are best stored at 2°C (refrigeration) to prevent deterioration. The column and spot samples are sub-sampled using clean apparatus (scalpel blade or spoon spatula), the sub-sample size varying between 1 ml (e.g. organic-rich sediment and peat) and 10 ml (e.g. soil and cave sediment). The pollen grains and spores can be extracted using a variety of procedures, however, the following is highly recommended because it reduces the chance of fragile grains being destroyed through laboratory procedures:

1 Dispersion in 1 per cent sodium pyrophosphate or hexametaphosphate.
2 Removal of coarse organic and mineral fraction by sieving through 90 μm, 125 μm, 150 μm or 200 μm mesh sizes (choice of mesh size depends on geographical location of site).
3 Removal of fine organic and mineral fraction by sieving through a 5 μm mesh size.
4 Separation of the organic fraction from the mineral fraction by heavy liquid flotation using sodium polytungstate at a specific gravity of 2.0 g/cm^3.
5 Removal of remaining mineral fraction (mixed with the organic fraction) by hydrofluoric acid (40 per cent) (optional).
6 Acetolysis of the organic fraction.
7 Staining of the sample in safranine and water (optional).
8 Mounting of the sample in glycerol jelly or silicon oil.

The pollen grains and spores are identified using a high magnification microscope (×400 and ×1000) using transmitted light, with phase and interference contrast.

3.3.2.2 Diatoms

Diatom samples should be stored at 2°C (refrigeration) for optimum preservation. The column and spot samples are sub-sampled using clean apparatus (scalpel blade or spoon spatula), with the sub-sample size 1 ml. The diatoms can be extracted using a variety of procedures, however, the following is highly recommended to help avoid destruction of fragile diatoms in the laboratory (Battarbee, 1986):

1 Disperse the sub-sample in hot 10 per cent hydrochloric acid (HCl) for 15 minutes; this procedure will remove carbonate, and metal salts and oxides.
2 Add 30 per cent hydrogen peroxide (H_2O_2); heat gently to remove organic matter.
3 Sieve through a 500 μm mesh to remove coarse organic and mineral materials (if necessary).
4 Check quality of extraction using a test slide (water mount); sonicate in an ultrasonic bath to separate frustules and colonies into single valves (if necessary).
5 Add distilled water to dilute the concentrate.
6 Add microspheres to establish the diatom concentration (if necessary).
7 Check quality of microsphere – diatom ratio using a test slide (water mount).
8 Transfer 0.2 ml of suspension to a coverslip using a graduated pipette; allow the water to evaporate at room temperature.
9 Mount the coverslip onto a glass slide using a resin of high refractive index (e.g. Naphrax).

The diatoms are examined using oil immersion and phase contrast at magnifications of ×400 and ×1000.

3.3.2.3 Ostracods and foraminifera

Ostracod samples should be stored at 2°C (refrigeration) to prevent deterioration. The core, column and spot samples are sub-sampled using clean apparatus (scalpel blade or spoon spatula), with a sub-sample size 5–20 ml. Once washed through a 50 μm or 20 μm mesh to extract the remains, the ostracods are stored in alcohol or dry. The ostracods are examined using a low magnification microscope (×40–50) with reflected light.

3.3.2.4 Phytoliths

The standard method for the extraction of phytoliths involves the following (Lentfer and Boyd, 1998):

1. Sub-sampling 5 ml of sediment.
2. Dispersion of the sub-sample in 5 per cent Calgon solution for 12 hours.
3. Removal of organic matter by simmering (70°C) the sub-sample in 15–30 per cent hydrogen peroxide for c. 20 minutes.
4. Removal of carbonates by simmering (70°C) the sub-sample in 15 per cent hydrochloric acid (HCl) for c. 20 minutes.
5. Sieving through 250 μm mesh.
6. Dispersion by simmering (70°C) the sub-sample in 5 per cent Calgon solution followed by repeated centrifuging at 2000 rpm for 1 minute and 30 seconds.
7. Heavy liquid separation using a mixture of cadmium and potassium iodides (CdI_2, KI).
8. Dehydration using 50 per cent and 100 per cent ethanol.
9. Mounting in benzyl benzoate.

Alternative methods for the extraction of phytoliths have been discussed by Lentfer and Boyd (1999), and include thermal oxidation and multiple centrifuging of the sample at 2000 rpm for 2 minutes to remove clay instead of sieving using a 5 μm mesh. In addition, Lentfer and Boyd (2000) have devised a scheme for the recovery of pollen, spores and phytoliths using a single method. The procedure involves dispersion of the sample in 5 per cent Calgon solution, alkali digestion using 10 per cent potassium hydroxide (KOH), removal of carbonates with 15 per cent HCl, sieving (250 μm mesh), removal of clay using 5 per cent Calgon solution, heavy liquid separation using a mixture of cadmium iodide and potassium iodide at a specific gravity of 2.35 g/cm^3, staining of the sample with safranin, dehydration with 50 per cent and 100 per cent ethanol and finally mounting in silicone oil.

3.3.2.5 Cladocera

Under optimal conditions, Cladocera occur in high concentrations in sediments (c. 100–100,000/cm^3). Extraction procedures vary depending on the nature of the sediments (e.g. calcareous or non-calcareous), but the following method provides a general insight (modified from Frey, 1986):

1. Disperse 1 cm^3 of sediment in 100 ml of 10 per cent potassium or sodium hydroxide (KOH or NaOH) by gentle heating (<100°C) and agitation.
2. Transfer to a centrifuge tube, centrifuge twice and wash in distilled water, decanting the supernatant.
3. Wash sample in 10 per cent hydrochloric acid (HCl) to remove calcareous substances (if necessary), centrifuge twice and wash in distilled water, decanting the supernatant.
4. Treat sample with 40 per cent hydrofluoric acid (HF) to remove mineral matter (if necessary), centrifuge three times and wash in distilled water, decanting the supernatant into a suitable PTFE container (neutralize using calcium carbonate).
5. Sieve the sample through either 55 μm or 37 μm mesh sizes to remove unwanted mineral and organic fractions.
6. Transfer the >55 μm or >37 μm fraction to a graduated centrifuge tube and add distilled water to a maximum volume of 10 cm^3 or 20 cm^3.
7. Mix with glycerol jelly (alternatively use glycerol or silicone oil) and safranin stain, and transfer aliquots of 0.05 ml to microscope slides.
8. Qualitative preparations to aid identification are made from residues, and involve scanning water-mounted material at low magnification and transferring specimens of interest to a microscope slide; these specimens are mounted using polyvinyl-lactophenol and stained with lignin pink.

3.3.2.6 Insects

Insects can be extracted using a variety of procedures, however, the following is highly recommended because it reduces the chance of fragile insects being destroyed in the laboratory process (modified from Coope, 1986):

1 Disperse the sample using 1 per cent sodium pyrophosphate; heating may be required.
2 Sieve the sample through a 300 μm mesh.
3 Mix the >300 μm sample residue with paraffin.
4 Add cold water and allow to stand for 30 minutes until the lighter fraction floats on the surface.
5 Decant the floating fraction into a 300 μm mesh.
6 Wash the sample using household detergent and hot water to remove the paraffin, and finally wash in alcohol.
7 Transfer the sample into a suitable plastic container using alcohol.

Insect samples are best stored at 2°C (refrigeration) to prevent deterioration. The sample size varies according to the nature of the geological or archaeological context. In optimal circumstances, a column of bulk sample should be obtained, with each bulk sample measuring 5 cm in thickness and with a maximum volume of 20 litres. However, there are many situations where this procedure is not possible, and spot samples of variable volume have to be obtained. In addition, the extraction procedure described above will vary according to the composition of the material and between analysts. For example, at the Neolithic lake settlement of Weier (Switzerland), bulk samples for insect analysis were taken from discrete stratigraphic layers (judgemental sampling strategy) within a byre. Each sample consisted of 200 ml of organic-rich material, which was subsequently sub-sampled (100 ml each). These were processed by wet sieving using two mesh sizes – 1.25 mm and 0.3 mm (300 μm) – and the >0.3 mm fraction retained for analysis (Overgaard Nielsen *et al.*, 2000). The recovery of mites, in particular, requires wet sieving through a 50 μm or 100 μm mesh, and flotation using either saturated magnesium sulphate or paraffin (Denford, 1978; Kenward *et al.*, 1980; Coope, 1986; Schelvis, 1997). Larger samples for insect analysis are often necessary from alluvial sediments (c. 10–15kg), and are processed by paraffin flotation (Smith and Howard, 2004). Archaeological contexts such as wells, ditches and basins at the La Tene and Gallo-Roman site of Touffréville were bulk sampled and 10kg processed (Ponel *et al.*, 2000).

Insect taxonomies follow: Western Europe – Lucht (1987), Koch (1989–1992); Scandinavia – Silfverberg (1979); Egypt – Alfieri (1976).

3.3.2.7 Mollusca

Molluscs can be extracted using a variety of procedures, however, the following is highly recommended (modified from Ložek, 1986):

1 Dispersal of air-dried sample in water and dilute hydrogen peroxide (5 per cent).
2 Sieving through a 500 μm mesh size.
3 Air drying the sample and storage in plastic bags.

Mollusc samples may be dry stored; with the sample size varying between 1 and 10 litres according to the nature of the geological or archaeological context. In optimal circumstances, a column of bulk samples should be obtained, with each bulk sample measuring 5–15 cm in thickness and with a maximum volume of 10 litres.

3.3.2.8 Waterlogged wood and charcoal

Both waterlogged wood and charcoal may be recovered as part of the flotation or wet sieving residue (described above). However, in many cases waterlogged wood is either selected as part of a judgemental sampling approach (spot samples) or randomly selected where large quantities exist. For example, a prehistoric wooden trackway may contain multiple phases of construction, and each phase will contain large quantities of worked and unworked wood. A sub-sampling strategy of 20 per cent of the total wood content is recommended for analysis, with each sample wrapped in plastic to maintain the moisture content. For charcoal, bulk samples of 40 litres are appropriate, with the charcoal extracted using the following procedure:

1 Disperse the air-dried sample in sodium pyrophosphate (1 per cent).

2 Sieve through 1 mm, 2 mm and 4 mm mesh sizes to remove fine mineral and organic fractions.
3 Air dry the sample and extract using forceps.

Concentrations of charcoal vary between contexts. Although a minimum number of 200–300 fragments is recommended (75–100 from each size fraction), samples with less than 200 fragments are sometimes examined depending upon the aims of the analysis (see Asouti, 2003). The charcoal can be identified by fracturing individual pieces and using a low magnification microscope (×10) with reflected light, while waterlogged wood can be thin-sectioned and identified using a high magnification microscope with transmitted light (×10–40). Both waterlogged wood and charcoal can be identified using characteristics noted in the tangential, radial and longitudinal sections (Schweingruber, 1978).

3.3.2.9 Faunal remains

At those archaeological sites where non-specific zooarchaeological remains are being recovered by wet sieving, dry sieving or flotation, one of the important considerations is the size of the sieve mesh being used. It is advisable that the smallest mesh size should be no less than 500 μm, and 1 mm is often acceptable. Experimental studies conducted on sites in the UK and North America confirm this approach, and also highlight the importance of obtaining faunal samples specifically for the purpose of faunal analysis and not for other bioarchaeological analyses. In particular, flotation should not be relied on for faunal samples since the recovery is often poor, increased fragmentation may occur and there are difficulties comparing the statistical results (bone counts and MNI) of the faunal analysis based upon remains recovered from dry/wet sieving and flotation (James, 1997).

3.4 Reconstructing environmental change and human subsistence: case studies

The final section of this chapter provides four examples of how an integrated, inter-disciplinary approach to bioarchaeological research can improve our knowledge and understanding of human economy and diet (microscale), human impact on the natural environment (mesoscale), natural and human induced environmental change (macroscale), and the adaptation of humans to new environments (megascale).

3.4.1 Microscale: bog bodies

In 1984, there was a remarkable archaeological find at Lindow Moss (Cheshire, England) – an almost complete body of a healthy young male, c. 25 years of age, c. 168 cm in height and at least 60kg in weight (at death). 'Lindow Man', preserved by peat and dated by radiocarbon assay to c. 1960 BP or earlier (Iron Age/Romano-British), had died from two blows to the head, garrotting and an incision to the neck (Fig. 3.36). The forensic evidence clearly indicated that he had been murdered, possibly a ritual sacrifice (Stead *et al.*, 1986). The body provided a unique opportunity to analyse plant micro- and macrofossil remains preserved within the stomach, duodenum and small intestine, with the aim of providing information on the 'past diet, past culinary practices, alimentary disease and the ancient uses of medicines and drugs' (Hillman, 1986: 99). The plant macrofossil component provided evidence for the presence of finely ground wheat bran (*Triticum dicoccum, T. spelta*) and/or rye bran (*Secale* sp.), and barley (*Hordeum* sp.) chaff, mixed with weed seeds from cultivated crops, to form wholemeal bread or something similar (e.g. griddlecakes, dumplings) (Holden, 1986).

The pollen record consisted of high concentrations of cereal and herbaceous pollen, interpreted as having entered the body as a component of food and/or drink ('trapped into the cereal inflorescences' [Scaife, 1986: 129]). This interpretation was possibly confirmed by the presence of mistletoe (*Viscum album*) pollen, noted for its medicinal properties and which may have been part of a beverage (Scaife, 1986). Large numbers of sub-fossil *Trichuris trichiura* (human whipworm) and *Ascaris lumbricoides* (maw worm) ova revealed the health of Lindow Man. According to Jones (1986), heavy infestation of *Trichuris* worms may cause prolapse of

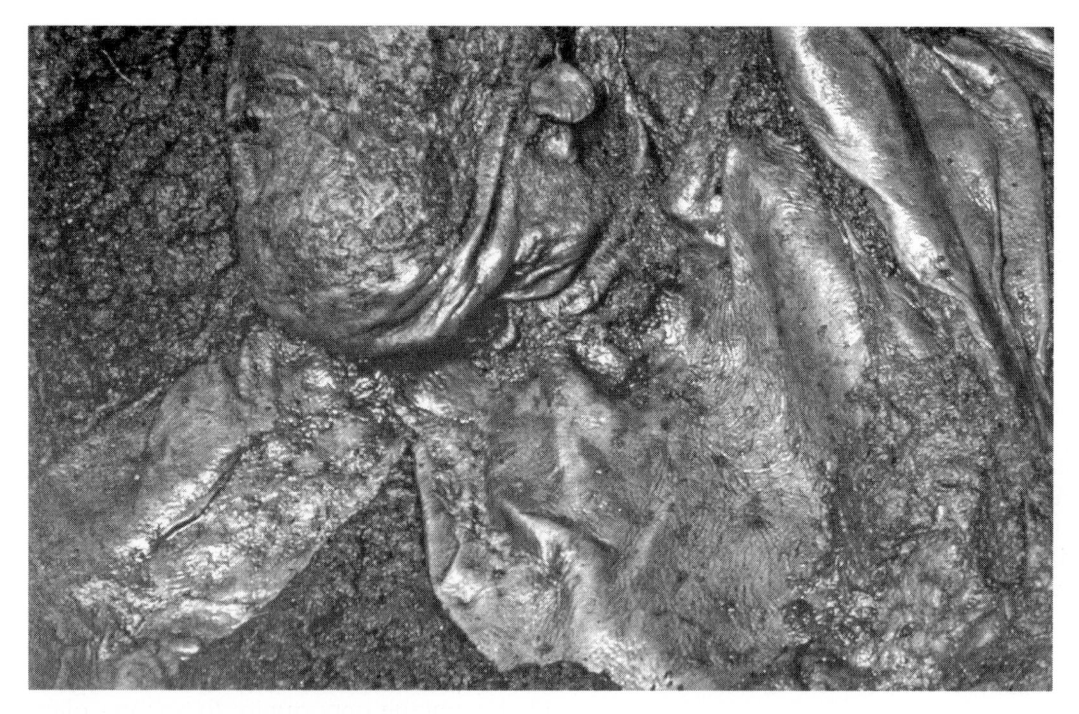

FIGURE 3.36 Lindow Man (Photo: Rick Turner).

the rectum, diarrhoea and blood in the faeces, while *Ascaris* causes digestive disorders, abdominal pain, vomiting and disturbed sleep.

The study of Lindow Man complemented previous work on Iron Age/Roman bog bodies in Denmark, namely 'Borremose' (Brandt, 1950), 'Tollund Man' and 'Grauballe Man' (Helbaek, 1950, 1958). Archaeobotanical evidence from these bodies for the consumption of wild seeds, cereal grain and cereal chaff probably reflects the gathering of wild plants (e.g. *Chenopodium album*), and the use of low quality products to supplement food supplies during periods of crop failure or the accidental addition of weed plants commonly found in cultivated fields to food, e.g. porridge (Hillman, 1986). Both Tollund Man and Grauballe Man also provided evidence for stomach and intestinal infections (*Trichuris* ova), comparable with those found in Lindow Man. One unusual aspect of the Tollund Man and Grauballe Man evidence that was not present in Lindow Man, was their ingestion of psychotrophic drugs, as indicated by the toxic sclerotia of *Claviceps purpurea* also known as ergot. Two main explanations were provided for its presence: that it was accidentally eaten as an unknown component of the crop, or deliberately eaten to induce a hallucinogenic state prior to ritual execution. These studies have all provided fascinating insights on Iron Age/Roman diet and food processing, and improved understanding of the composition of cultivated fields, and ritual practices.

3.4.2 Mesoscale: woodland clearance – the elm decline

Integrating bioarchaeological data at subregional (mesoscale) and regional (macroscale) scales often becomes a difficult challenge and success is entirely due to the spatial and temporal resolution of the available data (see McCorriston and Weisberg, 2002, Khabur Basin Project (Syria) for an excellent example of interdisciplinary mesoscale research). This problem is illustrated by the so-called 'elm (*Ulmus*) decline', which was a substantial pan-European decline in *Ulmus* pollen percentages occurring in Britain during a short period between about 6400 and 5300 cal BP (see Fig. 3.37). The broadly

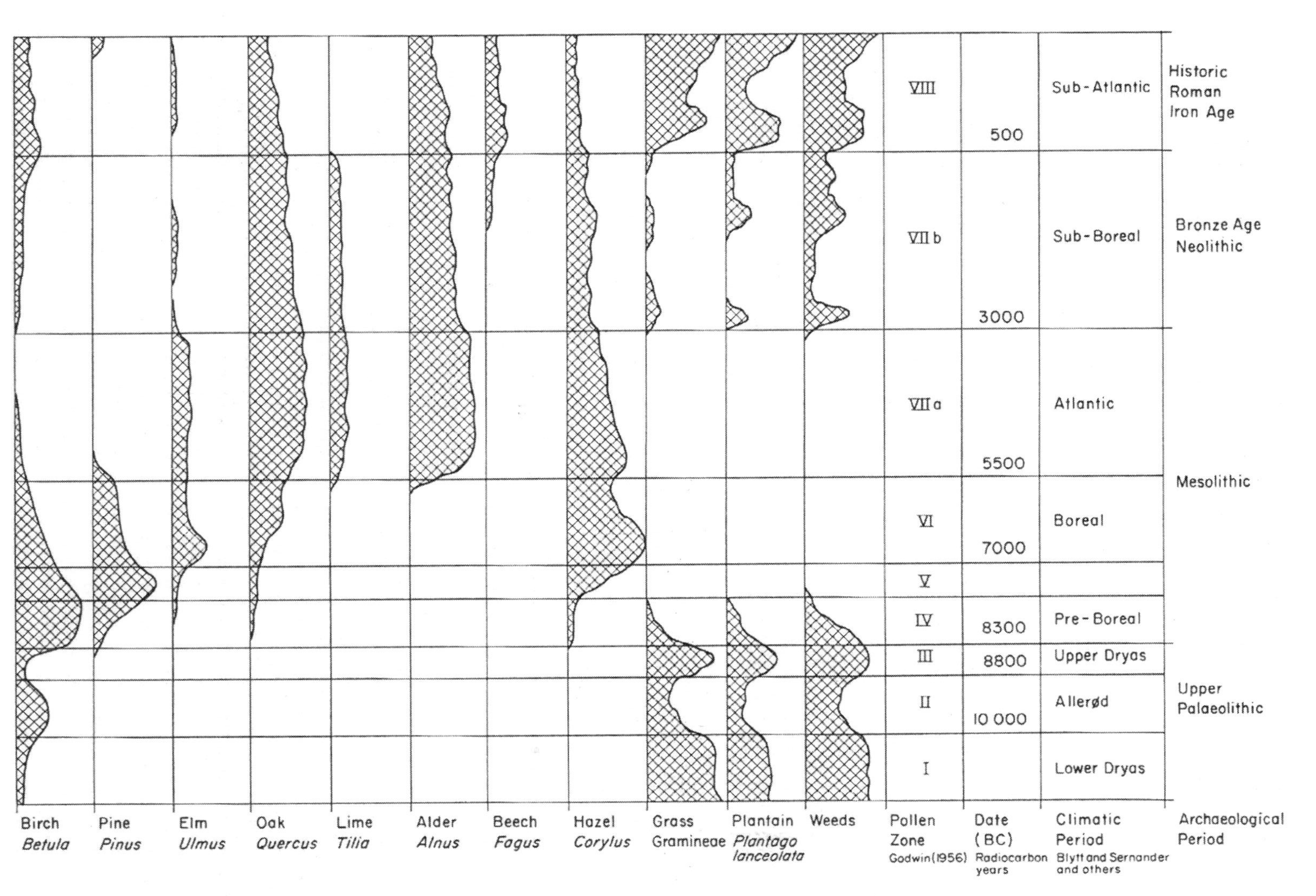

FIGURE 3.37 Diagrammatic representation of the main changes in vegetation in England during the last 14,000 years ((after Moore and Webb, 1978).

synchronous occurrence of this event in many radiocarbon dated pollen diagrams led to the suggestion that a change in climate (with higher summer temperatures) was the primary cause. Dismissal of this interpretation due to the absence of evidence for a long-term decline in other arboreal taxa has resulted in a debate centred on three themes:

- human activities;
- disease (so-called 'dutch elm disease');
- pollen taphonomy.

Three categories of data support arguments in favour of an anthropogenic cause for the elm decline:

1. Pollen-stratigraphical evidence from numerous mire basins indicating broadly contemporaneous woodland disturbance, in particular a short-term decline in woodland taxa in addition to elm, e.g. *Quercus* (oak), and an expansion of other taxa favouring open woodland, e.g. *Fraxinus* (ash).
2. Pollen evidence for cereal cultivation during the Early Neolithic cultural period.
3. Ethnographic evidence for the harvesting of elm, *Tilia* (lime) and *Hedera* (ivy) leaves which were regarded as highly nutritious and 'ideally suited to a pastoral economy in a heavily forested environment' (Scaife, 1988: 22).

These data suggest selective felling of elm in areas having soils suitable for farming (Mitchell, 1956). However, many pollen diagrams provide no direct evidence for simultaneous cultivation (cereal pollen) during the elm decline, and therefore the relationships between deforestation and human activities are often ambiguous. Accentuating this problem are mire basins located some distance from known centres of human occupation. Although the absence of cereal pollen may be attributed to animal husbandry practices, such as the collection of branches for leaf foddering, it is debatable whether selective gathering of elm leaves would cause a long-term reduction in elm woodland (Moe and Rackham, 1992). In conclusion, although the nutritional value of the leaves, bark and twigs for animal fodder is not in doubt (Moore, 1985a; Robinson and Rasmussen, 1989), much of the pollen-stratigraphy is ambiguous unless supported by direct evidence for land use or settlement activities that have resulted in the exploitation of *Ulmus* trees (Garbett, 1981; Rasmussen, 1989a, 1989b).

These issues led to a growing belief that the elm decline was simply an artefact of natural vegetation succession and pollen taphonomy (see Tauber, 1965). To test this hypothesis, an optimal site for investigating the relationship between the elm decline and human activity should be located proximal to centres of known Neolithic activity. One of the best examples is Gatcombe Withy Bed (Isle of Wight, UK), a mire basin that clearly demonstrates the effectiveness of using high-resolution pollen analysis (2 mm sub-sample intervals) to detect subtle changes in vegetation succession and human disturbance (Scaife, 1988) (see Fig. 3.38). Five distinct phases of human activity were recorded:

- pre-elm decline disturbance of woodland;
- elm decline;
- agricultural activity within woodland;
- shifting cultivation;
- woodland regeneration and pastoralism.

These data provide very persuasive evidence for an anthropogenic cause for the elm decline due to broadly contemporaneous localized settlement.

Arguments against an anthropogenic cause have primarily focused on disease. Three categories of data provide support:

1. A decline in elm populations during this relatively short period of time would have necessitated extraordinarily intensive and extensive efforts by human groups to remove woodland.
2. Zooarchaeological evidence from Hampstead Heath (UK) for the presence of *Scolytus scolytus* (elm bark beetle), the beetle that transmits the fungus *Ceratocystis ulmi* (Girling, 1988).
3. Modern records for the devastating effects of elm disease in the UK.

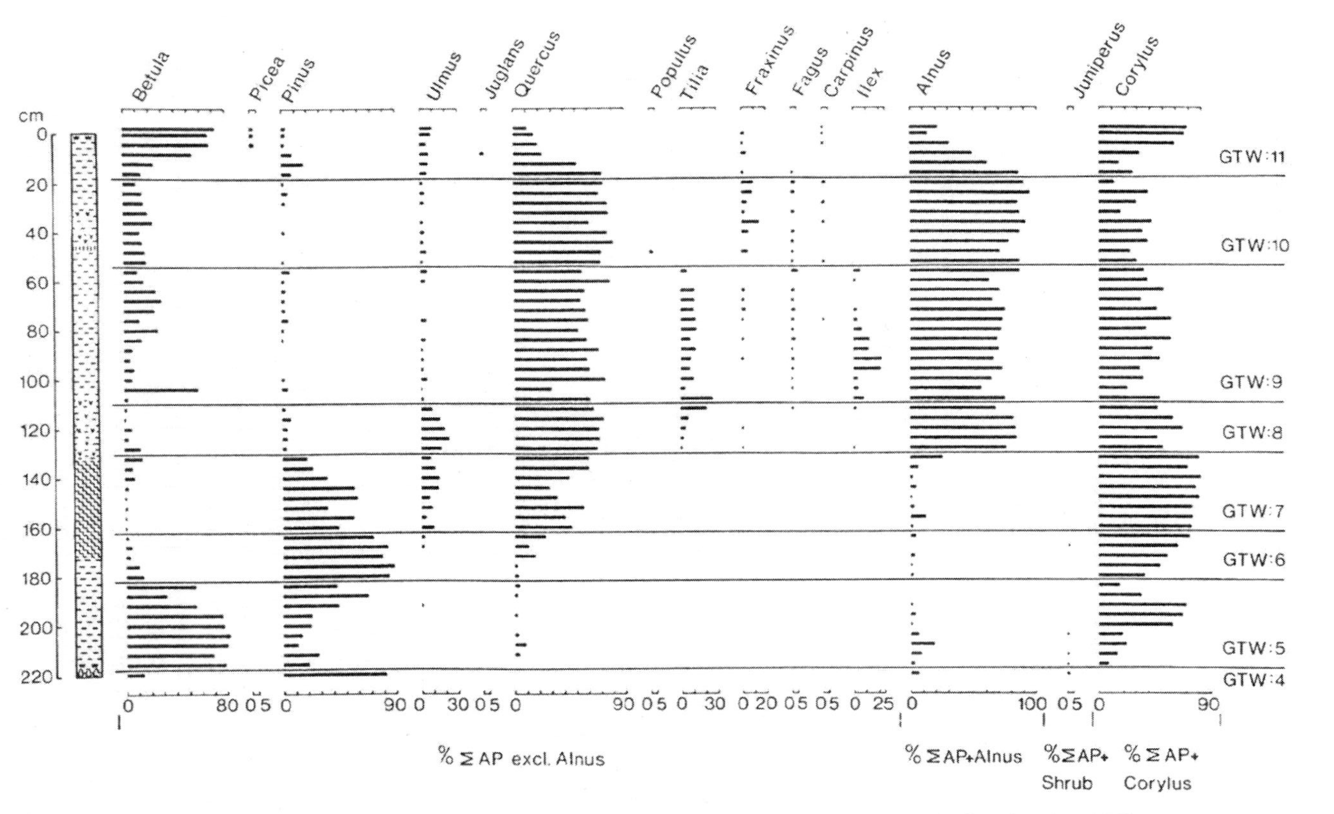

FIGURE 3.38 Percentage pollen diagram from Gatcombe Withy Bed, Isle of Wight, UK. (After Scaife, 1988)

The findings at Hampstead Heath for the elm bark beetle in an immediately *pre-elm decline* stratigraphic context, supported by pollen evidence from several sites for cereal cultivation during the same time period (Edwards and Hirons, 1984), provides compelling evidence for disease and human activity acting in unison to cause the elm decline (Girling, 1988). Whether farming facilitated the spread of the disease by creating opportunities for its easier transmission through woodland, or whether disease created woodland glades suitable for cultivation, or pastoralism, seems to have varied spatially. This is suggested by the records from London and the Isle of Wight, and may account for temporal variability in the timing of the event (over at least 500 years; Smith and Pilcher, 1973) as well as the presence of multiple elm declines, including short-term phases of woodland regeneration. This spatial variability in the nature of the elm decline may also account for the absence of a decline in elm pollen in some geographical areas due to the nature of human activities or the population levels of elm (e.g. Surrey and Kent, UK).

3.4.3 Macroscale: the environmental context of hunting, gathering and farming in the northern Mediterranean

Towards the end of the last cold stage (late Würm Lateglacial period; 11,000–10,000 BP), palaeoenvironmental records in the northern Mediterranean (Spain, France and Italy) indicate cold, dry climatic conditions (Guiot *et al.*, 1989; Lowe, 1992; Ponel and Lowe, 1992; Pons *et al.*, 1992; Reille *et al.*, 1992; Reille and Lowe, 1993; Watson, 1994; van Husen, 1997; Branch, 1999). Across biozones dominated by steppe-tundra and mixed coniferous woodland, e.g. *Abies* (fir) and *Pinus* (pine), zooarchaeological evidence indicates that human groups hunted red deer and steppe ass (*Equus asinus hydruntinus*). During the early postglacial (Holocene interglacial), climatic amelioration resulted in the expansion of arboreal vegetation, e.g. *Betula* (birch) and *Pinus*, followed by the succession of *Quercus* (oak), *Corylus* (hazel), *Ulmus* (elm), *Tilia* (lime) and *Acer* (maple) between 9000 and 8000 years BP (COHMAP Members, 1988; Rossignol-Strick *et al.*, 1992; Lowe and Watson, 1993; Magny and Ruffaldi, 1995; Watts *et al.*, 1996a, 1996b). Environmental archaeological evidence indicates that hunting (red deer, roe deer, cattle, pig and ibex), fishing (including shellfish) and gathering, e.g. wild barley, legumes (lentil, wild pea, vetch) and grape, replaced specialized hunting (Barker, 1973, 1975; Barker *et al.*, 1990). According to Barker (1985), in Italy 'one common system probably involved winter hunting and fishing on the coast, spring and summer hunting and fishing in the intermontane basins, ibex hunting in the hills in the early autumn, and cattle and deer hunting in the intermediate valleys in the late autumn and early spring' (p. 62).

Diachronous adoption of farming across the northern Mediterranean from 7000 BP is indicated by bioarchaeological evidence for the cultivation of emmer wheat, barley and legumes, animal husbandry (sheep and goats) and hunting (Dennell, 1983). The cause continues to be a major point of debate; four main models are proposed (Barker, 1985: 71–72):

1 Maritime colonists, who introduced and practised farming.
2 Colonists, who enabled indigenous populations to adopt farming.
3 Exchange among indigenous foragers.
4 Domestication by indigenous populations.

The balance of bioarchaeological evidence favours a combination of models (3) and (4), with initial domestication of indigenous einkorn wheat, barley and sheep followed by the introduction of domesticated wheat (emmer and bread wheat) from the Near East. Complex inter-regional variations in the timing and nature of early agricultural activities are clear from the available data. In the Ligurian Apennines/Alps of northern Italy, archaeological evidence indicates that the focus of human activity moved from the upland zone to lowland, coastal areas, and that the socio-economic structure was characterized by increased sedentism, and the exploitation of domesticated plants and animals. At the coastal cave of Arene Candide (Italy), c. 800 years of human activity during the Early Neolithic was characterized by the exploitation of both coniferous and deciduous woodland for animal fodder, such as *Quercus pubescens* and *Ulmus*, and the utilization of cereals (Branch 1997, 1999).

In the absence of direct palaeoecological evidence for human activities, identifying the impact of Neolithic activities on the landscape of the northern Mediterranean has been problematical. In the Ligurian Apennines/Alps, the diachronous regional invasion of *Fagus sylvatica* woodland, and the sudden decline of major woodland taxa (e.g. *Ulmus* and *Tilia*), may have been encouraged by human activities. These activities may have caused: (a) the creation of gaps in the woodland following clearance or felling; and (b) disturbance of the ground surface and soil structure enabling *Fagus* seed germination and growth. The impact of human activity on the vegetation may also account for changes in peatland hydrology that occurred during the colonization of *Fagus*, as well as the increase in pollen types indicative of a more open vegetation cover, such as *Corylus*, *Fraxinus*, *Hedera*, Apiaceae, *Rumex*, *Plantago lanceolata*, *Campanula*, *Artemisia* and *Cirsium*. These changes in woodland composition and structure are broadly supported by archaeobotanical evidence, which indicates the selective exploitation of *Ulmus* and other trees, such as *Fraxinus* and *Tilia* (e.g. Arene Candide).

However, an alternative explanation for the colonization of *Fagus* woodland, and possibly some of the changes in woodland composition mentioned above, is climate change. The evidence for increased precipitation during this period and its relationship with the adoption of farming has been discussed by Barker (1985) who believes that climate change and farming 'are a critical part of the equation' (Barker, 1985: 255), although the 'precise environmental data needed to test such an hypothesis properly are still rarely if ever available'. Since 1985, the database of palaeoenvironmental information has grown significantly, producing a body of evidence for the following:

1 The colonization of *Fagus* in Europe (Huntley *et al.*, 1989).
2 Sea level rise in the Mediterranean (van Andel and Shackleton, 1982).
3 Peat initiation in the northern Apennines (Branch, 1999).
4 Palaeoclimatic reconstructions from the Massif Central (Guiot *et al.*, 1989, 1993).
5 Sedimentological events from the eastern Pyrenees (Reille and Lowe, 1993) and Alps (Bortenschlager, 1982).
6 Pedogenetic developments in the Apennines (Cremaschi, 1990).
7 Lake level, peat-stratigraphical and glaciological records from the Jura mountains and north Alpine regions (Suc, 1984; Guiot, 1987; Scaife, 1987; Magny, 1992; Magny and Ruffaldi, 1995).

The period between 5000 and 3000 BP was characterized by demographic changes and the widespread adoption of farming in the northern Mediterranean. There are outstanding archaeobotanical and zooarchaeological records for human activity over a wide range of altitudes as part of a complex system of agricultural activity, involving cultivation and transhumant pastoralism. Pollen-stratigraphical records indicate further reductions in woodland, with decreases in arboreal pollen values (notably *Ulmus* and *Tilia*) and an increase in the diversity of light-loving herbaceous and fern taxa (notably *Plantago lanceolata*, *Artemisia*, *Filipendula*, *Rumex* and *Pteridium*). The abiotic records in many regional biozones also suggest that the changes in woodland cover resulted in erosion, colluviation and deposition of minerogenic sediments within mire basins and lakes. The expansion of Ericaceous and leguminous dwarf shrubs in upland and lowland areas indicates that deforestation and erosion also caused substantial soil degradation (Branch, 1999). In the Ligurian Apennines, the use of non-arboreal pollen as an indication of landscape openness suggests that non-arboreal pollen percentages of between 10 and 20 per cent may reflect the formation of woodland glades, and that values of 40–50 per cent indicate farmland (see also Aaby, 1994). These records are supported by pollen-stratigraphic evidence from northern and southern Italy (Schneider, 1985; Drescher-Schneider, 1994; Wick, 1994; Watts *et al.*, 1996a, 1996b), the Swiss Alps and southern French Alps (Pons, 1992), and the eastern Pyrenees (Reille and Lowe, 1993).

From the third millennium BP, pollen data from the northern Mediterranean identify the formation of new and distinctive vegetation communities, dominated by indicators of open woodland, heathland and grassland, as well as cereal cultivation. In the northern Apennines, for example, there is evidence for the widespread reduction of *Abies* woodland (Branch, 1999) and the formation of farmland and/or glades, and perhaps open country with farmland. In the Po valley, there is archaeobotanical evidence for mixed farming, involving millet, barley and wheat cultivation, with grapes, figs and legumes, and zooarchaeological records of animal husbandry involving transhumant pastoralism within intermontane basins and highland valleys. The reduction in woodland cover throughout the northern Mediterranean corresponds to an increase in herbaceous taxa represented by a distinctive group of pollen types. These signify a clear anthropogenic signal, and are broadly similar to indicator assemblages that have been recorded in similar investigations carried out in north-west Europe (Behre (ed.) 1986; Behre, 1990; Aaby, 1986, 1994; Turner, 1986; Birks, 1990; Gaillard *et al.*, 1992, 1994).

The distinctive pollen assemblage listed in Table 3.7 is likely to indicate the influence of woodland management, pastoralism and cereal cultivation. The impact of changing patterns of deforestation and land use was catastrophic, with evidence of multiple episodes of soil erosion and colluviation. These records are derived from studies in the northern Apennines, Po plain, Alpine margin and central Alps (Baffico *et al.*, 1987; Scaife, 1987; Macphail, 1988; Marziani *et al.*, 1991; Barker, 1994; Drescher-Schneider, 1994; Lowe *et al.*, 1994; Montanari *et al.*, 1994; Oeggl, 1994; Watson, 1996). In the northern Apennines, the scale of erosion and colluviation is reflected in an increase in the number of shallow mire basins, the deposition of minerogenic sediments in basins and an increase in lake water levels from c. 1700 years BP (c. 300–400AD). During later prehistory and into the historic periods, pollen evidence across the northern Mediterranean indicates widespread woodland clearance reflecting the increasing influence of human activity, probably as part of a broad-spectrum agricultural system involving woodland management (*Juglans* and *Castanea*), cultivation and animal husbandry. Although there is evidence for climatic oscillations in the northern Mediterranean during this period (Watts, 1985; Bottema, 1991; Magny, 1992; Magny and Ruffaldi, 1995; Yu and Harrison, 1995; Watts *et al.*, 1996a, 1996b; Branch, 1999), the precise relationships between these events and changes in settlement and subsistence practices remains uncertain (Barker, 1985).

TABLE 3.7 Anthropogenic indicators in pollen diagrams for the northern Mediterranean (Branch, 1999)

Juglans	*Castanea*	*Olea* type
Juniperus	*Erica* type	*Sinapis* type
Chenopodium type	*Plantago lanceolata* type	Lactuceae indet
Poaceae	*Rumex* type	*Urtica* type
Plantago media/major type	*Galium* type	*Bidens* type
Artemisia	*Serratula* type	

3.4.4 Megascale: colonization, hunting and farming in early prehistoric South America

Two broad research themes make South America arguably the most fascinating continent on Earth for environmental archaeological research. The first concerns the timing and nature of human colonization, the second focuses on the initiation of farming – when, where and why? The timing of the colonization of the Americas as a whole is a highly controversial subject, which was fuelled during the 1920s by the discovery of worked stone artefacts and zooarchaeological remains in Mexico – 'Folsom point' and 'Clovis point' (later radiocarbon dated to c. 11,000–10,000 BP and c. 11,500 BP respectively). These finds resulted in various theories, often lacking empirical data, which purported colonization of both North and South America prior to 12,000 BP (Lavallée, 2000).

Bearing in mind that some clearly remain questionable, the records of occupation and their radiocarbon dates are as follows:

1 Pedra Furada (Brazil), >45,000–14,000 BP, and again between 12,000 and 6000 BP.
2 Minas Gerais (Brazil), 20,000–14,500 BP and 12,000–11,000 BP.
3 Alice Boër (Brazil), >14,000 BP.
4 Muaco (Venezuela), c. 16,000–14,000 BP.
5 Taima-Taima (Venezuela), 14,000–13,000 BP.
6 El Abra (Columbia), 12,500–11,000 BP.
7 Pikimachay (Peru), 20,000 BP, and again between 14,000 and 13,000 BP (MacNeish *et al.*, 1980–83).
8 Monte Verde (Chile), 12,500 BP (Dillehay, 2000).
9 Los Toldos (Argentina), 12,600 BP.

Monte Verde, and the upper archaeological layers at Pikimachay (see Fig. 3.39) and Pedra Furada, fortunately provide widely accepted evidence for human occupation in Chile, Brazil and Peru prior to 12,000 BP (Dillehay, 2000; Lavallée, 2000). The migratory route into these diverse regions of South America, based upon genetic, skeletal and linguistic evidence, was probably across the Bering land bridge from Asia (due to a global reduction in relative sea level by 120 m during the last cold stage), and then overland (across the isthmus of Panama) and/or by sea down the Pacific coast. The environments early people entered were diverse and, in many places, inhospitable (Markgraf, 1989). Towards the end of the last glaciation, forests gradually replaced savannas in lowland areas by 13,000 BP, while upland areas below glacial limits (<3500 m) in Peru and Chile consisted of steppe-tundra and semi-desert. The onset of the present interglacial (c. 10,000 BP) resulted in the expansion of tropical forest in Brazil, desert vegetation in Chile, and puna grassland in upland areas of Peru, Bolivia and Argentina.

Various caves and rock-shelters in Peru, above altitudes suitable for cultivation (>3500 m), have provided excellent evidence for hunting and/or herding (pastoral economy) of animals. These include Telarmachay, Pachamachay and Uchkumachay, and collectively suggest that between 9000 and 6500/6000 cal BP there was a gradual change to specialized hunting of *Lama vicugna* (vicuña) and *Lama guanicoe* (guanaco), and reduced exploitation of *Hippocamelus antisensis* (Andean deer). After 6500/6000 cal BP, the faunal assemblages from Telarmachay record a higher proportion of young wild *Lama*, which is clearly at odds with normal hunting strategies that aim to protect young animals and females in order to create a sustainable food resource. These assemblages, interpreted as the negative effects of early attempts at animal husbandry, represent a transitional phase to a mixed economy based upon domesticated *Lama glama* (llama) and *Lama pacos* (alpaca), and to a lesser extent *Cavia porcellus* (guinea pig), alongside hunting after 4500 cal BP. The reason for this shift from a highly successful economic system based upon hunting to a rather more complex system combining hunting and herding is a major area of debate. Lavallée (2000) has ruled out changes in the natural environment, and instead suggests that knowledge and understanding of animal behaviour, and hence the potential benefits of greater control, was the key factor.

FIGURE 3.39 Pikimachay Cave, Ayacucho, Peru.

Plant remains are not ubiquitous on sites, although the presence of processing tools suggests that berries, nuts, roots, tubers and seeds were being utilized for food (e.g. *Opuntia, Prosopis*) and other purposes. After c. 8000 cal BP, the central Andean region witnessed major changes in its cultural history (Lavallée, 2000). This was due to the richness and accessibility of specific ecosystems that contained abundant marine resources, and plant and animal species suitable for domestication. The discovery of numerous coastal sites between Ecuador and Chile during the 1990s confirmed earlier theories that these areas became a focus for early human settlement due to the abundance of marine life, especially *Engraulis ringens* (anchovy). One site, called Paloma (Peru), became fully sedentary between 7000 and 4800 cal BP, and consisted of reed and grass/rush houses. The domestic rubbish contained the remains of anchovies, sardines, shellfish and marine mammals, suggesting that marine food rather than plants were the main component of the diet (Benfer, 1990). The evidence suggests that it was not until 5000 cal BP that a diverse range of domesticated plants appeared at these sites, including *Manihot esculenta* (manioc), *Gossypium barbadense* (cotton), *Ipomoea batatus* (sweet potato), *Solanum tuberosum* (potato) and *Chenopodium quinoa* (quinoa).

In contrast, and perhaps surprisingly, the upland areas provide the earliest evidence for horticulture. At Guitarrero cave (Peru), records of *Capsicum chinense* (pimento), *Phaseolus lunatus* and *P. vulgaris* (beans) between 10,600 and 7700 cal BP represent the oldest domesticated plants in America (Lavallée, 2000). *Cucurbita* sp. (gourd), *Lagenaria* sp. (calabash) and *Zea mays* (maize) at Guitarrero cave and Pikimachay (Peru) followed these taxa chronologically (MacNeish *et al.*, 1980–83). The evidence suggests that the appearance of these cultivars was diachronous, which implies that the adoption of farming was determined by the importance of hunted and gathered food in the diet in specific parts of the Andes. Where was the centre of domestication? Lavallée (2000) has proposed three hypotheses:

1 In Mexico, followed by diffusion into North and South America.
2 Tropical lowland savannas of the Amazon during the cool, dry climate of the early Holocene.
3 Mid-altitude Andean basins.

The bioarchaeological and geochronological evidence suggests that the third hypothesis is the most realistic, although the overall paucity of data, especially from the lowland tropics, means that further well-dated sequences are required to test the model proposed. Lynch (1973) has proposed that hunters simultaneously exploited both the highland valleys and lowland tropics, probably as part of a seasonal migratory pattern of resource exploitation of the diverse ecosystems (climate, soils, vegetation and animals) that existed in these areas. For this reason, domestication in both biozones seems highly likely, initially by accidental manipulation, and/or experimental horticulture, followed by deliberate cultivation. This gradual shift in plant and animal exploitation practices is attributed to climate change, population growth and pressure, or simply the consequence of long-term socio-technical developments, mainly in coastal areas (Lavallée, 2000).

REFERENCES

Aaby, B. (1986) Palaeoecological studies of mires. In (B.E. Berglund, ed.) *Handbook of Holocene palaeoecology and palaeohydrology*, 145–164. Sussex: Wiley.

Aaby, B. (1994) NAP percentages as an expression of cleared areas. In (B. Frenzel, ed.) *Evaluation of land surfaces cleared from forests in the Roman Iron Age and the time of migrating Germanic tribes based on regional pollen diagrams*, 13–28. Stuttgart: Gustav Fischer Verlag (European Climate and Man 7).

Alfieri, A. (1976) *The Coleoptera of Egypt*. Cairo: Atlas Press.

Allaby, R.G., Jones, M.K. and Brown, T.A. (1994) DNA in charred wheat grains from the Iron Age hill fort at Danebury, England. *Antiquity* **68**, 126–312.

Andersen, S.T. (1970) The relative pollen productivity and tree representation of north European trees and correlation factors for tree pollen spectra. *Danmarks Geologiske Undersolgelse* RII, 96.

Andersen, S.T. (1973) The differential pollen productivity of trees and its significance for the interpretation of a pollen diagram from a forested region. In (H.J.B. Birks and R.G. West, eds.) *Quaternary Plant Ecology*, 109–115.

Andersen, S.T. (1979) Identification of wild grasses and cereal pollen, *Danmarks Undersolgelse Arbog* 1978, 69–72.

Andersen, S.T. (1992) Pollen proxy data for human

impact on vegetation, based on methodological experiences. In (B. Frenzel, ed.) *Evaluation of land surfaces cleared from forests by prehistoric man in Early Neolithic times and the time of migrating Germanic tribes*, 1–12. European Palaeoclimate and Man, 3. Strasbourg: European Science Foundation.

Ashworth, A.C., Buckland, P.C. and Sadler, J.P. (eds.) (1997) *Studies in Quaternary Entomology – An Inordinate Fondness of Insects*. Quaternary Proceedings **5**. Chichester: Wiley.

Asouti, E. (2003) Woodland vegetation and fuel exploitation at the prehistoric campsite of Pinarbaşi, south-central Anatolia, Turkey: the evidence from the wood charcoal macro-remains. *Journal of Archaeological Science* **30**, 1185–1201.

Atkinson, T.C., Briffa, K.R., Coope, G.R., Joachim, M.J. and Perry, D.W. (1986) Climatic calibration of coleopteran data. In (B.E. Berglund, ed.) *Handbook of Holocene Palaeoecology and Palaeohydrology*, 851–858. Chichester: Wiley.

Auban, J.B., Barton, C.M. and Ripoll, M.P. (2001) A taphonomic perspective on Neolithic beginnings: theory, interpretation, and empirical data in the western Mediterranean. *Journal of Archaeological Science* **28**, 597–612.

Baffico, O., Cruise, G.M., Macphail, R.I., Maggi, R. and Nisbet, R. (1987) Monte Aiona – Prato Mollo. *Archeologia in Liguria III.1, Scavi e Scoperte 1982–1986*, 57–66.

Barker, G. (1973) Cultural and economic change in the prehistory of central history. In (C. Renfrew, ed.) *The Explanation of Cultural Change – Models in Prehistory*, 359–370. London: Duckworth.

Barker, G. (1975) Prehistoric territories and economies in central Italy. In (E.S. Higgs, ed.) *Palaeoeconomy*, 111–175. Cambridge: Cambridge University Press.

Barker, G. (1985) *Prehistoric Farming in Europe*. Cambridge: Cambridge University Press.

Barker, G. (1994) The exploitation of the Matese mountain and upper Biferno valley from prehistoric times to the present day: environment, economy and society. In (P. Biagi and J. Nandris, eds.) *Highland zone exploitation in southern Europe*, 205–220. Brescia: Museo Civico Di Scienze Naturali Di Brescia (Monografie di Natura Bresciana 20).

Barker, G., Biagi, P., Clark, G., Maggi, R. and Nisbet, R. (1990) From hunting to herding in the Val Pennavaira (Liguria – Northern Italy). In (P. Biagi, ed.) *The Neolithisation of the Alpine region*, 99–121. Brescia: Museo Civico Di Scienze Naturali Di Brescia (Monografie di Natura Bresciana 13).

Barton, H., Torrence, R. and Fullagar, R. (1998) Clues to stone tool function re-examined: comparing starch grain frequencies on used and unused obsidian artefacts. *Journal of Archaeological Science* **25**, 1231–1238.

Battarbee, R.W. (1986) Diatom analysis, In (B.E. Berglund, ed.) *Handbook of Holocene Palaeoecology and Palaeohydrology*, 527–570. Chichester: Wiley.

Battarbee, R.W. (1988) The use of diatom analysis in archaeology: a review. *Journal of Archaeological Science* **15**, 621–644.

Behre, K-E., ed. (1986) *Anthropogenic Indicators in Pollen Diagrams*. Rotterdam: Balkema.

Behre, K-E. (1990) Some general remarks on human impact in pollen diagrams, In (D. Moe and S. Hicks, eds.) Impact of prehistoric and Medieval man on the vegetation: man at the forest limit: 31–33. *Journal of the European Study Group on Physical, Chemical, Mathematical and Biological Techniques Applied to Archaeology* **31**, 31–33.

Bell, M.G. (1982) The effects of land-use and climate on valley sedimentation. In (A.F. Harding, ed.) *Climatic Change in Later Prehistory*, 127–142. Edinburgh: Edinburgh University Press.

Bell, M.G. (1983) Valley sediments as evidence of prehistoric land-use on the South Downs. *Proceedings of the Prehistoric Society* **49**, 119–150.

Bell, M.G. (1986) Archaeological evidence for the date, cause and extent of soil erosion on the Chalk. In (C.P. Burnham and J.I. Pitman, eds.) *Soil Erosion*, 72–83. (SEESOIL 3).

Bell, M.G. and Walker, M.J.C. (1992) *Late Quaternary Environmental Change: Human and Physical Perspectives*. Harlow: Longman.

Benfer, R.A. (1990) The preceramic period site of Paloma, Peru: bioindications of improved adaptation to sedentism. *Latin American Antiquity* **1**, 284–318.

Berglund, B.E. (1994) Methods for quantifying prehistoric deforestation. In (B. Frenzel, ed.) *Evaluation of Land Surfaces Cleared from Forests in the Roman Iron Age and the Time of Migrating Germanic Tribes based on Regional Pollen Diagrams*, 5–12. European Palaeoclimate and Man 7. Strasbourg: European Science Foundation.

Berggren, G. (1981) *Atlas of seeds 3: Salicaceae to Cruciferae*. Stockholm: Swedish Natural Science Research Council.

Bird, D.W., Richardson, J.L., Veth, P.M. and Barham, A.J. (2002) Explaining shellfish variability in middens on the Meriam Island, Torres Strait, Australia. *Journal of Archaeological Science* **29**, 457–469.

Birks, H.J.B. (1990) Indicator values of pollen types from post-6000 BP, pollen assemblages from southern England and southern Sweden. *Quaternary Studies in Poland* **10**, 21–31.

Bogaard, A., Palmer, C., Jones, G., Charles, M. and Hodgson, J.G. (1999) A FIBS approach to the use of weed ecology for the archaeobotanical recognition of crop rotation regimes. *Journal of Archaeological Science* **26**, 1211–1224.

Bogaard, A., Jones, G., Charles, M. and Hodgson, J.G. (2001) On the archaeobotanical inference of crop sowing time using the FIBS method. *Journal of Archaeological Science* **28**, 1171–1183.

Bortenschlager, S. (1982) Chronostratigraphic subdivisions of the Holocene in the Alps. *Striae* **16**, 75–79.

Bottema, S. (1991) Pollen proxy data from southeastern Europe and the Near East. In (B. Frenzel, ed.) *Evaluation of climate proxy data in relation to the European Holocene*, 63–79. Stuttgart: Gustav Fischer Verlag (European Palaeoclimate and Man 1).

Bradshaw, R.H.W. (1981) Quantitative reconstruction of local woodland vegetation using pollen analysis from a small basin in Norfolk, England. *Journal of Ecology* **69**, 941–955.

Bradshaw, R.H.W. (1991) Spatial scale in the pollen record. In (D.R. Harris and K.D. Thomas, eds.) *Modelling Ecological Change*, 41–52. London: Institute of Archaeology.

Bradshaw, R.H.W. and Millar, N.G. (1988) Recent successional processes investigated by pollen analysis of closed canopy forest sites. *Vegetatio* **76**, 45–54.

Branch, N.P. (1997) Palynological study of the early and middle Neolithic cave deposits of Arene Candide: preliminary results. In (R. Maggi, ed.) *Arene Candide: a functional and environmental assessment of the Holocene sequence*, 89–102. Rome: Il Calamo.

Branch, N.P. (1999) *Vegetation History and Human Activity in the Ligurian Apennines and Alps during the last 14,000 years.* Unpublished PhD thesis, University of London.

Branch, N.P. (2002) L'analisi palinologica per lo studio della vegetazione e del suo uso. In (N. Campana and R. Maggi, eds.) *Archeologia in Valle Lagorara, Diecimila anni di storia intorno a una cava di diaspro*, 339–352. Firenze: Istituto Italiano di Preistoria e Protostoria.

Branch, N.P. (2003) *Pollen analysis of modern human faecal material, and its implications for archaeological and forensic science.* ArchaeoScape, Department of Geography, Royal Holloway University of London, Technical Report.

Branch, N.P. and Canti, M.G. (1994) Research proposal for testing pollen translocation rates in coarse mineral substrates. *Journal of the South-East England Soils Discussion Group* **10**, 73–80.

Branch, N.P., Athersuch, J., Bates, M.R., Green, C.P., Rackham, C. and Swindle, G.E. (2003) *Environmental Archaeological Assessment: A249 Road Improvement Scheme, Isle of Sheppey, Kent.* ArchaeoScape, Royal Holloway University of London, Unpublished Report.

Brandt, J. (1950) Plant remains in an early Iron Age Body. *Arboger for Nordisk Oldkyndighed og Historie*, 347–351.

Brasier, M.D. (1980) *Microfossils.* London: George Allen and Unwin.

Briggs, D.J., Gilbertson, D.D. and Harris, A.L. (1990) Molluscan taphonomy in a braided river environment and its implications for studies of Quaternary cold-stage river deposits. *Journal of Biogeography* **17**, 623–637.

Brochier, J.E. (1983) Bergeries et feux néolithiques dans le Midi de la France, caractérisation et incidence sur le raisonnement sédimentologique. *Quatar* **33/34**, 181–193.

Brochier, J.E. (1990) Des techniques géo-archéologiques au service de l'étude des paysages et de leur exploitation. *Archéologie et espaces: X_e Rencontres Internationales d'Archaeologie et d'Histoire d'Antibes, 1989*, 453–472. Juan-les-Pins: Editions APDCA.

Brochier, J.E. (1996) Feuilles ou fumiers? Observations sur le rôle des poussières sphérolitiques dans l'interpretatation des dépôts archéologiques holocènes. *Anthropozoologica* **24**, 19–30.

Brochier, J.E., Villa, P., Giacomarra, M. and Tagliacozzo, A. (1992) Shepherds and sediments: Geo-ethnoarchaeology of pastoral sites. *Journal of Anthropological Archaeology* **11**, 47–102.

Brothwell, D.R. (1981) *Digging Up Bones.* Cornell University Press.

Buckland, P.C., McGovern, T.H., Sadler, J.P. and Skidmore, P. (1994) Twig layers, floors and middens. In (B. Ambrosiani and H. Clarke, eds.) *Developments around the Baltic and North Sea in the Viking Age*, 132–143. Stockholm: Birka Studies 3.

Buckland, P.C., Amorosi, T., Barlow, L.K., Dugmore, A.J., Mayewski, P.A., McGovern, T.H., Ogilvie, A.E.J., Sadler, J.P. and Skidmore, P. (1996) Bioarchaeological and climatological evidence for the fate of Norse farmers in medieval Greenland. *Antiquity* **70**, 88–96.

Bunting, M.J., Tipping, R. and Downes, J. (2001) 'Anthropogenic' pollen assemblages from a Bronze Age cemetery at Ling Fiold, West Mainland, Orkney. *Journal of Archaeological Science* **28**, 487–500.

Butzer, K.W. (1982) *Archaeology as Human Ecology.* Cambridge: Cambridge University Press.

Butzer, K.W. and Freeman, L.G. (1968) Pollen analysis at the Cueva del Toll, Catalonia: a critical re-appraisal. *Geologie en Mijnbouw* **47**, 116–120.

Caballero, M., Ortega, B., Valadez, F., Metcalfe, S., Macias, J.L. and Sugiura, Y. (2002) Sta. Cruz Atizapan: a 22-ka lake level record and climatic implications for the late Holocene human occupation in the Upper Lerma Basin, Central Mexico. *Palaeogeography, Palaeoclimatology, Palaeoecology* **186**, 217–235.

Cabral, J.P. and Silva, da A.C.F. (2003) Morphometric analysis of limpets from an Iron-Age shell midden found in northwest Portugal. *Journal of Archaeological Science* **30**, 817–829.

Canti, M.G. (1997) An investigation into microscopic

calcareous spherulites from herbivore dungs. *Journal of Archaeological Science* **24**, 219–231.

Canti, M.G. (1998a) The micromorphological identification of faecal spherulites from archaeological and modern materials. *Journal of Archaeological Science* **25**, 435–444.

Canti, M.G. (1998b) Origin of calcium carbonate granules found in buried soils and Quaternary deposits. *Boreas* **27**, 275–288.

Canti, M.G. (1999) The production and preservation of faecal spherulites: animals, environment and taphonomy. *Journal of Archaeological Science* **26**, 251–258.

Canti, M.G. (2003) Earthworm activity and archaeological stratigraphy: a review of products and processes. *Journal of Archaeological Science* **30**, 135–148.

Canti, M.G. and Piearce, T. (2003) Morphology and dynamics of calcium carbonate granules produced by different earthworm species. *Pedobiologia* **47**, 511–521.

Carrión, J.S. (1992) Late Quaternary pollen sequence from Carihuela Cave, southeastern Spain, *Review of Palaeobotany and Palynology* **71**, 37–77.

Carrión, J.S. and Dupré, M. (1994) Pollen data from Mousterian sites in southeastern Spain, *AASP Contributions Series* **29**, 17–26.

Carrión, J.S., Dupré, M., Pilar Fumanal, M. and Montes, R. (1995) A palaeoenvironmental study in semi-arid southeastern Spain: the palynological and sedimentological sequence at Perneras Cave (Lorca, Murcia). *Journal of Archaeological Science* **22**, 355–367.

Carrión, J.S., Munuera, M. and Navarro, C. (1998) The palaeoenvironment of Cariheula Cave (Granada, Spain): a reconstruction on the basis of palynological investigations of cave sediments. *Review of Palaeobotany and Palynology* **99**, 317–340.

Carrott, J. and Kenward, H. (2001) Species associations among insect remains from urban archaeological deposits and their significance in reconstructing the past human environment. *Journal of Archaeological Science* **28**, 887–905.

Carruthers, W. (2000) The mineralised plant remains. In (A. Lawson and C. Gingell, eds.) *Potterne 1982–1985: Animal Husbandry in Later Prehistoric Wiltshire* (Wessex Archaeology Report 17).

Charles, M., Jones, G. and Hodgson, J.G. (1997) FIBS in archaeobotany: functional interpretation of weed floras in relation to husbandry practices. *Journal of Archaeological Science* **24**, 1151–1161.

Clark, J.S. (1988) Particle motion and the theory of charcoal analysis: source area, transport, deposition and sampling. *Quaternary Science* **30**, 67–80.

Clark, R.L. (1982) Point-count estimation of charcoal in pollen preparations and thin sections of sediments. *Pollen et Spores* **24**, 523–535.

Clarke, C. (1999) Palynological investigations of a Bronze Age cist burial from Whitsome, Scottish Borders, Scotland. *Journal of Archaeological Science* **26**, 553–560.

COHMAP Members (1988) Climatic changes of the last 18,000 years: Observations and model simulations. *Science* **241**, 1043–1052.

Coil, J., Korstanje, M.A., Archer, S. and Hastorf, C.A. (2003) Laboratory goals and considerations for multiple microfossil extraction in archaeology. *Journal of Archaeological Science* **30**, 991–1008.

Coles, G.M., Gilbertson, D.D., Hunt, C.O. and Jenkinson, R.D.S. (1989) Taphonomy and the palynology of cave deposits. *Cave Science* **16**, 83–90.

Coles, G.M. and Gilbertson, D.D. (1994) The airfall-pollen budget of archaeologically important caves: Creswell Crags, England. *Journal of Archaeological Science* **21**, 735–755.

Coope, G.R. (1986) Coleoptera analysis. In (B.E. Berglund, ed.) *Handbook of Holocene Palaeoecology and Palaeohydrology*, 703–713. Chichester: Wiley.

Cremaschi, M. (1990) Pedogenesi Medio Olocenica ed uso dei suoli durante il Neolitico in Italia Settentrionale. In (P. Biagi, ed.) *The Neolithisation of the Alpine Region*, 71–90. Brescia: Museo Civico Di Scienze Naturali Di Brescia (Monografie di Natura Bresciana 13).

Cushing, E.J. (1967) Evidence for differential pollen preservation in late-Quaternary sediments from Minnesota. *Review of Palaeobotany and Palynology* **4**, 87–101.

Cwynar, L.C. (1987) Fire and the forest history of the North Cascade Range. *Ecology* **68**, 791–802.

Darwin, C. (1881). *The Formation of Vegetable Mould through the Action of Worms, with Observations on their Habits*. London: John Murray.

Davis, S.J.M. (1987) *The Archaeology of Animals*. London: Batsford.

Davies, P., Gale, C.H. and Lees, M. (1996) Quantitative study of modern wet-ground molluscan faunas from Bossington, Hampshire. *Journal of Biogeography* **23**, 371–377.

Day, S.P. (1996) Dogs, deer and diet at Star Carr: a reconstruction of C-isotope evidence from Early Mesolithic dog remains from the Vale of Pickering, Yorkshire, England. *Journal of Archaeological Science* **23**, 783–787.

Delcourt, P.A., Delcourt, H.R., Ison, C.R., Sharp, W.E. and Gremillion, K.J. (1998) Prehistoric human use of fire, the eastern agricultural complex, and Appalachian oak-chestnut forests: palaeoecology of Cliff Palace Pond, Kentucky. *American Antiquity* **63**, 89–100.

Denford, S. (1978) Mites and their potential use in Archaeology. In (D.R. Brothwell, K.D. Thomas and J. Clutton-Brock, eds.) *Research Problems in Zooarchaeology*, 77–83. Institute of Archaeology (UCL): Occasional Publication 3.

Dennell, R.W. (1983) *European Economic Prehistory*. London: Academic Press.

Devine, J.D., Sigurdsson, M. and Davies, A.N. (1984) Estimates of sulfur and chlorine yield to the atmosphere from volcanic eruptions and potential climatic effects. *Journal of Geophysical Research* **89**, 6309–6325.

Dickson, J.H. (1978) Bronze age mead. *Antiquity* **52**, 108–113.

Dillehay, T.D. (2000) *The Settlement of the Americas*. New York: Basic Books.

Dimbleby, G.W. (1957) Pollen analysis of terrestrial soils. *New Phytologist* **56**, 12–28.

Dimbleby, G.W. (1961a) Soil pollen analysis. *Journal of Soil Science* **12**, 1–11.

Dimbleby, G.W. (1961b) Transported material in the soil profile. *Journal of Soil Science* **12**, 12–22.

Dimbleby, G.W. (1962) *The Development of British Heathlands and their Soils*. Oxford Forestry Memoirs, 23.

Dimbleby, G.W. (1977) *Ecology and Archaeology*. London: Arnold.

Dimbleby, G.W. (1985) *The Palynology of Archaeological Sites*. London: Academic Press.

Drescher-Schneider, R. (1994) Forest, forest clearance and open land during the time of the Roman Empire in northern Italy (the botanical record). In (B. Frenzel, ed.) *Evaluation of land surfaces cleared from forests in the Mediterranean region during the time of the Roman Empire*, 45–58. Stuttgart: Gustav Fischer Verlag (European Palaeoclimate and Man 1).

Edwards, K.J. (1979) Palynological and temporal inference in the context of prehistory, with special reference to the evidence from lake and peat deposits. *Journal of Archaeological Science* **6**, 255–270.

Edwards, K.J. (1982) Man, space and the woodland edge – speculation on the detection and interpretation of human impact in pollen profiles. In (S. Limbrey and M. Bell, eds.) *Archaeological Aspects of Woodland Ecology*, 5–22. British Archaeological Reports International Series, S146.

Edwards, K.J. (1983) Quaternary palynology: multiple profile studies and pollen variability. *Progress in Physical Geography* **7**, 687–709.

Edwards, K.J. (1989) The cereal pollen record and early agriculture. In (A. Milles, D. Williams and N. Gardner, eds.) *The Beginnings of Agriculture*, 113–135. British Archaeological Reports International Series, S496.

Edwards, K.J. (1991) Using space in cultural palynology: the value of the off-site pollen record. In (D.R. Harris and K.D. Thomas, eds.) *Modelling Ecological Change*, 61–73. London: Institute of Archaeology.

Edwards, K.J. (1998) Detection of human impact on the natural environment: palynological views. In (J. Bayley, ed.) *Science in Archaeology: an Agenda for the Future*, 69–88. London: English Heritage.

Edwards, K.J. and Hirons, K.R. (1984) Cereal pollen grains in pre-elm decline deposits: implications for the earliest agriculture in Britain and Ireland. *Journal of Archaeological Science* **11**, 71–78.

Elias, S.A. (1994) *Quaternary Insects and their Environments*. London: Smithsonian Institution Press.

Evans, J.G. (1972) *Land Snails in Archaeology*. London: Seminar Press.

Evershed, R.P., Turner-Walker, G., Hedges, R.E.M., Tuross, N. and Leyden, A. (1995) Preliminary results for the analysis of lipids in ancient bone. *Journal of Archaeological Science* **22**, 277–290.

Frey, D.G. (1986) Cladocera analysis. In (B.E. Berglund, ed.) *Handbook of Holocene Palaeoecology and Palaeohydrology*, 667–686. Chichester: Wiley.

Gaillard, M.-J., Birks, H.J.B., Emanuelsson, U. and Berglund, B.E. (1992) Modern pollen/land-use relationships as an aid in the reconstruction of past land-uses and cultural landscapes: an example from south Sweden. *Vegetation History and Archaeobotany* **1**, 3–17.

Gaillard, M.-J., Birks, H.J.B., Emanuelsson, U., Karlsson, P., Lageros, P. and Olausson, D. (1994) Application of modern pollen/land-use relationships to the interpretation of pollen diagrams-reconstructions of land-use history in southern Sweden, 3000–0 BP. *Review of Palaeobotany and Palynology* **82**, 47–73.

Gale, S.J. and Hunt, C.O. (1985) The stratigraphy of Kirkhead Cave, an Upper Palaeolithic site in northern England. *Proceedings of the Prehistoric Society* **51**, 283–304.

Garbett, G.G. (1981) The elm decline: the depletion of a resource. *New Phytologist* **88**, 573–585.

Gautier, A. (1993) Trace fossils in archaeozoology. *Journal of Archaeological Science* **20**, 511–523.

Gilbert, R.I. and Mielke, J.H. (eds.) (1985) *The Analysis of Prehistoric Diets*. Orlando: Academic Press.

Girling, M.A. (1988) The bark beetle Scolytus scolytus (Fabricius) and the possible role of elm disease in the early Neolithic. In (M. Jones, ed.) *Archaeology and the Flora of the British Isles*, Oxford University Committee for Archaeology, 14, 34–38. Oxford: Oxford University Committee for Archaeology.

Gobalet, K.W. (2001) A critique of faunal analysis; inconsistency among experts in blind tests. *Journal of Archaeological Science* **28**, 377–386.

Grave, P. and Kealhofer, L. (1999) Assessing bioturbation in archaeological sediments using soil morphology and phytolith analysis. *Journal of Archaeological Science* **26**, 1239–1248.

Greenfield, H.J. (1999) The origin of metallurgy: distinguishing stone from metal cut-marks on bones from archaeological sites. *Journal of Archaeological Science* **26**, 797–808.

Guilizzoni, P., Lami, A., Marchetto, A., Jones, V., Manca, M. and Bettinetti, R. (2002) Palaeoproductivity and environmental changes during the Holocene in central

Italy as recorded in two crater lakes (Albano and Nemi). *Quaternary International* **88**, 57–68.

Guiot, J. (1987) Late Quaternary climatic change in France estimated from multivariate pollen time series. *Quaternary Research* **28**, 100–118.

Guiot, J., Pons, A., De Beaulieu, J-L. and Reille, M. (1989) A 140,000 year climatic reconstruction from two European pollen records. *Nature* **338**, 309–313.

Guiot, J., Harrison, S.P. and Prentice, I.C. (1993) Reconstuction of Holocene precipitation patterns in Europe using pollen and lake-level data. *Quaternary Research* **40**, 139–149.

Gustafsson, S. (2000) Carbonized cereal grains and weed seeds in prehistoric houses – an experimental perspective. *Journal of Archaeological Science* **27**, 65–70.

Haldane, C. (1991) Recovery and analysis of plant remains from some Mediterranean shipwreck sites. In (J.M. Renfrew, ed.) *New Light on Early Farming, Recent Developments in Palaeoethnobotany*, 213–223. Edinburgh: Edinburgh University Press.

Haldane, C. (1992) Direct evidence for organic cargoes in the Late Bronze Age. *World Archaeology* **24**, 348–360.

Hale, D.N. and Noel, M.J. (1991) Some quantitative pollen extraction tests leading to a modified technique for cave sediments. In (P. Budd, B. Chapman, C. Jackson, R. Janaway and B. Ottaway, eds.) *Archaeological Sciences*, 319–324. Proceedings of the Conference on the Application of Scientific Techniques in Archaeology, 1989 (Oxbow Monograph 9).

Halkett, D., Hart, T., Yates, R., Volman, T.P., Parkington, J.E., Orton, J., Klein, R.G., Cruz-Uribe, K. and Avery, G. (2003) First excavation of intact Middle Stone Age layers at Ysterfontein, Western Cape Province, South Africa: implications for Middle Stone Age ecology. *Journal of Archaeological Science* **30**, 955–971.

Hall, A.R. and Kenward, H.K. (1990) Environmental evidence from the Colonia: General Accident and Rougier Street. *The Archaeology of York*, **14**, 289–434. London: Council for British Archaeology.

Havinga, A.J. (1964) Investigation into the differential corrosion susceptibility of pollen and spores. *Pollen et Spores* **6**, 621–635.

Havinga, A.J. (1967) Palynology and pollen preservation. *Review of Palaeobotany and Palynology* **2**, 81–98.

Helbaek, H. (1950) Botanical study of the stomach of Tollund Man. *Arboger for Nordisk Oldkyndighed og Historie*, 329–341.

Helbaek, H. (1958) The last meal of Grauballe Man: an analysis of food remains in the stomach. *Kuml*, 83–116.

Hendey, N.I. (1964) *An introductory account of the smaller algae of British coastal waters. Part V: Bacillariophyceae (Diatoms).* Fisheries Investigations Series IV. London: HMSO.

Hellqvist, M. and Lemdahl, G. (1996) Insect assemblages and local environment in the Mediaeval town of Uppsala, Sweden. *Journal of Archaeological Science* **23**, 873–881.

Hillman, G.C. (1984) Interpretation of archaeological plant remains: the applications of ethnographic models from Turkey. In (W. van Zeist and W.A. Casparie, eds.) *Plants and Ancient Man*, 1–41. Rotterdam: Balkema.

Hillman, G.C. (1986) Plant foods in ancient diet: the archaeological role of palaeofaeces in general and Lindow Man's gut contents in particular. In (I.M. Stead, J.B. Bourke and D. Brothwell, eds.), *Lindow Man The Body in the Bog*, 99–115. London: British Museum Publications.

Hillman, G.C. (1991) Phytosociology and ancient weed floras: taking account of taphonomy and changes in cultivation methods. In (D.R. Harris and K.D. Thomas, eds.) *Modelling Ecological Change*, 27–40. London: Institute of Archaeology, University College London.

Hillman, G.C., Mason, S., de Moulins, D. and Nesbitt, M. (1996) Identification of archaeological remains of wheat: the 1992 London workshop. *Circaea* **12**, 195–209.

Hofmann, W. (2000) Response of the Chydorid faunas to rapid climatic changes in four alpine lakes at different altitudes. *Palaeogeography, Palaeoclimatology, Palaeoecology* **159**, 281–292.

Holden, T.G. (1986) Preliminary report on the detailed analyses of the macroscopic remains from the gut of Lindow Man. In (I.M. Stead, J.B. Bourke and D. Brothwell, eds.) *Lindow Man: The Body in the Bog*, 116–125. London: British Museum Publications.

Holden, T.G. and Núnez, L. (1993) An analysis of the gut contents of five well-preserved human bodies from Tarapacá, northern Chile. *Journal of Archaeological Science* **20**, 595–611.

Horrocks, M., Irwin, G.J., McGlone, M.S., Nichol, S.L. and Williams, L.J. (2003) Pollen, phytoliths and diatoms in prehistoric coprolites from Kohika, Bay of Plenty, New Zealand. *Journal of Archaeological Science* **30**, 13–20.

Hunt, C.O. (1993) Mollusc taphonomy in caves: a conceptual model. *Cave Science* **20**, 45–49.

Huntley, B., Bartein, P.J. and Prentice, I.C. (1989) Climatic control of the distribution and abundance of Beech (*Fagus* L.) in Europe and North America. *Journal of Biogeography* **16**, 551–560.

Innes, J.B. and Simmons, I.G. (1988) Disturbance and diversity: floristic changes associated with pre-elm decline woodland recession in north-east Yorkshire. In (M. Jones, ed.) *Archaeology and the Flora of the British Isles*, 7–20. Oxford: Oxford University Committee for Archaeology Monograph 14.

Innes, J.B. and Blackford, J.J. (2003) The ecology of Late Mesolithic woodland disturbances: model testing with fungal spore assemblage data. *Journal of Archaeological Science* **30**, 185–194.

Ioannidou, E. (2003) Taphonomy of animal bones: species, sex, age and breed variability of sheep, cattle and pig bone density. *Journal of Archaeological Science* **30**, 355–365.

Iriarte, J. (2003) Assessing the feasibility of identifying maize through the analysis of cross-shaped size and three-dimensional morphology of phytoliths in the grasslands of southeastern South America. *Journal of Archaeological Science* **30**, 1085–1094.

Jacobson, G.L. and Bradshaw, R.H.W. (1981) The selection of sites for palaeovegetation studies. *Quaternary Research* **16**, 80–96.

James, S.R. (1997) Methodological issues concerning screen size recovery rates and their effects on archaeofaunal interpretations. *Journal of Archaeological Science* **24**, 385–397.

Janssen, C.R. (1986) The use of local pollen indicators and of the contrast between regional and local pollen values in the assessment of the human impact on vegetation. In (K-E. Behre, ed.) *Anthropogenic Indicators in Pollen Diagrams*, 203–209. Rotterdam: Balkema.

Jans, M.M.E., Nielson, C.M., Smith, C.I., Collins, M.J. and Kars, H. (2004) Characterisation of microbial attack on archaeological bone. *Journal of Archaeological Science* **31**, 87–95.

Jeppesen, E., Leavitt, P., De Meester, L. and Jensen, J.P. (2001) Functional ecology and palaeolimnology: using cladoceran remains to reconstruct anthropogenic impact. *Trends in Ecology and Evolution* **16**, 191–198.

Jones, A.K.G. (1986) Parasitological investigations on Lindow Man. In (I.M. Stead, J.B. Bourke and D. Brothwell, eds.) *Lindow Man The Body in the Bog*, 136–139. London: British Museum Publications.

Jones, G.E.M. (1984) Interpretation of archaeological plant remains; ethnographic models from Greece. In (W. van Zeist and W.A. Casparie, eds.) *Plants and Ancient Man*, 43–61. Rotterdam: Balkema.

Jones, G., Bogaard, A., Charles, M. and Hodgson, J.G. (2000) Distinguishing the effects of agricultural practices relating to fertility and disturbance: a functional ecological approach in archaeobotany. *Journal of Archaeological Science* **27**, 1073–1084.

Kenward, H.K. (1976) Reconstructing ancient ecological conditions from insect remains: some problems and an experimental approach. *Ecological Entomology* **1**, 7–17.

Kenward, H.K. (1978) The value of insect remains as evidence of ecological conditions on archaeological sites. In (D.R. Brothwell, K.D. Thomas and J. Clutton-Brock, eds.) *Research Problems in Zooarchaeology*, 25–38. Institute of Archaeology (UCL): Occasional Publication 3.

Kenward, H.K. (1985) Outdoors-indoors? The outdoor component of archaeological insect assemblages. In (N.R.J. Fieller, D.D. Gilbertson and N.G.A Ralph, eds.) *Palaeobiological Investigations: Research Design, Methods and Data Analysis*, 97–104. British Archaeological Reports International Series, 266.

Kenward, H.K., Hall, A. and Jones, A.K.G. (1980) A tested set of techniques for the extraction of plant and animal macrofossils from waterlogged archaeological deposits. *Science and Archaeology* **22**, 3–15.

Kenward, H.K. and Hall, A. (1997) Enhancing bioarchaeological interpretation using indicator groups: stable manure as a paradigm. *Journal of Archaeological Science* **24**, 663–673.

Kerney, M.P. (1963) Late-glacial deposits on the chalk of south-east England. *Philosophical Transactions of the Royal Society of London* B730, 203–254.

Kerney, M.P., Preece, R.C. and Turner, C. (1980) Molluscan and plant biostratigraphy of some Late Devensian and Flandrian deposits in Kent. *Philosophical Transactions of the Royal Society of London* Series B, 291, 1–43.

Koch, K. (1989–1992) *Die Käfer Mitteleuropas, ökologie 1, 2, 3*. Krefeld: Goecke and Evers.

Kos, A.M. (2003) Characterisation of post-depositional taphonomic processes in the accumulation of mammals in a pitfall cave deposit from southeastern Australia. *Journal of Archaeological Science* **30**, 781–796.

Lageard, J.G.A., Chambers, F.M. and Thomas, P.A. (1995) Recording and reconstruction of wood macrofossils in three-dimensions. *Journal of Archaeological Science* **22**, 561–567.

Lam, Y.M., Chen, X., Marean, C.W. and Frey, C.J. (1998) Bone density and long bone representation in archaeological faunas: comparing results from CT and photon densitometry. *Journal of Archaeological Science* **25**, 559–570.

Lavallée, D. (2000) *The First South Americans* (translated by P.G. Bahn). Utah: The University of Utah Press.

Laville, H. (1976) Deposits in calcareous rockshelters: analytical methods and climatic interpretation. In (D.A. Davidson and M.L. Shackley, eds.) *Geoarchaeology*, 137–155. London: Duckworth.

Lee, G., Davis, A.M., Smith, D.G. and McAndrews, J.H. (2004) Identifying fossil wild rice (*Zizania*) pollen from Cootes Paradise, Ontario: a new approach using scanning electron microscopy. *Journal of Archaeological Science* **31**, 411–421.

Legge, A.J. (1972) Cave climates. In (E.S. Higgs, ed.) *Papers in Economic Prehistory*, 97–103. Cambridge: Cambridge University Press.

Legge, A.J. and Rowley-Conwy, P.A. (1988) *Star Carr Revisited*. London: Birkbeck College.

Lentfer, C.J. and Boyd, W.E. (1998) A comparison of three methods for the extraction of phytoliths from sediments. *Journal of Archaeological Science* **25**, 1159–1183.

Lentfer, C.J. and Boyd, W.E. (1999) An assessment of

techniques for the deflocculation and removal of clays from sediments used in phytolith analysis. *Journal of Archaeological Science* **26**, 31–44.

Lentfer, C.J. and Boyd, W.E. (2000) Simultaneous extraction of phytoliths, pollen and spores from sediments. *Journal of Archaeological Science* **27**, 363–372.

Lowe, J.J. (1982) Three Flandrian pollen profiles from the Teith valley, Perthshire, Scotland: II, analysis of deteriorated pollen. *New Phytologist* **90**, 371–385.

Lowe, J.J. (1992) Lateglacial and early Holocene lake sediments from the northern Apennines, Italy – pollen stratigraphy and radiocarbon dating. *Boreas* **21**, 193–208.

Lowe, J.J., Branch, N. and Watson, C. (1994) The chronology of human disturbance of the vegetation of the northern Apennines during the Holocene. In (P. Biagi and J. Nandris, eds.) *Highland zone exploitation in southern Europe*, 171–189. Brescia: Museo Civico Di Scienze Naturali Di Brescia (Monografie di Natura Bresciana 20).

Lowe, J.J. and Walker, M.J.C. (1984) *Reconstructing Quaternary Environments*. Harlow: Longman.

Lowe, J.J. and Watson, C. (1993) Lateglacial and Early Holocene pollen stratigraphy of the Northern Apennines, Italy. *Quaternary Science Reviews* **12**, 727–738.

Loy, T. (1994) Methods in the analysis of starch residues on prehistoric stone tools, In (J. Hather, ed.) *Tropical Archaeobotany*, 86–114. London: Routledge.

Ložek, V. (1986) Mollusca analysis. In (B.E. Berglund, ed.) *Handbook of Holocene Palaeoecology and Palaeohydrology*, 729–735. Chichester: Wiley.

Lucht, W.H. (1987) *Die Kafer Mitteleuropas, Katalog*. Krefeld: Goecke and Evers.

Lupo, K.D. (1994) Butchering marks and carcass acquisition strategies: distinguishing hunting from scavenging in archaeological contexts. *Journal of Archaeological Science* **21**, 827–837.

Lynch, T.F. (1973) Harvest timing, transhumance, and the process of domestication. *American Anthropologist* **75**, 1254–1259.

MacNeish, R.S. ***et al.*** (1980–83) *Prehistory of the Ayacucho Basin, Peru*. R.S. Peabody Foundation for Archaeology, University of Michigan Press.

Macphail, G. (1988) *Pollen stratigraphy of Holocene peat sites in eastern Liguria, Northern Italy*. Thesis presented for the degree of Doctor of Philosophy (CNAA) City of London Polytechnic.

Madrigal, T.C. and Capaldo, S.D. (1999) White-tailed deer marrow yields and Late Archaic hunter-gatherers. *Journal of Archaeological Science* **26**, 241–249.

Magny, M. (1992) Holocene lake-level fluctuations in Jura and the northern subalpine ranges, France: regional pattern and climatic implications. *Boreas* **21**, 319–334.

Magny, M. and Ruffaldi, P. (1995) Younger Dryas and early Holocene lake-level fluctuations in the Jura mountains, France. *Boreas* **24**, 155–172.

Magnin, F., Tatoni, T., Roche, P. and Baudry, J. (1995) Gastropod communities, vegetation dynamics and landscape changes an old-field succession in Provence, France. *Landscape and Urban Planning* **31**, 249–257.

Maier, U. (1996) Morphological studies of free-threshing wheat ears from a Neolithic site in southwest Germany, and the history of the naked wheats. *Vegetation History and Archaeobotany* **5**, 39–55.

Mangafa, M. and Kotsakis, K. (1996) A new method for the identification of wild and cultivated charred grape seeds. *Journal of Archaeological Science* **23**, 409–418.

Mannino, M.A. and Thomas, K.D. (2001) Intensive Mesolithic exploitation of coastal resources? Evidence from a shell deposit on the Isle of Portland (southern England) for the impact of human foraging on populations of intertidal rocky shore molluscs. *Journal of Archaeological Science* **28**, 1101–1114.

Mannino, M.A., Spiro, B.F. and Thomas, K.D. (2003) Sampling shells for seasonality: oxygen isotope analysis on shell carbonates of the inter-tidal gastropod Monodonta lineate (da Costa) from populations across its modern range and from a Mesolithic site in southern Britain. *Journal of Archaeological Science* **30**, 667–679.

Markgraf, V. (1989) Palaeoclimates in Central and South America since 18,000 BP based on pollen and lake-level records. *Quaternary Science Reviews* **8**, 1–24.

Marziani, G.P., Iannone, A., Patrignani, G. and Schiattareggia, A. (1991) Reconstruction of the tree vegetation near a Bronze Age site in Northern Italy based on the analysis of charcoal fragments. *Review of Palaeobotany and Palynology* **70**, 241–246.

Mays, S.A. (1997) Carbon stable isotope ratios in mediaeval and later human skeletons from Northern England. *Journal of Archaeological Science* **24**, 561–567.

McCobb, L.M.E., Briggs, D.E.G., Evershed, R.P., Hall, A.R. and Hall, R.A. (2001) Preservation of fossil seeds from a 10th century AD cess pit at Coppergate, York. *Journal of Archaeological Science* **28**, 929–940.

McCobb, L.M.E., Briggs, D.E.G., Carruthers, W.J. and Evershed, R.P. (2003) Phosphatisation of seeds and roots in a Late Bronze Age deposit at Potterne, Wiltshire, UK. *Journal of Archaeological Science* **30**, 1269–1281.

McCorriston, J. and Weisberg, S. (2002) Spatial and temporal variation in Mesopotamian agricultural practices in the Khabur Basin, Syrian Jazira. *Journal of Archaeological Science* **29**, 485–498.

Mellars, P.A. and Wilkinson, M.R. (1980) Fish otoliths as indicators of seasonality in prehistoric shell middens: the evidence from Oronsay (Inner Hebrides). *Proceedings of the Prehistoric Society* **46**, 19–44.

Mitchell, G.F. (1956) Post-Boreal pollen diagrams from Irish raised bogs. *Proceedings of the Royal Irish Academy* **B57**, 185–251.

Moe, D. and Rackham, O. (1992) Pollarding and a possible explanation of the neolithic elmfall. *Vegetation History and Archaeobotany* **1**, 63–68.

Montanari, C., Guido, M.-A., Boccaccio, A. and Rametta, M. (1994) Paleoecologia olocenica del Monte Gottero (Parma) nel quadro della storia del popolamento vegetale dell'Appennino Ligure. *Il Quaternario* **7**, 373–380.

Moore, J. (2000) Forest fire and human interaction in the early Holocene woodlands of Britain. *Palaeoecology, Palaeoclimatology, Palaeoecology* **164**, 125–137.

Moore, P.D. (1980) Resolution limits of pollen analysis as applied to archaeology. *MASCA Journal* **1**, 118–120.

Moore, P.D. (1983) Palynological evidence of human involvement in certain palaeohydrological events. *Quaternary Studies in Poland* **4**, 97–105.

Moore, P.D. (1985a) The death of the elm. *Nature* **107**, 32–34.

Moore, P.D. (1985b) Forests, man and water. *Journal of Environmental Studies* **25**, 159–166.

Moore, P.D. (1986a) Unravelling human effects. *Nature* **321**, 204.

Moore, P.D. (1986b) *Hydrological changes in mires*. In (B.E. Berglund, ed.) *Handbook of Holocene Palaeoecology and Palaeohydrology*, 91–107. Chichester: Wiley.

Moore, P.D. and Webb, J.A. (1978) *Pollen Analysis*. London: Hodder and Stoughton.

Moreno-Garcia, M., Orton, C. and Rackham, J. (1996) A new statistical tool for comparing animal bone assemblages. *Journal of Archaeological Science* **23**, 437–453.

Morrison, K.D. (1994) Monitoring regional fire history through size-specific analysis of microscopic charcoal: the last 6000 years in South India. *Journal of Archaeological Science* **21**, 675–685.

Munson, P.J. and Garniewicz, R.C. (2003) Age-mediated survivorship of ungulate mandibles and teeth in canid-ravaged faunal assemblages. *Journal of Archaeological Science* **30**, 405–416.

Navarro Camacho, C., Carrión, J.S., Navarro, J., Munuera, M. and Prieto, A.R. (2000) An experimental approach to the palynology of cave deposits. *Journal of Quaternary Science* **15**, 603–619.

Nesbitt, M. and Samuel, D. (1996) From staple crop to extinction? The archaeology and history of the hulled wheats. In (S. Padulosi, K. Hammer and J. Heller, eds.) *Hulled Wheats*, 41–100. Rome: International Plant Genetic Resources Institute.

Newman, M.E., Parboosingh, J.S., Bridge, P.J. and Ceri, H. (2002) Identification of archaeological animal bone by PCR/DNA analysis. *Journal of Archaeological Science* **29**, 77–84.

Nicholson, R.A. (1996) Bone degradation, burial medium and species representation: debunking the myths, an experimental approach. *Journal of Archaeological Science* **23**, 513–533.

Nicholson, R.A. (1998) Bone degradation in a compost heap. *Journal of Archaeological Science* **25**, 393–403.

Nisbet, R. (1997) Arene Candide: charcoal remains and prehistoric woodland use, 103–112. In (R. Maggi, ed.) *Arene Candide: a functional and environmental assessment of the Holocene sequence*, Rome: Il Calamo.

Oeggl, K. (1994) The palynological record of human impact on highland zone ecosystem. In (P. Biagi and J. Nandris, eds.) *Highland zone exploitation in southern Europe*, 107–122. Brescia: Museo Civico Di Scienze Naturali Di Brescia (Monografie di Natura Bresciana 20).

Overgaard Nielsen, B., Mahler, V. and Rasmussen, P. (2000) An arthropod assemblage and the ecological conditions in a byre at the Neolithic settlement of Weier, Switzerland. *Journal of Archaeological Science* **27**, 209–218.

Panagiotakopulu, E. (1999) An examination of biological materials from coprolites from XVIII dynasty Amarna, Egypt. *Journal of Archaeological Science* **26**, 547–551.

Panagiotakopulu, E. (2001) New records for ancient pests: archaeoentomology in Egypt. *Journal of Archaeological Science* **28**, 1235–1246.

Payne, S. and Munson, P.J. (1985) Ruby and how many squirrels? The destruction of bones by dogs. In (N.R.J. Fieller, D.D. Gilbertson and N.G.A. Ralph, eds.) *Palaeobiological Investigations: Research Design, Methods and Data Analysis*. Oxford: British Archaeological Reports International Series, 266, 31–39.

Pearsall, D.M., Chandler-Ezell, K. and Chandler-Ezell, A. (2003a) Identifying maize in neotropical sediments and soils using cob phytoliths. *Journal of Archaeological Science* **30**, 611–627.

Pearsall, D.M., Chandler-Ezell, K. and Zeidler, J.A. (2003b) Maize in ancient Ecuador: results of residue analysis of stone tools from the Real Alto site. *Journal of Archaeological Science* **31**, 423–442.

Pearsall, D.M., Chandler-Ezell, K. and Chandler-Ezell, A. (2004) Maize can *still* be identified using phytoliths: response to Rovner. *Journal of Archaeological Science* **31**, 1029–1038.

Piperno, M. and Giacobini, G. (1992) A taphonomic study of the paleosurface of Guattari Cave (Monte Circeo, Latina, Italy). *Quaternaria Nova* **1**, 143–161.

Piperno, D.R. and Holst, I. (1998) The presence of starch grains on prehistoric stone tools from the humid Neotropics: indications of early tuber use and agriculture in Panama. *Journal of Archaeological Science* **25**, 765–776.

Piperno, D.R., Ranere, A.J., Holst, I. and Hansell, P. (2000) Starch grains reveal early root crop horticulture in the Panamanian tropical forest. *Nature* **407**, 894–897.

Plug, I. (1997) Late Pleistocene and Holocene hunter-gatherers in the Eastern Highlands of South Africa and Lesotho: a faunal interpretation. *Journal of Archaeological Science* **24**, 715–727.

Polet, C. and Katzenberg, M.A. (2003) Reconstruction of the diet in a medieval monastic community from the coast of Belgium. *Journal of Archaeological Science* **30**, 525–533.

Ponel, P. and Lowe, J.J. (1992) Coleoptera, pollen and radiocarbon evidence from the Prato Spilla 'D' succession, N. Italy. *Comptes Rendus de L'Academie de Sciences Paris* **315**, 1425–1431.

Ponel, P., Matterne, V., Coulthard, N. and Yvinec, J-H. (2000) La Tene and Gallo-Roman natural environments and human impact at the Touffréville rural settlement, reconstructed from coleopteran and plant macroremains (Calvados, France). *Journal of Archaeological Science* **27**, 1055–1072.

Pons, A. (1992) Modern vegetation in Provence and in the southern Alps, Lateglacial-Holocene history. *Cahiers de Micropaleontologie* **7**, 259–281.

Pons, A., Guiot, J., De Beaulieu, J-L. and Reille, M. (1992) Recent contributions to the climatology of the last Glacial/Interglacial cycle based on French pollen sequences. *Quaternary Science Reviews* **11**, 439–448.

Preece, R.C. (1981) The value of shell microsculpture as a guide to the identification of land Mollusca from Quaternary deposits. *Journal of the Conchological Society of London* **30**, 331–337.

Preece, R.C. and Robinson, J.E. (1984) Late Devensian and Flandrian environmental history of the Ancholme Valley, Lincolnshire: Molluscan and Ostracod evidence. *Journal of Biogeography* **11**, 319–352.

Preece, R.C., Kemp, R.A. & Hutchinson, J.N. (1995). A late-glacial colluvial sequence at Watcombe Bottom, Ventnor, Isle of Wight, England. *Journal of Quaternary Science* **10**, 107–121.

Preece, R.C., Bridgland, D.R. & Sharp, M.J. (1998). Stratigraphical investigations. In (R.C. Preece and D.R. Bridgland, eds.) *Late Quaternary Environmental Change in North-West Europe: Excavations at Holywell Coombe, South-east England*, 33–68. London: Chapman & Hall.

Rackham, D.J. (1986) Assessing the relative frequencies of species by the application of a stochastic model to a zoo-archaeological database. In (L.H. van Wijngaarden-Bakker, ed.) *Database Management and Zooarchaeology*. Journal of the European Study Group of Physical, Chemical, Biological and Mathematical techniques applied to Archaeology, Research Volume 40.

Rasmussen, P. (1993) Analysis of goat/sheep faeces from Egolzwil 3, Switzerland: evidence for branch and twig foddering of livestock in the Neolithic. *Journal of Archaeological Science* **20**, 479–502.

Rasmussen, P. (1989a) Leaf-foddering of livestock in the Neolithic: archaeobotanical evidence from Weier, Switzerland. *Journal of Danish Archaeology* **8**, 51–71.

Rasmussen, P. (1989b) Leaf foddering in the earliest Neolithic agriculture. *Acta Archaeologica* **60**, 71–86.

Reille, M., Pons, A. and De Beaulieu, J-L. (1992) Late and postglacial vegetation, climate and human action in the French Massif Central. *Cahiers de Micropaleontologie* **7**, 93–106.

Reille, M. and Lowe, J.J. (1993) A re-evaluation of the vegetation history of the eastern Pyrenees (France) from the end of the last glacial to the present. *Quaternary Science Reviews* **12**, 47–77.

Rhode, D. (2003) Coprolites from Hidden Cave, re-visited: evidence for site occupation history, diet and sex of occupants. *Journal of Archaeological Science* **30**, 909–922.

Richard, H. (1993) Palynological micro-analysis in Neolithic lake dwellings. *Journal of Archaeological Science* **20**, 241–262.

Richards, M.P. and Mellars, P.A. (1998) Stable isotopes and the seasonality of the Oronsay middens. *Antiquity* **72**, 178–184.

Rick, T.C. and Erlandson, J.M. (2000) Early Holocene fishing strategies on the California coast: evidence from CA-SBA-2057. *Journal of Archaeological Science* **27**, 621–633.

Robertson, J.D. (1936). The function of the calciferous glands of earthworms. *Journal of Experimental Biology* **13**, 279–297.

Robinson, M.A. (1981) The use of ecological groupings of Coleoptera for comparing sites. In (M. Jones and G. Dimbleby, eds.) *The Environment of Man: The Iron Age to the Anglo-Saxon Period*, 279–286. British Archaeological Reports British Series, 87.

Robinson, M.A. (1983) Arable/pastoral ratios from insects? In (M. Jones, ed.) *Interpreting the Subsistence Economy*, 19–53. British Archaeological Reports International Series, 181.

Robinson, D. and Rasmussen, P. (1989) Botanical investigations at the Neolithic lake village at Weier, north east Switzerland: leaf hay cereals as animal fodder. In (A. Milles, D. Williams and N. Gardner, eds.) *The Beginnings of Agriculture*, 149–163. Oxford: British Archaeological Publications (British Series 496).

Robinson, S., Nicholson, R.A., Pollard, M. and O'Connor, T.P. (2003) An evaluation of nitrogen porosimetry as a technique for predicting taphonomic durability in animal bone. *Journal of Archaeological Science* **30**, 391–403.

Rogers, A.R. (2000) Analysis of bone counts by maximum likelihood. *Journal of Archaeological Science* **27**, 111–125.

Rogers, A.R. and Broughton, J.M. (2001) Selective

transport of animal parts by ancient hunters: a new statistical method and an application to the Emeryville Shellmound Fauna. *Journal of Archaeological Science* **28**, 763–773.

Rossignol-Strick, M., Planchais, N., Paterne, M. and Duzer, D. (1992) Vegetation dynamics and climate during the deglaciation in the south Adriatic basin from a marine record. *Quaternary Science Reviews* **11**, 415–423.

Rovner, I. (2004) On transparent blindfolds: Comments on the identifying maize in Neotropical sediments and soils using cob phytoliths. *Journal of Archaeological Science* **31**, 815–819.

Scaife, R.G. (1986) Pollen in human palaeofaeces; and a preliminary investigation of the stomach and gut contents of Lindow Man. In (I.M. Stead, J.B. Bourke and D. Brothwell, eds.) *Lindow Man: The Body in the Bog*, 126–135. London: British Museum Publications.

Scaife, R.G. (1987) Pollen analysis and the later prehistoric vegetational changes of the Val Chisone. In (R. Nisbet and P. Biagi, eds.) Balm' Chanto: un riparo sottoroccia dell'Eta del Rame nelle Alpi Cozie. *Archeologia dell'Italia Settentrionale* **4**, 89–101. Como: New Press.

Scaife, R.G. (1988) The elm decline in the pollen record of south east England and its relationship to early agriculture. In (M. Jones, ed.) *Archaeology and the Flora of the British Isles*, Oxford University Committee for Archaeology, **14**, 21–33. Oxford: Oxford University Committee for Archaeology.

Schneider, R. (1985) Palynologie research in the southern and southeastern Alps between Torino and Trieste – a review of investigations concerning the last 15,000 years. *Dissertationes Botanicae* **87**, 83–103.

Schelvis, J. (1997) Mites in the background, use and origin of remains of mites in Quaternary deposits. In (A.C. Ashworth, P.C. Buckland and J.P. Sadler, eds.) *Studies in Quaternary Entomology – An Inordinate Fondness of Insects*, 233–236. Quaternary Proceedings 5. Chichester: Wiley.

Schlumbaum, A., Jacomet, S. and Neuhaus, J-M. (1998) Coexistence of tetraploid and hexaploid naked wheat in a Neolithic lake dwelling of central Europe: evidence from morphology and ancient DNA. *Journal of Archaeological Science* **25**, 1111–1118.

Schmid, E. (1969) Cave sediments and prehistory. In (D. Brothwell and E. Higgs, eds.) *Science in Archaeology*, 151–166. London: Thames and Hudson.

Schweingruber, F.H. (1978) *Microscopic wood anatomy*. Teufen: Fluck-Wirth.

Sept, J.M. (1994) Bone distribution in a semi-arid riverine habitat in eastern Zaire: implications for the interpretation of faunal assemblages at early archaeological sites. *Journal of Archaeological Science* **21**, 217–235.

Silfverberg, H. (1979) *Enumeratio Coleopterorum Fennoscandia et Daniae*. Helsingfors: Helsingfors Entomologiska Bytesförening.

Simmons, I.G. (1969a) Pollen diagrams from the North York Moors. *New Phytologist* **69**, 807–827.

Simmons, I.G. (1969b) Evidence for vegetation changes associated with Mesolithic man in Britain. In (P.J. Ucko and G.W. Dimbleby, eds.) *The Domestication and Exploitation of Plants and Animals*, 111–119. London: Duckworth.

Simmons, I.G. (1975a) Towards and ecology of Mesolithic man in the uplands of Great Britain. *Journal of Archaeological Science* **2**, 1–15.

Simmons, I.G. (1975b) The ecological setting of Mesolithic man in the highland zone. In (J.G. Evans, S. Limbrey and H. Cleere, eds.) *The Effect of Man on the Landscape: the Highland Zone*, 57–63. Oxford: Council for British Archaeology Research Report, 11.

Simmons, I.G. (1993) Vegetation change during the Mesolithic in the British Isles: some amplifications. In (F.M. Chambers, ed.) *Climate Change and Human Impact on the Landscape: Essays in Honour of A.G. Smith*, 109–118. London: Chapman and Hall.

Simmons, I.G. and Innes, J.B. (1987) Mid-Holocene adaptations and later Mesolithic forest disturbance in northern England. *Journal of Archaeological Science* **14**, 385–403.

Simmons, I.G. and Innes, J.B. (1996a) Prehistoric charcoal in peat profiles at North Gill, North Yorkshire Moors, England. *Journal of Archaeological Science* **23**, 193–197.

Simmons, I.G. and Innes, J.B. (1996b) The ecology of an episode of prehistoric cereal cultivation on the North York Moors, England. *Journal of Archaeological Science* **23**, 613–618.

Smart, T.L. and Hoffman, E.S. (1988) Environmental interpretation of archaeological charcoal. In (H. Hastorf and V. Popper, eds.) *Current Palaeoethnobotany*, 167–25. Chicago: University of Chicago Press.

Smith, A.G. and Pilcher, A.R. (1973) Radiocarbon dates and the vegetation history of the British Isles. *New Phytologist* **72**, 903–914.

Smith, D.N. and Howard, A.J. (2004) Identifying changing fluvial conditions in low gradient alluvial archaeological landscapes: can coleopteran provide insights into changing discharge rates and floodplain evolution? *Journal of Archaeological Science* **31**, 109–120.

Smith, H. and Jones, G. (1990) Experiments on the effects of charring on cultivated grape seeds. *Journal of Archaeological Science* **17**, 317–327.

Sparks, B.W. and West, R.G. (1972) *The Ice Age in Britain*. London: Methuen.

Stead, I.M., Bourke, J.B. and Brothwell, D. (1986) *Lindow Man: The Body in the Bog*. London: British Museum Publications.

Stephan, E. (2000) Oxygen isotope analysis of animal bone phosphate: method refinement, influence of consolidants, and reconstruction of palaeotemperatures for Holocene sites. *Journal of Archaeological Science* **27**, 523–535.

Stockmarr, J. (1971) Tablets with spores used in absolute pollen analysis. *Pollen et Spores* **13**, 615–621.

Stott, A.W., Evershed, R.P., Jim, S., Jones, V., Rogers, J.M., Tuross, N. and Ambrose, S. (1999) Cholesterol as a new source of palaeodietary information: experimental approaches and archaeological applications. *Journal of Archaeological Science* **26**, 705–716.

Suc, J.-P. (1984) Origin and evolution of the Mediterranean vegetation and climate in Europe. *Nature* **307**, 429–432.

Szeroczyńska, K. (1998) Palaeolimnological investigations in Poland based on Cladocera (Crustacea). *Palaeogeography, Palaeoclimatology, Palaeoecology* **140**, 335–345.

Szeroczyńska, K. (2002) Human impact on lakes recorded in the remains of Cladocera (Crustacea). *Quaternary International* **95–96**, 165–174.

Tauber, H. (1965) Differential pollen dispersion and the interpretation of pollen diagrams. *Danmarks Geologiske Undersøgelse* **89**, 1–69.

Tauber, H. (1967) Investigation of the mode of pollen transfer in forested areas. *Review of Palaeobotany and Palynology* **3**, 277–287.

Tipping, R. (1994) 'Ritual' floral tributes in the Scottish Bronze Age – palynological evidence. *Journal of Archaeological Science* **21**, 133–139.

Tipping, R., Carter, S. and Johnston, D. (1994) Soil pollen and soil micromorphological analyses of old ground surfaces on Biggar Common, Borders Region, Scotland. *Journal of Archaeological Science* **21**, 387–401.

Tolonen, K. (1986) Charred particle analysis. In (B.E. Berglund, ed.) *Handbook of Holocene Palaeoecology and Palaeohydrology*, 485–496. Sussex: Wiley.

Triat-Laval, H. (1978) *Contribution pollen analytique a l'histoire tardi- et postglaciaire de la vegetation de la basse vallee du Rhone*. These presentee a l'Universite d'Aix-Marseille III.

Turner, C. (1985) Problems and pitfalls in the application of palynology to Pleistocene archaeological sites in western Europe. In (J. Renault-Miskovsky, A. Bui-Thi-Mai and M. Girard, eds.) *Palynologie Archeologique*, 347–373. C.N.R.S. Centre de Recherches Archeologiques Notes et Monographies Techniques, 17.

Turner, J. (1986) Principal components analysis of pollen data with special reference to anthropogenic indicators. In (K-E. Behre, ed.) *Anthropogenic Indicators in Pollen Diagrams*, 221–232. Rotterdam: Balkema.

van Andel, T.H. and Shackleton, J.C. (1982) Late Palaeolithic and Mesolithic coastlines of Greece and the Aegean. *Journal of Field Archaeology* **9**, 445–454.

van Husen, D. (1997) LGM and Late-glacial fluctuations in the eastern Alps. *Quaternary International* **38/39**, 109–118.

Vaughan-Williams, A. (2004) *Plant remains from Southwark Cathedral, London, UK*. ArchaeoScape, Royal Holloway Department of Geography, Internal Report.

Waller, M.P. (1998) An investigation into the palynological properties of fen peat through multiple pollen profiles from south-eastern England. *Journal of Archaeological Science* **25**, 631–642.

Watson, C. (1994) *The Late Weichselian and Holocene vegetational history of the northern Apennines, Italy*. Thesis presented for the Degree of Doctor of Philosophy, Department of Geography, Royal Holloway, University of London, February 1994.

Watson, C. (1996) The vegetational history of the northern Apennines, Italy: information from three new sequences and a review of regional vegetational changes. *Journal of Biogeography* **23**, 805–841.

Watts, W.A. (1985) A long pollen record from Laghi di Monticchio, southern Italy: a preliminary account. *Journal of the Geological Society of London* **142**, 491–499.

Watts, W.A., Allen, J.R.M., Huntley, B. and Fritz, S.C. (1996a) Vegetation history and climate of the last 15,000 years at Laghi di Monticchio, southern Italy. *Quaternary Science Reviews* **15**, 113–132.

Watts, W.A., Allen, J.R.M. and Huntley, B. (1996b) Vegetation History and Palaeoclimate Of The Last Glacial Period At Lago Grande Di Monticchio, Southern Italy. *Quaternary Science Reviews* **15**, 133–154.

Weiss, E. and Kislev, M.E. (2004) Plant remains as indicators of economic activity: a case study from Iron Age Ashkelon. *Journal of Archaeological Science* **31**, 1–13.

Welter-Schultes, F.W. (2001) Land snails from an ancient shipwreck: the need to detect wreck-independent finds in excavation analysis. *Journal of Archaeological Science* **28**, 19–27.

Whittington, G. (1993) Palynological investigations at two Bronze Age burial sites in Fife. *Proceedings of the Society of Antiquaries of Scotland* **123**, 211–213.

Wick, L. (1994) Vegetation development and human impact at the forest limit: palaeoecological studies in the Splugen Pass area (northern Italy). In (P. Biagi and J. Nandris, eds.) *Highland zone exploitation in southern Europe*, 123–132. Brescia: Museo Civico Di Scienze Naturali Di Brescia (Monografie di Natura Bresciana 20).

Williams, D. (1973) Flotation at Siraf. *Antiquity* **47**, 288–292.

Wilson, D.G. (1984) The carbonization of weed seeds and their representation in macrofossil assemblages. In (W. van Zeist and W.A. Casparie, eds.) *Plants and Ancient Man: Studies in Palaeo-ethnobotany*, 39–43. Rotterdam: Balkema.

Wright, P. (2003) Preservation or destruction of plant remains by carbonization? *Journal of Archaeological Science* **30**, 577–583.

Yu, G. and Harrison, S.P. (1995) Holocene changes in atmospheric circulation patterns as shown by lake status changes in northern Europe. *Boreas* **24**, 260–268.

Zeder, M.A. (2001) A metrical analysis of a collection of modern goats (*Capra hircus aegargus* and *C. h. hircus*) from Iran and Iraq: implications for the study of caprine domestication. *Journal of Archaeological Science* **28**, 61–79.

Zheng, Y., Dong, Y., Matsui, A., Udatsu, T. and Fujiwara, H. (2003) Molecular genetic basis of determining subspecies of ancient rice using the shape of phytoliths. *Journal of Archaeological Science* **30**, 1215–1221.

Zohary, D. and Hopf, M. (1994) *Domestication of Plants in the Old World*. Oxford: Clarendon Press.

Zohary, D., Tchernov, E. and Kolska Horwitz, L.R. (1998) The role of unconscious selection in the domestication of sheep and goats. *Journal of Zoology* **245**, 129–135.

4

Dating and numerical analysis: the age and significance of environmental evidence

4.0 Chapter summary

This chapter introduces some of the numerical methods environmental archaeologists use in data analysis, including the basic (e.g. normal distribution and errors) and more advanced (e.g. ordination) approaches that are crucial to fully exploit measured data. Relevant dating methods are introduced, with discussions of relative, age-equivalent, incremental and radiometric techniques. Due to limitations of space these must be considered just an introduction to the topics and readers requiring further information are directed to more specialized texts given at the beginning of the relevant section. Stress is constantly placed on the value of multi-method approaches to dating, and examples are given where these have been used to maximum effect.

4.1 Numerical principles

4.1.1 General background

Many of the principles outlined below can be found in most introductory statistical texts. Good examples are Haynes (1982), Sanders (1995) and Mann (1998).

4.1.1.1 Quantification

Environmental archaeologists commonly measure a large number of variables (e.g. soil depth, rainfall, temperature), producing data known statistically as observations. These observations can take many different forms, the details of which have been discussed in the previous chapters, including presence or absence, percentage values and ranked scales (e.g. cold, warm and hot). One of the key challenges of environmental archaeology is to know how to analyse and describe these observations most efficiently. A central precursor to this analysis is the extent to which such data are normally distributed.

4.1.1.2 Normal distribution

Statistically we use the concept of a normal (or Gaussian) distribution. If sample size, money and time were no object (a situation unfortunately rare), we could undertake an infinite number of analyses on a single sample. Each result would be slightly different due to instrumentation differences, sampling errors etc., but the values would cluster around a central point with fewer and fewer occurring away from that central mass of values.

Many datasets from environmental archaeology produce a frequency plot of values that approximates to this normal distribution (Fig. 4.1). Instead of taking an infinite number of

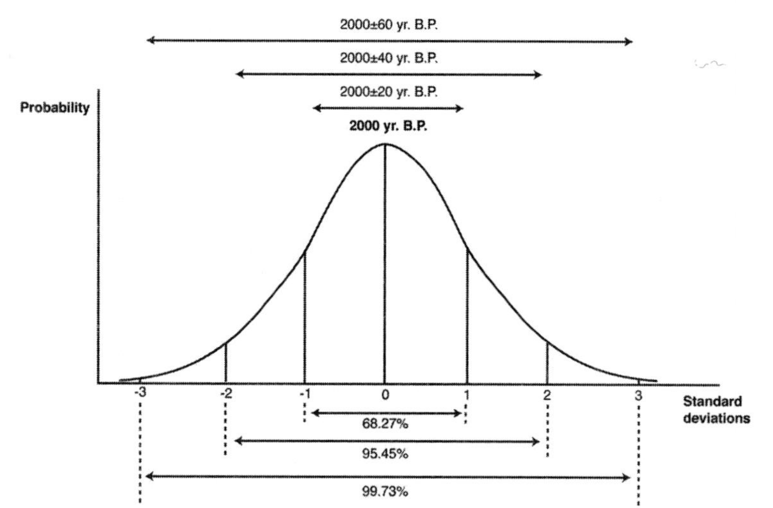

FIGURE 4.1 A generalized normal (Gaussian) distribution with standard deviations. A radiometric age of 2000 ± 20 years is shown as an example.

readings, we can model such a curve and save the expense, time and material. When using a frequency plot, we often refer to the *average* or *mean* value. This is the arithmetic mean – a measure of the central tendency of a dataset produced by dividing the sum of all the data by the number of observations. The problem with the arithmetic mean as a measure of central tendency is that outlier values can shift it to an unrepresentative value. The median can be used instead of the mean in some cases; this is the middle value of all the observations once they are arranged in order. Alternatively, the *mode* is the value that occurs most frequently. In real life these measures may all be different, but if data is normally distributed, all three will be identical.

4.1.1.3 Standard deviation

Using averages only tells part of the story. A single value, for example a mean radiometric date, could be misleading. Assuming that the dating method produced a normal distribution of measured values, then a significant number of them will differ from the mean value by varying amounts. This variation is measured using standard deviation (or σ), which effectively quantifies the differences from the mean (the area under the curve) and thus gives a measure of the degree of uncertainty. Depending on what dating technique is used, a different standard deviation is commonly given. For instance, radiocarbon is routinely quoted as being to 1σ or 68.27 per cent confidence limits. Thus, if we had a modern age report of 2000 ± 20 AD, we would say that we were 68.27 per cent confident that the true age fell between the years 1980 and 2020 (Fig. 4.1). If we wanted to quote to 2σ (or 3σ) we would have error margins of, respectively, ±40 (and ±60), allowing us 95.45 per cent confidence that the age fell between 1960 and 2040 (and 99.73 per cent confidence that the age lies somewhere between 1940 and 2060).

4.1.1.4 Tests of significance

Two or more datasets can be statistically analysed to see whether they are truly different, using a t-test. To calculate t, we need the *degrees of freedom* (the number of values that are capable of varying) denoted for each group of values as n_1, n_2 etc.

If we have two means ($\bar{x}_1$, and $\bar{x}_2$) that were measured with s_1 and s_2 standard deviations, and n_1 and n_2 degrees of freedom respectively, then the t value can be calculated as follows:

$$t = (\bar{x}_1 - \bar{x}_2) / \sqrt{\left(\frac{\sigma_1^2}{n_1 - 1} + \frac{\sigma_2^2}{n_2 - 1}\right)}$$

The value thus determined can then be compared against critical t-values published in most books of standard statistical tables.

If the calculated value is smaller than the critical t-value, then there is no significant difference between the datasets. Generally speaking, for most tests we are concerned with assessing to a level of significance of 0.05, i.e. a 5 per cent chance that the ages may be different. One problem with the t-test is that it assumes that the values are normally distributed. This is not always the case when small sample sizes have been used. The larger the number of samples, the more robust the statistical analysis.

Alternatively, we can use the χ^2-test (chi-square). Here, we identify whether the differences measured in a dataset are due to chance and are therefore normally distributed and statistically indistinguishable. The χ^2 value can be calculated as follows:

$$\chi^2 = \sum_{i=1}^{i=n} \frac{(\bar{x}_i - \bar{x})}{\sigma_i^2}$$

Again, the value determined can be compared against a critical χ^2 value at the 0.05 significance level. If the calculated value is smaller than the critical, the dataset comprises a randomly distributed range of points, i.e. they follow a normal distribution.

4.1.1.5 Precision and accuracy

Precision and accuracy often causes confusion but understanding the difference is more than academic. A precise value is one where the error term associated with a value is relatively small. An accurate value is where the number reported is correct, though this may not necessarily be precise. An excellent analogy can be seen in Fig. 4.2 where the value being reported can be compared to a set of fairground targets.

The first target (A) shows a cluster of shots that are off-centre. The fact that the shots (values) are tightly clustered indicates they are precise but as they are significantly off the bullseye, they are inaccurate. In the second target (B), the shots are approximately around the centre but are not nearly so tightly clustered as in Target A. Here, the values are accurate (because they are near the bullseye) but would be said to be imprecise. Finally, the last target (C) has a tight cluster of shots around the centre – the best of both worlds! These values are both accurate and precise.

4.2 Advanced numerical analyses

4.2.1 Ordination methods

Analyses of data are significantly easier if we have only a few variables. Typically, however, as environmental archaeologists, we will have numerous datasets each measuring different parameters (multivariate analyses). Trying to summarize this data qualitatively is problematic and inherently subjective. When there are three or more variables, we can consider the data as occupying three-dimensional space, much as two variables would be represented on a two-dimensional plot. Ideally, we require a method that allows us to represent the data in a low (typically two or three) dimension space that best characterizes the original variables. Ordination (also referred to as mapping or factor) methods are such an approach, reducing the number of variables but maintaining as much as possible the structure of the data in the original space. Here we shall restrict ourselves

A.

B.

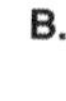

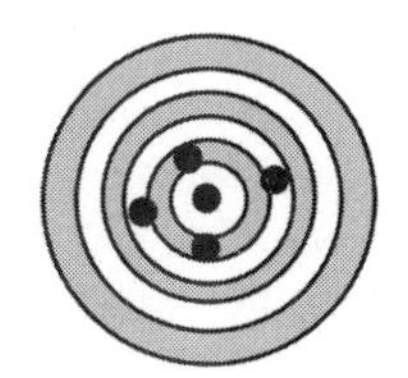

C.

FIGURE 4.2 Targets illustrating differences between precision and accuracy: (A) Precise but inaccurate, (B) Accurate but imprecise, (C) Accurate and precise.

to analysis of data that have no known environmental controls but for which we are trying to provide a summary and detect underlying gradients of variation. For this type of problem we use a family of techniques known as *indirect gradient analysis*. They comprise Principal Components Analysis (PCA), Correspondence Analysis (CA) and Detrended Correspondence Analysis (DCA). Excellent overviews can be found in Everitt (1978) and ter Braak (1987). Examples of software available for these methods include CANOCO™, DECORANA™ NT-SYS™, Statistica™ and R™ (http://www.r-project.org/).

PCA utilizes a linear response model to produce a line of best fit, by least-squares regression to the multivariate dataset. Also known as Empirical Orthogonal Function (EOF) analysis, a new variable (factor) is generated by maximizing (extracting) the variance described by the regression while minimizing the variance about it. This regression explains the greatest amount of the variance of the data and can be used to explore a controlling mechanism. Once the first variable has been extracted, a second variable is identified by driving a new linear regression against the remaining data. In this instance, the remaining data is the sum of the squares between the values and the fitted line of the first new variable. Eigenvalues are generated by the analysis to describe the goodness of the fit and are expressed as a percentage of the total variance in the data. Once the amount of variation described by the new variables describes little extra (the so-called 'scree test'; Cattell, 1966) no further analysis is required. Typically, two dimensions/axes will encompass the majority of the variance. PCA must be used on data that have the same degree of variance. Relatively large variance in data will dominate analysis, and must therefore be transformed by one of several methods, e.g. standardization, square root or log.

CA (also known as reciprocal averaging) recognizes that data generated for particular variables, e.g. species abundance, often have a bell-shaped response curve to an environmental variable, with optimal conditions allowing maximum representation, but more extreme conditions restrict their abundance. Thus, whereas PCA deals with a linear response, CA allows us to explore a unimodal response. If we consider a species, we can identify the optimum environmental conditions for that species, i.e. where the maximum number of that species occurs (referred to as the species score). CA generates a new variable (the first axis) that maximizes the dispersion of the species scores with further axes being generated that are not correlated to the previous axis. The advantage of CA is that it can be applied to both presence-absence data and abundance data.

A complication with CA is that the extremes of the axes can be compressed relative to the middle of the axes and the second axis often shows a quadratic (arch) relationship with the first axis (Hill and Gauch, 1980). In DCA, the former is corrected by nonlinearly rescaling the axis, the practical upshot of which is that the average widths are equal. By detrending the second axis by either polynomials or segments (ter Braak and Smilauer, 1998) the mean value of the site scores on subsequent axes can be made approximately zero and therefore be statistically more robust.

4.2.2 Time series

In many datasets where we have a record taken through time, e.g. in a stratigraphic sequence, we often observe (or subsequently identify) a systematic or periodic pattern over time. Spectral analyses decompose a time series into different contributions of variability from frequencies associated with different time scales (Schulz *et al.*, 2002) allowing us not only to provide a summary description of the data, but to look at potential controls, extract a signal despite the noise in the dataset and in exceptional circumstances use it as a predictive tool. An excellent introduction to the topic is Kendall and Ord (1990). Various statistical packages are available for analysis and these include AnalySeries™, Arand™, PAVE™ and Statistica™.

Most time series analysis techniques involve some form of filtering to remove noise and to isolate the recurring signal. Typically, there are four components: (a) a long-term movement (commonly referred to as a *trend*); (b) a seasonal element repeating itself at systematic periods over time; (c) a cyclic component which has a longer duration that varies from cycle to cycle; and (d) a random/irregular component. Numerous techniques can be used for separating trend, seasonal and/or cyclic variability from the irregular components. These techniques include classical seasonal decomposition (Census 1) which isolates the different components and Census X-11 seasonal decomposition which is a modification of the classical method (Makridakis *et al.*, 1998). To detect and isolate the different components, additive or multiplicative models utilize some form of smoothing (usually a moving average). The moving average can use any number of adjacent values, and utilize the mean or median values, the latter being more robust where a dataset may have outliers. Conversely, with no outliers, median smoothing may produce sharper-edged curves and does not allow for weighting. More sophisticated methods allow the user to test how observations are influenced by previous observation points or specific events (e.g. Auto Regressive Integrated Moving Average or ARIMA; Box and Jenkins, 1976), though these require considerable statistical experience.

Where the seasonal fluctuations are not known, spectral analysis can be used to explore cyclical patterns of data. Such methods are largely based on Fourier transformations and allow the user to identify the wavelength of a cycle (sine or cosine) as the number of cycles per unit time (frequency) while also identifying the period as the length of time required for one full cycle. For datasets where the frequency content changes over time, Fourier analysis of overlapping subsets of the time series allows the user to explore the changing frequency content. Alternatively, through decomposing the time series into time-frequency space, wavelet analysis allows the user to identify the dominant modes of variability and how these vary over time (Torrence and Compo, 1998). Interactive wavelet analysis software is available at http://paos.colorado.edu/research/wavelets/. For unevenly spaced data, the program SPECTRUM is available at http://www.palmod.uni-bremen.de/~mschulz/#software (Schulz and Stattegger, 1997).

4.3 Dating methods

4.3.1 General background

In this section we will attempt to outline some of the key geochronological tools that are available to environmental archaeologists for the Quaternary period (the last 2 million years). The information is by no means exhaustive and, in addition to references cited in the relevant sections, it is suggested readers consult the detailed reviews given by Aitken (1990), Noller *et al.* (2000) and Grün *et al.* (forthcoming) for specific details. The purpose of this chapter is to give workers a feel for the issues to be considered when selecting a particular dating technique and the sampling option(s).

4.3.1.1 Terminology

Within this chapter we will be making a distinction between the use of the terms *age* and *date*, following Colman *et al.* (1987). A *date* is a specific point in time that few dating methods can achieve. The exceptions are historical records or where the dating method has no uncertainty, such as dendrochronology. The more appropriate term the environmental archaeologist will commonly use is *age*, which is used to describe a unit of time measured from the present. Here, an uncertainty is associated with the time quoted. We are therefore not dealing with an absolute timescale that a date suggests. It should be noted, however, that the verb *date* is firmly entrenched in the literature when used in the context of describing the process by which an age estimate is produced and will be used as such in this chapter.

For historical calendar dates, the time should be expressed as the year AD or BC. With regards to describing ages, thousands of years ago will be abbreviated to *ka*. As a measure of time from present is implied in *ka* there is no need to express them with the addition of Before Present (BP). The exception to the latter is radiocarbon, which is quoted as *ka* (or *yr*) *BP* to indicate the measure of age has been taken from 1950 AD. As a radiocarbon age does not equate precisely with a sidereal/calendar age, a calibrated radiocarbon age should have the additional designation *cal* i.e. *cal ka* (or *yr*) *BP*. Wherever possible, the dating technique used to derive the age estimate should also be given to provide an additional degree of confidence, e.g. *luminescence* ka.

4.3.1.2 Selection of dating methods

Prior to sending samples for measurement, the most appropriate method for dating a deposit must be considered. For experienced environmental archaeologists this is often intuitive. The following factors should be considered:

Applicability

We will be covering only a relatively small number of the significant range of dating methods available to the environmental archaeologist. No single method is appropriate for all deposit types and it is therefore essential to consider the type of material and applicable dating methods (Fig. 4.3). For instance, radiocarbon dating is superb for dating organic materials but will not provide an age on volcanic deposits or inorganic sediments. Luminescence dating can date sediments, but not organic matter. In ideal circumstances, several dating methods should be applied if a suitable range of different materials is available.

Time range

A key consideration is the likely age of the deposit to be dated and whether the technique best suited to that type will extend to the period of interest (Fig. 4.3). For instance, while radiocarbon can be used to date samples back to 50 ka BP, recent historical samples are problematic. If we wish to go further back, luminescence dating may provide ages back to 300 ka. In some instances the age of the deposits is unknown and, if so, the cheapest method should be applied, e.g. palynostratigraphy, to provide a rangefinder age estimate.

Cost and time

The environmental archaeologist is consistently working within tight financial budgets and strict deadlines. One of the primary concerns is, therefore, affordability and processing time. Due to ever changing technological advances and currency exchange rates, this book will not attempt to provide costs of different dating methods. It is therefore strongly recommended that workers directly approach laboratories to enquire about costs and processing time.

Precision and accuracy

Different dating methods have different capabilities in their degree of precision and accuracy, a factor of considerable importance when we use methods that do not provide absolute dated chronologies (see Section 4.1.1.5 for definitions). A summary of the main methods, with their associated precision is given in Fig. 4.3.

The extent to which particular dating techniques can achieve the degrees of accuracy and precision required by particular branches of environmental archaeology depends on the qustions being posed.

Technique	Range, ka	Precision (1σ)	Ocean Cores	Corals	Volcanics	Sediments	Tufa	Speleothem	Shells	Bones	Teeth	Wood & Plant Residues
Typology	0->2000	<20%	—	—	—	🙂	—	—	—	😐	😐	—
Palyno-stratigraphy	0->2000	<10%	—	—	—	🙂	—	—	—	—	—	🙂
Amino acid racemization	0-700 (<2000)	<25%	—	?	—	—	—	—	😐	🙁	😐	?
Palaeomagnetism	0->2000	<1%	😐	—	😐	😐	—	🙁	—	—	—	—
Tephrochronology	0->2000	Instant event	—	—	🙂	—	—	—	—	—	—	—
Dendro-chronology.	0-12 (>2000)	Annual	—	—	—	—	—	—	—	—	—	🙂
Varve chronology	0->2000	Annual?	—	—	—	🙂	—	—	—	—	—	—
Radiocarbon (^{14}C)	0-40 (60)	0.1%	🙂	🙂	—	—	🙁	🙁	🙂	🙂	🙂	🙂
Luminescence	0-300 (800)	3-5%	🙁	🙂	😐	🙂	?	🙂	—	—	—	—
Electron spin Resonance	0.5-2000	2-20%	🙁	🙂	🙂	🙁	😐	🙂	😐	🙁	🙂	—

FIGURE 4.3 Summary of dating methods available to the environmental archaeologist, with approximate age ranges, levels of current precision and suitability to different material types.

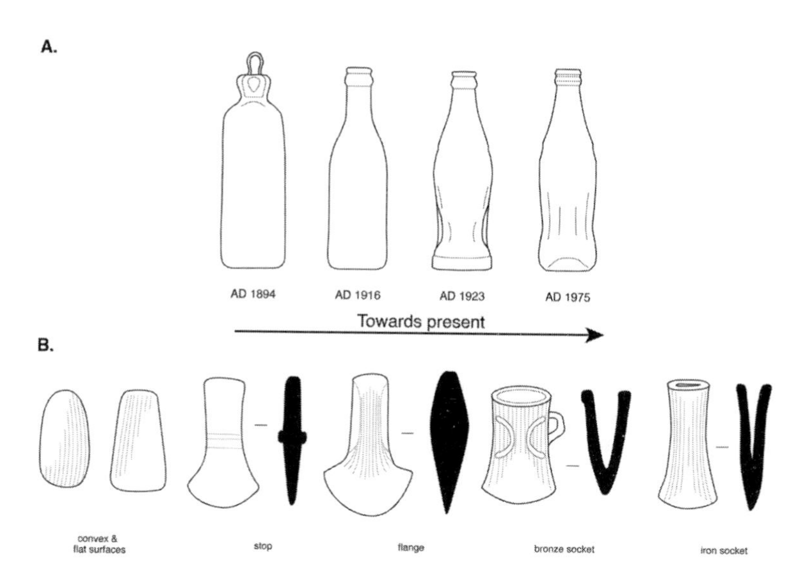

FIGURE 4.4 Typological development of (A) Coca-Cola bottles compared to (B) Western European Neolithic and Bronze Age axes. ('Coca-Cola' is a registered trademark of The Coca-Cola Company.)

4.3.1.3 Combining ages

Where ages are made on the same sample or can be considered statistically indistinguishable, the results can be combined to reduce the overall error and provide a more precise age. The equation for combining a series of ages $\bar{x}_i$ to derive a mean ($\bar{x}_m$) weights the results to those age estimates with the smaller error terms and is as follows:

$$\bar{x}_m = \frac{\Sigma \bar{x}_i/\sigma_i^2}{\Sigma 1/\sigma_i^2}$$

The error on the pooled mean is given by:

$$\sigma_m = \sqrt{\left(\frac{1}{\Sigma 1/\sigma_i^2}\right)}$$

4.3.2 Relative dating

Relative dating methods are those that provide an age sequence rather than an actual age, sometimes with a suggestion of the magnitude of difference between members of a sequence. Although numerous methods of this type exist, the ones of most consequence to the environmental archaeologist are typology, palynostratigraphy and aminostratigraphy.

4.3.2.1 Typology

Principles

Unlike the other methods discussed in this chapter, typology does not require expensive laboratory analyses. During the course of our lives, most of us would have noticed changes in the design of general household items and goods. An excellent example is that shown by the evolution in the design of Coca-Cola® bottles over the last two centuries (Fig. 4.4). There is an immediately recognisable pattern of development, from the first curved forms through to the more familiar shape of plastic bottles we see today. In much the same way, typology exploits known developments in artefacts through the ages, looking at changes in the shape and design. Artefacts can be defined by material, shape and decoration to characterize types that can be used to recognize distinct periods.

Various different methodologies exist that will be briefly introduced here but two basic principles underlie the general approach. The first is that an artefact of any given age will have a recognizable form for a place. A trained typologist (often in the field) will be able to recognize artefacts by this form and then place it within a relative timescale. Secondly, the changes in design are often very slow and tend to be evolutionary. Artefacts through the ages will show similar characteristics as they evolve over time, e.g. the obvious progression in the shape of axe heads over the last few thousand years, with a general development (though not always) towards technologically more advanced designs (Fig. 4.4).

In reality, typology has been used for several centuries. Greek and Roman architectural and artistic styles were utilized by Renaissance scholars, and coins (themselves a form of typology) have been used since the fourteenth century by authors such as the Italian Petrarch (AD 1304–1374), with published collections available from the sixteenth century (e.g. Joseph von Eckhel *Doctrina numorum veterum* – AD 1782–1798). It was not until the nineteenth century, however, that typology as we know it, came into its own. The first real recognition of the use of evolving designs was proposed by Augustus Lane-Fox (also known as Pitt Rivers, AD 1827–1900) who, in 1875, used Darwinian theory to order the complexity of collections he had made as a British Grenadier Guard. Although Lane-Fox made some attempt to place the changing designs into chronological time, it was the Swedish scholar Oscar Montelius (AD 1843–1921) who really developed a firm footing for the typological approach. In the 1800s, Montelius recognized distinctive styles and changes in design of Bronze Age weapons and tools collected by enthusiasts and museums over the years. He developed numerous local chronologies based on the premise that the simplest designs were the oldest, arranging artefacts and types chronologically on the basis of degrees of similarity and dissimilarities. As a result he was able to form local relative Bronze Age chronologies using tools and weapons, which could be tested against stratigraphic sequences where the simplest artefacts should be the deepest, i.e. the oldest. In 1885, he published the classic text *Dating in the Bronze Age, with special reference to Scandinavia,* dividing the Bronze Age in Scandinavia and northern Germany into six main periods. Despite subsequent finds, the system has remained remarkably robust, with similar schemes being developed throughout the world in ceramics (e.g. Nelson, 1916), gravestone designs (e.g. Dethlefson and Deetz, 1966), and even megalithic monuments (e.g. Boujot and Cassen, 1993), sometimes with relatively high precision (Harrington, 1954). Many such schemes are now available on the web as an interactive resource (e.g. Roman amphorae in Britain at www.intarch.ac.uk/journal/issue1/tyers/).

It is worth noting that the evolution of design depends very much on the needs of the original user. For instance, throughout the Palaeolithic, stone tools were notoriously slow to evolve and cannot, therefore, provide a particularly sensitive indicator of the passage of time. Conversely, in most eras, styles of pottery change rapidly and can provide a relatively precise age control for a deposit, e.g. some Mycenaean pottery styles change within as little as 20 years. In some instances, changes in style have been linked to extensive and detrimental environmental changes (e.g. Bicknell, 2000). An absolute timescale can be provided by independent age control, such as cross-correlating to datable objects within a historical framework.

The typology of individual artefacts can be taken a step further by the use of characteristics on the full range of materials found in a deposit. The two versions of this approach are frequency and contextual seriation.

Frequency seriation

Frequency seriation measures changes in the proportions of styles and was pioneered at the turn of the century by Kroeber (1916) working on the Zuñi culture in New Mexico. Here, it was recognized that in a geographically limited area, older style pottery was more frequently associated with dilapidated ruins, implying that buildings and their occupation overlapped, allowing him to order the sites independent of stratigraphic control. Subsequent work in the 1940s by archaeologists working on Mayan sites in the Yucatán, Mexico (Brainerd, 1951; Robinson, 1951) developed the ideas further on a statistically robust footing. As in New Mexico, an impressive range of ceramics had been excavated from individual buildings and monuments but the relative relationship in their construction and use was unknown. To get around this apparent impasse, measures in the proportional abundance (or frequency) of ceramic design found at these sites allowed the development of a relative chronology.

Two basic assumptions are necessary in the above approach. Firstly, a particular ceramic style must become popular, reach a peak in popularity and then fade away. If plotted as a frequency diagram, the shape is much like an aerial view of a battleship, thus the use of the term 'battleship curves'. Secondly, at a given time, a style of ceramic has to be equally popular in different areas (so that the artefact forms a similar proportion of the total finds).

Contextual seriation (sequence dating)
Unlike frequency seriation, the contextual approach exploits the length of time that different artefact designs are popular and was developed by Flinders Petrie at the Upper Egypt site of Diospolis Parva, reported in 1899. Here, numerous graves had been excavated and much like the Mayan buildings, the relative relationships between their construction and use were unknown. Because the graves were Pre-Dynastic Neolithic in age, they could not be linked to the historical king-lists. To place them in relative order, Petrie made an inventory of all the grave goods on separate sheets, listing all the artefacts found. He then placed all the slips in a column and rearranged their order until the greatest number of individual artefacts had the least number of slips.

Potential problems
Despite these advantages, there are several problems inherent with the method that the user should be aware of. A key constraint is how to describe artefacts objectively to allow quantitative comparisons. For some artefacts, universally applied definitions on usage and style can be problematic. Furthermore, beyond historical dates of >3000 BC, no absolute chronology can be provided by the method. Indeed, in some instances, the historical dating control itself may be called into question. Objects imported from distant sources can be used within a culture for a significant period and thus delayed prior to burial, superficially similar artefacts found in different areas may in fact be unconnected, and not all artefacts show evolving styles (e.g. British harpoon points; Smith, 1992). In addition, users should be aware that scarcity of resources may lead to the reutilization and reshaping of artefacts (Bettinger and Eerkens, 1999), and that artefacts once buried may be susceptible to post-depositional disturbance, particularly in relatively fine-grained sediments (Cahen and Moeyersons, 1977). Where types have limited age control, chronological interpretations should be treated with caution (Raban, 2000).

Nevertheless, despite the above problems, the method is remarkably robust in numerous situations and can be a rapid, powerful tool for developing a chronological framework for the environmental archaeologist.

4.3.2.2 Palynostratigraphy

Principles
Palynology utilizes identifiable pollen grains and spores from plants to reconstruct past vegetation cover. Comprehensive discussion of the technique can be found in Chapter 3, and useful online resources are available via the University of Arizona web site (http://www.geo.arizona.edu/palynology/polonweb.html).

Once a database of knowledge about vegetation history has built up, those same pollen and spores can be used as a relative dating tool. Palynostratigraphy has proved a reliable, cheap method for deriving the age of sequences in terrestrial, marine and, in some exceptional circumstances, ice deposits. The first applications of the technique can be traced back to Lennart von Post, who, at the start of the twentieth century, recognized distinct periods of postglacial vegetation development in Swedish sediments. Since then, the method has become extremely popular and widely adopted throughout the world (e.g. Guiot *et al.*, 1989; van der Knaap *et al.*, 2000). In remote locations, using changes in 'exotic' pollen types has provided correlation (and sometimes ages) from well-described palynostratigraphic changes developed from upwind sites (McGlone *et al.*, 2000). Further details on this method can be found in Birks and Birks (1980), Moore *et al.* (1991) and Lowe and Walker (1997).

Potential problems

Palynostratigraphy as a dating tool must be carried out with a clear awareness of the possible problems. In some instances, relatively coarse sediments and/or high pH (>6) significantly limits pollen preservation (Dimbleby, 1957). Also, dispersal is generally undertaken by a range of agents including water, insects and birds, and wind. As a result, not all plant types deliver similar quantities of pollen to the sediments around them. The dispersion and proximity relationships lead to significant variations between the pollen observed in a sediment and the vegetation of the source area. Tyldesley (1973) reported finding contemporary pine and birch pollen in northern Scotland and the Shetland Islands. The latter can be characterized as a treeless landscape and it seems most likely that the source was actually Scandinavia, an area with different vegetation dynamics and not representative of the northern British Isles.

In addition, microclimates and dispersal times mean not all changes in vegetation can be considered truly time-parallel. For instance, at the end of the last ice age, vegetation appears to have recovered relatively rapidly (within decades) in central Europe (e.g. Wick, 2000), allowing the development of precise correlative schemes. By contrast, at high latitudes (e.g. Lowe and Turney, 1997), vegetation change appears to have significantly lagged behind climate by up to several centuries, making any precise correlations more difficult. Also, different sediments can affect pollen accumulation in different ways. In bogs, for example, pollen tends to be trapped effectively on the surface, while in lake and marine sediments, wind and

Box 4.1 Correlating sequences in tropical Queensland, Australia

Individual pollen diagrams can be amalgamated to produce regional pollen assemblage zones, within restricted geographical areas. Vegetation changes can then be considered broadly synchronous, allowing the development of time-parallel horizons. As a result, principle changes in vegetation cover through any time period can often be characterized. An excellent example comes from the Atherton Tablelands in north-east Australia (Fig. 4.5). The area has over 60 volcanoes, many of which have infilled with pollen-rich sediments that can be used to reconstruct changes in the dominance of wet, dense rainforest (currently restricted to eastern areas) and dry-loving sclerophyll, such as eucalypts (that form a more open landscape to the west). Immediately offshore, a marine core (ODP-820; Moss and Kershaw, 2000) provides a regional summary of these changes (albeit at relatively low resolution), within which the records from the Atherton can be correlated and an alternative chronology provided by $\delta^{18}O$ (Martinson *et al.*, 1987). What is immediately apparent is the difference in age of the past eruptions in this area. For instance, Lynch's Crater clearly spans at least the last 190 ka (Kershaw, 1986) while Lake Barrine (Kershaw, 1983) appears to have taken place at the onset of the Holocene. Intriguingly, the local Aborigine tribe, the Ngadjonji, have a folk memory of the eruption of Barany (Lake Barrine) that appears to be the oldest human record of volcanic activity (Bindon and Raynal, 1998):

> Two men broke a taboo and angered the rainbow serpent, a major spirit of the area. The earth roared like thunder and the winds blew like a cyclone. The ground began to twist and crack and there were red clouds in the sky that had never been seen before. People ran from side to side but were swallowed by a crack which opened in the earth.

The Ngadjonji describe the area at the time of the eruption of Lake Barrine as scrub, consistent with the pollen assemblage at the base of the record at around 10 ka BP. Prior to any palynostratigraphic work their story was largely discounted, since the dominant natural vegetation in the area is now rainforest.

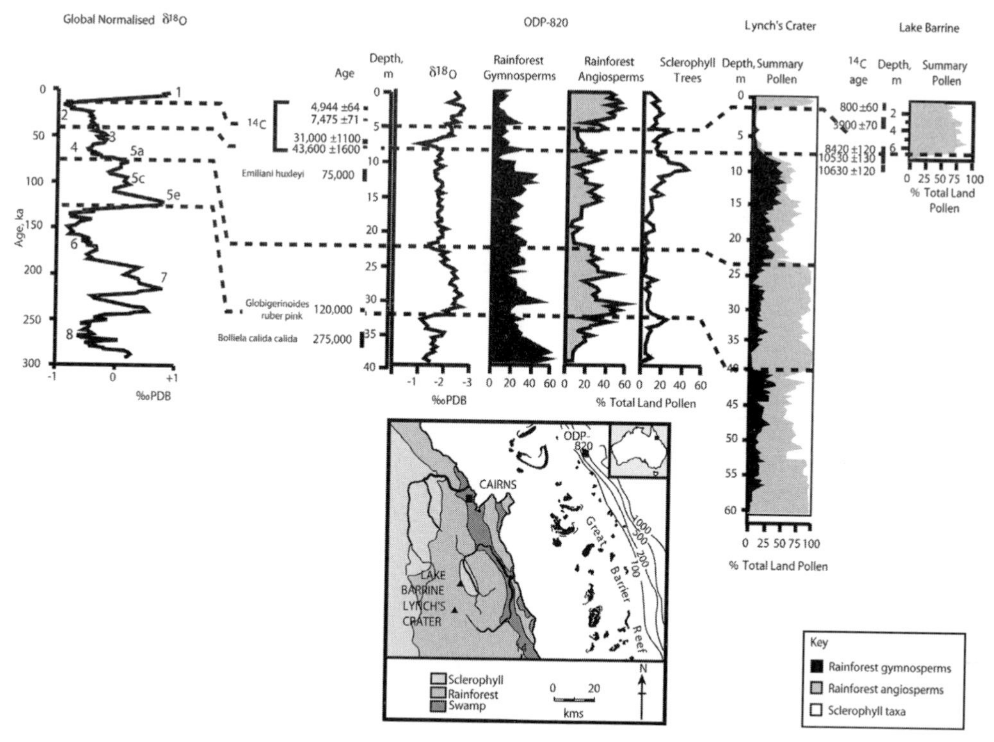

FIGURE 4.5 Summary pollen records from northern Queensland, Australia. (Modified from Kershaw, 1986; Hiscock and Kershaw, 1992; Moss and Kershaw, 2000) and the global δ^{18}O record. (Martinson *et al.*, 1987)

water circulation will influence grains depending on their morphology, making comparison between records from different environments problematic.

4.3.2.3 Aminostratigraphy

Principles

Aminostratigraphy is a chemical dating technique that exploits the alteration of amino acids within biological materials over time; it can be used on shell, bone and wood within a range of depositional environments. The first work in this area was undertaken in the 1950s by Abelson (e.g. 1954, 1956) who investigated proteins and amino acids in geological samples. Subsequent work by Hare (1963) on Quaternary marine material illustrated the potential of the method as a chronological tool. Research since the 1970s has seen the development of aminostratigraphies within a wide range of depositional contexts throughout the world, including British Quaternary stratigraphy (e.g. Bowen *et al.*, 1989), evolution of modern humans (e.g. Miller *et al.*, 1999a) and the extinction of megafauna in Australia (e.g. Miller *et al.*, 1999b). Detailed reviews can be found in McCoy (1987), Goodfriend *et al.* (1998) and Wehmiller and Miller (2000).

The principle is relatively simple, though the interpretation can sometimes prove to be contentious. Proteins within organic materials comprise structural amino acids that, following the death and burial of an individual (plant or animal), begin to naturally hydrolyse to lower molecular weight peptides and free amino acids. Although a significant proportion of the original amino acid population is lost during early diagenesis (Powell *et al.*, 1989), typically anywhere between 40 and 60 per cent remains in Quaternary fossils (Wehmiller and Miller, 2000). In general, around ten amino acids can be found in fossils, and most of these have one central carbon atom bonded to four different

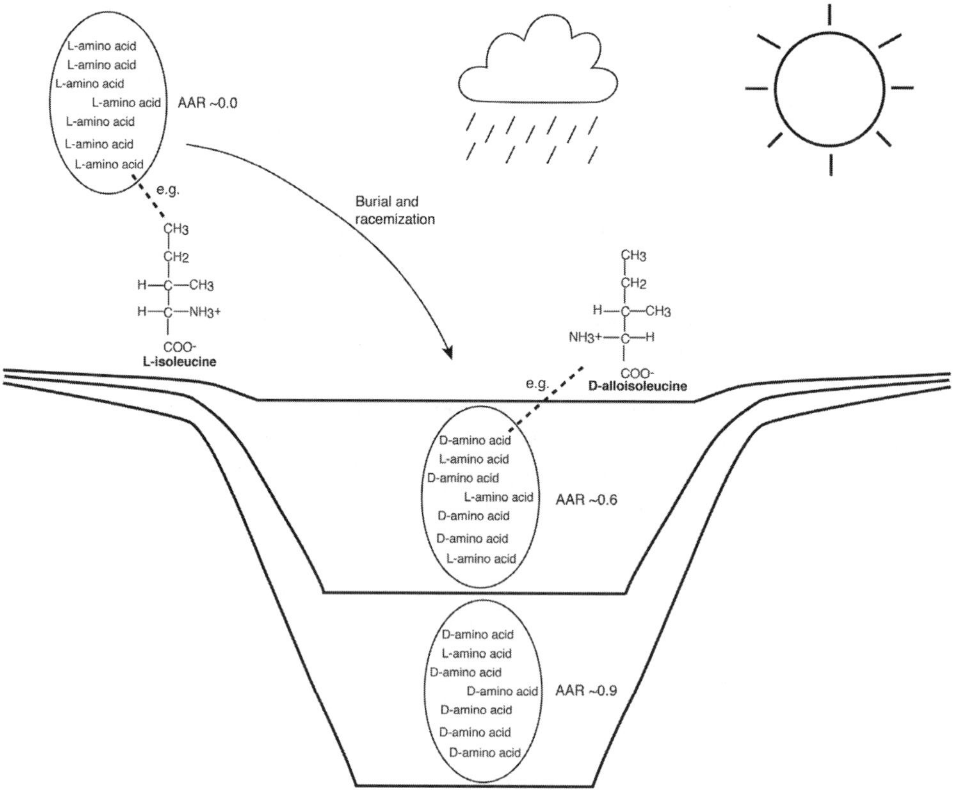

FIGURE 4.6 Schematic showing racemization of D/L ratios in eggshells over time.

atoms (Fig. 4.6). These molecules can exist in two geometrical configurations which are mirror images of one another. Given a sufficiently closed system, racemization takes place converting the amino acids from one form to the mirror image. In the case of the amino acid isoleucine, there are two adjacent centres of asymmetry. The racemization of the isoleucine around the carbon atom bonded to hydrogen, carboxyl and amino groups is referred to as epimerization (Fig. 4.6). Collectively, the term Amino Acid Racemization (AAR) is given to the measurement of these changes. Fortunately for the environmental geochronologist, the racemization and epimerization rates of the two most commonly investigated proteins (L-aspartic acid and L-isoleucine) are sufficiently slow to be of use as a chronological tool. As such, there is no specific dating limit of the method because of the range of factors that control the rate of racemization (and epimerization). In high-latitude regions, where temperatures are frequently below −10°C, equilibrium may take up to 5 Ma. In contrast, molluscs from tropical locations may reach equilibrium within 150 ka. Typically, the AAR signal is calibrated against one or more other dating techniques. Although analytical precision is relatively high, uncertainties (including the thermal history) may mean the overall level of precision can be as poor as 25 per cent at 1σ.

Methods

Racemization involves the conversion of the L-amino acid (left-handed) to the D-amino acid (right-handed) form. In modern samples, the amount of D-amino acids has been found to be negligible, thus the D/L ratio for racemization should increase from 0 (living sample) up to a value of around 1.1 (i.e. in equilibrium) in a sample of infinite age. For epimerization, the D/L ratio may get as high as 1.3 (L-isoleucine

(L-Ile) converting to D-alloisoleucine (D-alle)). To maintain comparability, it is crucial to take samples from as far within an outcrop as possible and avoid surface material (due to the latter's more extreme temperature history) (e.g. Miller and Brigham-Grette, 1989). Furthermore, taxonomic selection is crucial, with racemization rates being different between species (e.g. Miller and Brigham-Grette, 1989). Thus, AAR values obtained from different samples within the same stratigraphic unit will rarely be the same. Since it is unusual for a single species to be present throughout relatively long sequences, a range of different species is commonly investigated.

To measure the ratio of different amino acids, chromatographic methods are employed (Wehmiller and Miller, 2000). Numerous variations exist but the most common include ion-exchange high-pressure liquid chromatography (HPLC) and gas chromatography (GC) (e.g. Murray-Wallace, 1993). Typical sample sizes range from 0.1 and 1.0 g and samples are physically cleaned, then dissolved and hydrolysed in HCl. Generally speaking, HPLC is less labour intensive but GC allows a greater range of amino acids to be investigated. More recently, reversed phase liquid chromatography (RPLC) has also been applied with promising results (Kaufman and Manley, 1998), providing a combination of many of the best aspects of HPLC and GC (Wehmiller and Miller, 2000). GC will take at least a week for a batch of 3–4 samples. For HPLC, typically a batch of samples will take a week to go through the entire pretreatment and measurement procedure. Multiple AAR values can be obtained, providing a considerably faster and cheaper method than most dating techniques available to the environmental archaeologist.

Potential problems

The conversion of the amino acids is heavily dependent on the temperature history which can thus significantly affect the dating range of this technique. Temperature can have a significant effect on the D/L ratio, with a rise of 4°C approximately doubling the rate of epimerization. If the exact temperature changes over time are not known, using D/L ratios from the same area (i.e. samples that have experienced similar temperatures) can bypass the problem, as all samples will have experienced the same thermal history (e.g. Miller *et al.*, 1999a). Furthermore, caution should be applied to certain sample types. For instance, within bones, the porosity of the material may allow external chemical interference (e.g. leaching, bacterial contamination), providing AAR ages up to an order of magnitude different compared to other independent methods (Hare *et al.*, 1997; Bada *et al.*, 1999). Within mollusc shells, the amino acid composition may vary significantly between structural layers of the same sample, resulting in different AAR values being obtained (Miller and Brigham-Grette, 1989). Over glacial-interglacial samples, different oxygen isotope stages have been reported (e.g. Bowen *et al.*, 1989), though there has been considerable debate as to the selection of AAR values and the degree to which there is overlap between the different stages (McCarroll, 2002). To date, the greatest success has been the application of AAR to ratite eggshells which can generally be considered closed systems (Miller *et al.*, 1999a) and therefore not susceptible to chemical interference.

4.3.3 Age-equivalent dating methods

Age-equivalent methods do not provide an age per se. Instead they provide time-parallel marker horizons against which archaeological information can be discussed and correlated. The methods outlined here for the environmental archaeologist are palaeomagnetism and tephrochronology.

4.3.3.1 Palaeomagnetism

Principles

Palaeomagnetism (including archaeomagnetism) exploits the fact that many geological and archaeological materials can preserve the Earth's magnetic field strength and polarity at the time of formation/deposition. As the magnetic field changes temporally and spatially, it is possible to use these variations to provide chronological control. If the variations are known and dated, an absolute age is in theory possible using this method. Even if no specific age is known for the geomagnetic parameters, the properties can still be compared. Overviews of the method are given by Thompson and Oldfield (1986), Tarling (1991) and Wagner (1998).

The Earth's magnetic field comprises three distinct components:

1 *Declination* – the angle between true and magnetic north.
2 *Inclination* – the dip relative to the horizontal which at the equator is 0° while at the poles is 90°.
3 *Intensity* – the strength of the field.

If the Earth's magnetic field were purely dipolar (which can for all practical purposes be thought of as a simple bar magnet) and if this were orientated along the rotation axis of the planet, then under the present (normal) polarity state, the declinations would align themselves precisely to north and only the inclinations would be latitude. In practice, about 80 per cent of the field can be described by the dipole component, the rest is the non-dipole component. The latter (the remaining 20 per cent of the field) is not constant, changing as a result of currents in the Earth's core (partially controlled by variations in the Earth's orbital parameters; Yamazaki and Oda, 2002), resulting in variations in the declination and inclination values throughout the world (secular variation). In the late seventeenth century, Halley was the first to recognize that secular variations existed. Early studies concluded that not only did volcanic rocks acquire their magnetization on cooling, but that the direction paralleled the modern geomagnetic field for younger volcanics, while some older volcanics were magnetized in the opposite direction. Since then, palaeomagnetism has become global in application with the recognition of geomagnetic field reversals that can take place over longer intervals than the historical period.

Many materials when exposed to a magnetic field become magnetized in the same direction as the applied field, with an intensity proportional to the strength of the applied field. Palaeomagnetism aims to identify and isolate the original direction of the magnetic components preserved within those materials. For instance, rocks and sediments magnetized during formation, align themselves to the field, recording the natural remanent magnetism (NRM). In addition, the ferromagnetic crystals in molten lava can also align themselves to the magnetic field, resulting in thermoremanent magnetism (TRM). This form of remanence is lost if the same material is subsequently reheated to high temperatures (e.g. magnetite >570°C, haematite >675°C). The temperature at which the remanence is lost is the Curie temperature. Such features are not restricted to geological settings, however. Ferromagnetic particles in sediments will align themselves to the magnetic field as they settle out, hence detrital remanent magnetism (DRM). Grains of around ~1 μm in size are orientated by the magnetic field, being sufficiently large to preserve this signal against Brownian motion while small enough not to be entirely controlled by gravitational forces. Once deposited, such grains tend to occur within intergranular spaces, allowing retention of the alignment.

Determination of the above properties has led to the recognition of significant variations in the Earth's magnetic field, operating on a variety of different temporal scales, and therefore offering chronological control at various levels. Direct measurement of declination, inclination and intensity in major centres of population allows historical secular variations (>AD 1600) to be determined, e.g. Paris, London, Boston and Rome (Aitken, 1974; Thellier, 1981). Prior to the historical period, the datasets have been augmented by measurements made on archaeological materials of known age. Thus, heating of clays above the Curie temperature, results in TRM being imprinted within ceramics and bricks. Such finds allow master curves to be developed at a regional level (e.g. Thellier, 1981; Tarling, 1991; Bowles *et al.*, 2002). As a result, such schemes can provide a framework of up to 5000 years, with a precision of between 25 and 50 years (Tarling, 1991). Archaeomagnetic dating has been used, for example, to suggest temporal differences between late Minoan destruction sites across Crete (Downey and Tarling, 1984).

Longer records can be derived from lacustrine and loess depositional environments, where palaeosecular variations can be reconstructed on the basis of the DRM signal. Independent dating of the sediments allows these variations to be placed within a chronological framework (e.g. Saarinen, 1999; Snowball and Sandgren, 2002), though with greatly increased age uncertainty (dependent on the dating control). In addition, global palaeointensity curves can be produced from the marine environment where continuous sedimentation could in theory provide records encompassing the entire Quaternary. One of the most comprehensive is the stacked NAPIS-75 record from the North Atlantic (Laj *et al.*, 2000). Such records have identified geomagnetic field intensity minima (10^4–10^5 yrs) at 42–41 ka, and 34–32 ka, known as the Mono Lake and Laschamp excursions respectively (Voelker *et al.*, 2000). The potential exists that such excursions may be precisely correlated to Greenland and Antarctica through continuous published records of ^{10}Be recorded in the ice (e.g. Raisbeck *et al.*, 1987; Yiou *et al.*, 1997), acting as global time-parallel marker horizons independent of uncertainties in individual chronological schemes. Outside such excursions, where sedimentation rates of >10 cm/1000 years exist, a precision of 500 years is theoretically possible (Sarnthein *et al.*, 2002).

On the geological timescale, polarity reversals are also known to take place, i.e. a 180° shift in the geomagnetic field. Initial work focused on volcanic rocks from which K-Ar dating provided chronological control. Detailed records have now also been made from marine sediments, which have the advantage of providing continuous palaeomagnetic data (e.g. Biswas *et al.*, 1999), from which $\delta^{18}O$ measurements can also be made, providing simultaneous chronological control. Repeated and worldwide synchronous reversals provide the basis of a global chronological scheme, with two periods of uniform polarity (chrons) in the Quaternary: the Brunhes (normal) and Matuyama (reversed), the transition of which (the B-M boundary) has been dated at 779 ± 2 ka using K-Ar dating. These chrons are interrupted by shorter events (10^3–10^5 years) of opposite polarity to the main period. Where uncertainty exists as to their global significance, such events are termed excursions. Polarity reversals are of great significance if detected in sequences, and have provided comprehensive chronological control for early hominin sequences (Partridge *et al.*, 1999; Oms *et al.*, 2000; Zhu *et al.*, 2001).

Methods

In most studies, ideally all three magnetic components are measured. Thus, the direction of true North must be noted (or another known direction) and the angle of a flat surface relative to the horizontal. Where only intensity is being measured, samples can be taken from a structure or object without noting orientation. A variety of magnetometers can be used for the determination of direction and intensity of remanence. For samples that are strongly magnetized, a spinner magnetometer is usually sufficient, whereby the samples are exposed to fast rotation. Spinning induces a voltage in a coil that is proportional to the magnetized components perpendicular to the axis of rotation. The third component is measured by setting the sample in the magnetometer at a different orientation. Alternatively, if the signal is relatively weak, a cryogen magnetometer can be used. Here, superconducting loops are utilized within detectors set to measure one component of magnetism, i.e. that perpendicular to the loops. Depending on the model, single measurements can be made or multiple analyses must be undertaken (requiring subsequent reorientation of the sample). The precision on both machine types is approximately the same. A list of laboratories where analyses are undertaken can be found at http://www.geo.umn.edu/orgs/irm/irm.html.

Potential problems

A major limitation of palaeomagnetism can be the availability of suitable material encompassing a sufficient period of time to recognize variations. Sherds and bricks are rarely found in the position they were fired in. Pots, however, are usually fired in an upright position and bricks lie on their sides. The azimuth is no longer known and the declination component cannot be used, but the inclination and intensity values can still be extracted. The kilns, themselves, are often found *in situ*, but attention must be paid to whether the kiln has physically altered in any way since heating. For DRM in lacustrine environments, variations in pH result in changes to the interaction between clay particles which can greatly reduce the strength of the preserved geomagnetic field and must be considered at the sampling stage (Katari and Tauxe, 2000).

4.3.3.2 Tephrochronology

Principles

Tephrochronology provides time-parallel marker horizons that allow precise correlation between environmental and climatic records of the recent geological past (Westgate and Gorton, 1981; Sarna-Wojcicki, 2000). Few (if any) geochronological techniques can provide the precision offered by tephrochronology. The virtually instantaneous atmospheric deposition of tephra following an eruption can often lead to clear layers in a wide range of depositional environments, including, for example, lakes, mires, peats, soils, loess, marine sediments and glacier ice masses. The term tephra encompasses a wide range of airborne pyroclastic material ejected during a volcanic eruption, including blocks and bombs (>64 mm), lapilli (2–64 mm) and ash (<2 mm) (Thorarinsson, 1944). The distribution of active volcanoes throughout the world is widespread and associated with plate margins or intra-plate hotspot activity. Most of these have been active during the Quaternary, though the activity of many can be traced over a much longer geological period. There has therefore been a long, complex history of tephra ejection and deposition and, as a result, many areas of the world have the potential for the application of tephrochronological studies. The first attempts to utilize tephra horizons for correlation and geochronological control were based in New Zealand (Berry, 1928; Grange, 1931; Oliver, 1931). Once its potential was established, tephrochronology was rapidly adopted as a key tool for the correlation and dating of stratigraphic horizons. Regional tephrochronological frameworks (the recognition of one or more 'marker' tephra layers of known provenance and/or unique geochemistry) have been developed for many parts of the world (Turney and Lowe, 2001).

The geographical distribution of particles deposited on the Earth's surface following a volcanic eruption depends upon: (a) the size of the eruption; (b) the relative proportion of different types and sizes of ejecta; and (c) the direction and strength of the wind during the time the tephra remains in the atmosphere. The most extensive tephra plumes and associated deposits are generally those associated with the most explosive (plinian and ignimbrite) eruptions of large magnitude (Sparks *et al.*, 1997). Typically, tephra deposits show a decrease in horizon thickness as well as in mean grain size with increasing distance from eruptive sources (Pyle, 1989; Sparks *et al.*, 1992), though this is not always the case, as, for example, is illustrated by the tephra deposited by the Toba volcanic eruption ca. 75 ka (Rose and Chesner, 1987). The larger components of pyroclastic material (i.e. blocks, bombs and lapilli) tend to follow a ballistic trajectory, staying proximal to the eruption site and blanketing the immediate landscape. Conversely, ash particles are frequently erupted much higher into the atmosphere and have a much wider spread. This may account for a greater proportion of the overall tephra fallout with increasing distance from the volcano, as is the case, for example, in tephra deposited after the Mount St Helens eruption of 1980 (Carey and Sigurdsson, 1982).

The deposition of coarse tephra will typically occur within hours to days following an eruption, largely through density settling (Sparks *et al.*, 1997). In the case of finer ash particles, however, localized meteorological precipitation plays an increasingly important role with increasing distance from the volcano (Dugmore *et al.*, 1995). Rain droplets remove light particles from the air column, and this leads to a much more patchy pattern of ash deposition, with greater concentrations found in areas experiencing heavier and/or more persistent rain. For all practical purposes, therefore, tephra horizons can be regarded as instantaneous time-parallel marker horizons in archaeological contexts. An excellent example is the Kaharoa Tephra (Newnham *et al.*, 1998) which has been

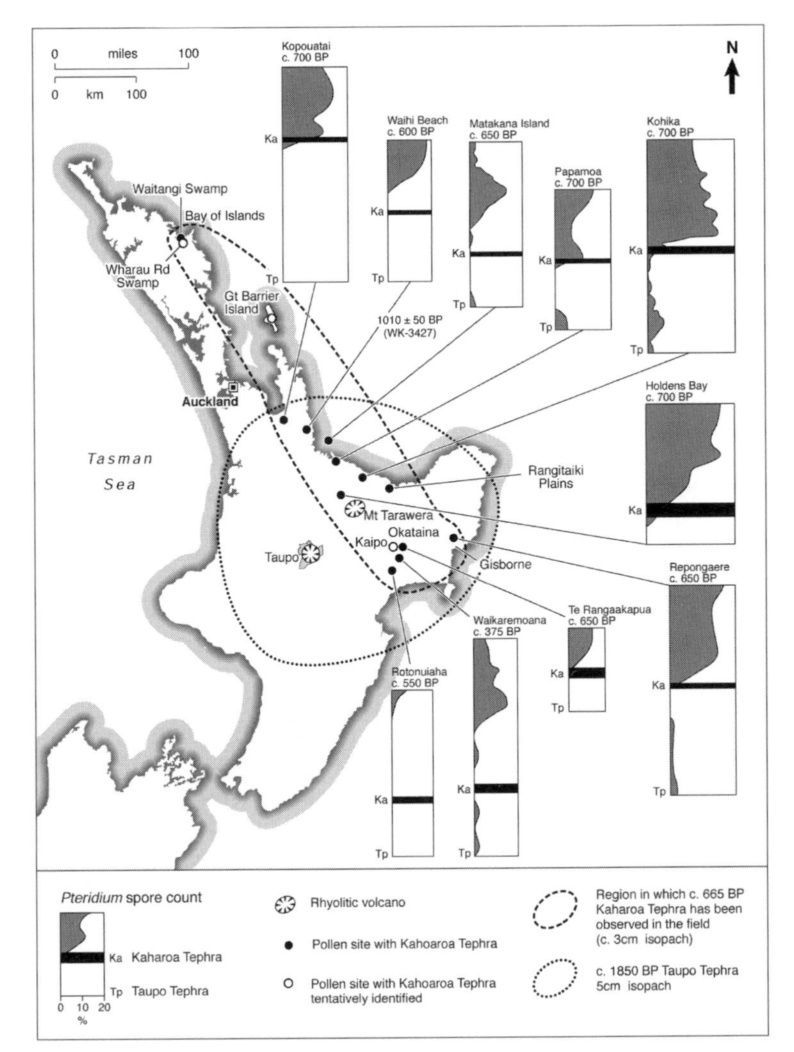

FIGURE 4.7 Pteridium (bracken) spore profiles across North Island, New Zealand containing the Kaharoa Tephra. (Modified from Newnham *et al.* (1998). The Kaharoa Tephra as a critical datum for earliest human impact in northern New Zealand. *Journal of Archaeological Science*, **25**, 533–544 (with permission from Elsevier)

dated to 665 ± 15 ^{14}C BP and found throughout North Island, New Zealand, in ombrogenous mires (Fig. 4.7). The earliest inferred human impacts occurred at the time of the eruption or after (as shown by the increase in bracken as a result of widespread deforestation).

Methods

If located sufficiently proximal to an eruption site, lake sediments may contain several distinct tephra horizons that can be observed by the naked eye. A preliminary tephrochronological framework can therefore be developed quickly by measures or observations of the varying grain size, colour and thickness of different tephra layers. As an approximate rule of thumb, basaltic horizons tend to be black, those of andesitic origin are grey, and rhyolitic material is predominantly white in colour. The colour of ash can, however, be misleading, and geochemical analyses or other diagnostic laboratory tests should be carried out to confirm the nature and composition of tephra layers (Westgate and Gorton, 1981). In the case of ash deposited more distally from the point of origin, the tephra particles may not be visible to the naked eye and may be extremely low in concentration within lake sediments. More detailed laboratory techniques are therefore required to detect and characterize these 'micro-tephra' horizons (Dugmore *et al.*, 1995; Pilcher and Hall, 1996; Pilcher *et al.*, 1996; Turney *et al.*, 1997), for which some researchers have suggested the term *cryptotephra* to be more appropriate (Lowe and Hunt, 2001; Turney *et al.*, 2004).

A large number of laboratory techniques are available for detecting the occurrence of cryptotephra particles in lake sediments. Non-destructive detection methods (i.e. methods which leave the host sediment undisturbed) include analysis of magnetic properties (e.g. Oldfield *et al.*, 1980; Guerrero *et al.*, 2000), X-ray analysis (e.g. Dugmore and Newton, 1992) and measurements of light reflectance from the host sediment (e.g. Caseldine *et al.*, 1999). Destructive detection methods (those which disturb the host sediments and/or remove, alter or even destroy the tephra particles) include the preparation of thin sections to reveal microscopically-thin tephra bands (e.g. Merkt *et al.*, 1993), analysis of the bulk geochemistry of the sediments in which the tephra particles occur (e.g. Lowe and Turney, 1997), magnetic separation (Mackie *et al.*, 2002), ashing of organic sediments (Pilcher and Hall, 1992), organic silica digestion (Rose *et al.*, 1996) and flotation of tephra shards (Turney, 1998). In archaeological contexts, pumice fragments associated with specific eruption events have actually been identified mixed with clay in pottery, allowing precise correlations over a relatively large area (Cioni *et al.*, 2000), though such examples are quite rare.

Although one or more of the above methods may reveal the presence of cryptotephra within deposits, the concentration of tephra shards is often significantly diluted by contemporaneous deposition of mineral sediments of non-volcanic origin, particularly within a limnological context. The detection of tephra in predominantly minerogenic sediments can therefore be difficult, unless recourse is made to flotation and/or magnetic methods (Turney, 1998; Mackie *et al.*, 2002). The flotants can be examined for the presence of glass microshards using a polarizing microscope, the shards being identified on the basis of their distinctive morphology and optical characteristics (including polarization, vesicularity and refractive index). An online workshop of these methods can be found at http://mediator.ads.qub.ac.uk/ms/virtualconference/.

The characterization of the mineral assemblage (e.g. relative abundance of ferromagnesian minerals) of a tephra horizon proximal to the eruption centre can be a powerful correlative tool (Westgate and Gorton, 1981; Shane, 2000). In the search for a more diagnostic basis for distinguishing between tephras of similar mineralogy, therefore, scientists have turned to geochemical techniques, an approach made increasingly more precise and accessible by the development of high-precision analytical machines within the last two decades or so. The electron microprobe, for example, uses Energy Dispersive Spectrometry (EDS) or Wavelength Dispersive Spectrometry (WDS) to measure variations in the concentration of the oxides of the major elements within individual tephra shards. WDS is the preferred option because it enables the relative importance of each major oxide to be monitored during analysis. This is particularly significant with respect to the oxides of alkali metals, which are vulnerable to mobilization through the crystal lattice of the tephra shards during analysis.

For all results obtained using electron microprobes, the values are reported as percentages of sample weight. It is rare for the total analysis to reach 100 per cent, because of small uncertainties introduced during sample preparation, variations in shard characteristics (e.g. thickness and structure of shard walls, which alter susceptibility to impurities, including the mounting medium) and the water content of the shards (Hunt and Hill, 1993). To facilitate comparisons between different sets of analyses produced by different laboratories, therefore, it has been recommended that all geochemical data be normalized to 100 per cent (Froggatt, 1992). This procedure has been criticized by Hunt and Hill (1993), however, who argue that no adjustment of the original measures should be undertaken but that all samples yielding oxide measure totals of less than 95 per cent be rejected. Comparisons of electron microprobe data obtained from different ash samples are typically based on simple bivariate or ternary plots (Hunt and Hill, 1996) or numerical analysis of the data using either similarity coefficients (e.g. Borchardt *et al.*, 1972) or discriminant function analysis (Stokes and Lowe, 1988; Charman and Grattan, 1999).

In recent years, attention has switched from the analysis of major elements to that of variations in the relative proportions or concentrations of minor, trace and rare earth elements. This has become possible through the development of more sophisticated element analysers, such as laser ablation ICP-MS, which can derive a comprehensive suite of geochemical data from extremely small surface areas on individual glass shards (e.g. Westgate *et al.*, 1994; Eastwood *et al.*, 1998; Pearce *et al.*, 1999). Measures of these indices may prove to be more discriminating than measures of major element content, and thus lead to a more reliable 'fingerprinting' methodology. Numerous laboratories undertake tephrochronological work and a comprehensive list of active researchers can be found at http://www.gris.cri.nz/inquatephra/.

Potential problems

In visible tephra horizons, characterization of mineral assemblages does have problems. Dissolution of mafic minerals in acidic environments (Hodder *et al.*, 1991) and preferential settling of tephra with distance from the centre (Sparks *et al.*, 1997) can lead to significant variations in mineralogical composition over relatively short distances. Such mineralogical assessments also have other limitations, because tephras derived from the same centre, or from several centres with similar geological characteristics and histories, will tend to have the same mineralogical composition. In order to isolate sufficient material for successful geochemical analysis, the shards are usually extracted from lake sediment samples using the acid digestion method described by Dugmore (1989). Although the ashing of sediment samples at furnace temperatures of 550°C should not melt glass shards (Swanson and Begét, 1994), it can result in changes in glass chemistry through alteration of alkali concentration (Dugmore *et al.*, 1995) as well as chemical reactions between glass surfaces and organic sediment components (Pilcher and Hall, 1992) and should be avoided at all times.

Although electron microprobe analysis of individual glass shards is a relatively sophisticated and rigorous technique, there are problems associated with this approach. Interlaboratory comparisons have so far shown that more rigorous standardization procedures need to be undertaken to facilitate and improve the accuracy of comparative data (Hunt and Hill, 1996), as some variation in results is introduced through differences in laboratory procedures, in machine precision and by operator error. Furthermore, post-depositional reworking (Boygle, 1999) and alteration of glass shards can occur, the latter through hydration and alkali exchange (Shane, 2000), processes which are dependent upon the duration of any subaerial exposure and differences in depositional environment (Dugmore *et al.*, 1992). Finally, even the detailed examination of major oxide content by electron microprobe may fail to produce confident distinctions between successive tephra layers which have emanated from the same volcanic source, despite the application of powerful discriminant statistical methods (e.g. Stokes *et al.*, 1992).

4.3.4 Incremental dating methods

4.3.4.1 Dendrochronology

Principles

The recognition that trees lay down annual growth rings goes back to Theophrastus (c. 300 BC), a student of Aristotle. By the time of the Rennaisance, Leonardo da Vinci had suggested there was a relationship between ring width and moisture availability and used this to reconstruct past climate. The recognition of the importance of tree rings for past studies continued, and by AD 1837, Babbage had suggested that the patterns of tree rings could be overlapped ('cross-dated') to produce continuous, long chronologies into the past (Baillie, 2000). The father of dendrochronology, however, must be considered to be Andrew Douglass who generated the first dendrochronological framework in Arizona, USA.

Trained as an astronomer, Douglass believed long-term changes in tree ring growth in ponderosa pines could be attributed to the 11-year solar cycle (Douglass, 1919). This work went beyond living trees in AD 1914 when archaeologists working on Native American sites such as Pueblo Bonito (Chaco Canyon) and Aztec (New Mexico) identified ancient timbers in excavations. From these Douglass developed a continuous master chronology over 1000 years in length. For many years there was a gap of unknown length between the timbers being excavated ('floating') and the living material. Numerous expeditions were organized, using the typological knowledge obtained from the pottery at sites to infer the likeliest spots where that gap could be bridged. In AD 1929, a buried charred beam was excavated that linked the two datasets, placing the sites in the American Southwest on an absolute calendar time.

In most temperate and many tropical environments there is sufficient seasonal climate change to allow the identification of growth rings. During periods of growth, an annual layer is produced (the cambium) on the inside of the bark. The cambium comprises two parts: the phloem cells that transport sugars and other photosynthetic products throughout the tree (later to become bark); and the xylem cells that transport water from the roots up the trunk (to become the building blocks of tree rings). The xylem cells typically form two parts. The inner part is termed the 'early' wood, generating relatively large cells, and is formed at the start of the growing season, when the growth factors that can be limiting (e.g. nutrients, temperature and moisture) are at an optimum. In gymnosperms (non-flowering seed plants such as conifers), these xylem cells are called tracheids. In angiosperms (flowering plants such as oaks, beech and ash), the xylem cells are referred to as vessels. Later in the growing season, the above factors can start to play a key role in limiting growth, generating relatively thin xylem cells that are thick walled and appear darker in cross-section. Over time, climatic and/or environmental conditions can change, generating extreme narrow and wide rings, depending on whether the conditions are detrimental or conducive to growth. As a result, it is possible to recognize annually-resolved periods of growth that can be used to derive a powerful, absolute timescale in sidereal years with no error. Dendrochronology has thus been used to identify the timing of climatic change (Baillie, 1999), volcanic eruptions (Jacoby *et al.*, 1999), earthquakes (e.g. Yamaguchi *et al.*, 1997), the construction of artefacts and buildings (e.g. Baillie *et al.*, 1985; Tercier *et al.*, 1996; Topham and McCormick, 2000) and fires (e.g. Lageard *et al.*, 1999).

Chronologies now exist throughout the world. In the British Isles, using the Belfast bog oak dataset, wood samples can be dated with absolute precision (using a minimum of 100 rings) back to 7.4 ka (Pilcher *et al.*, 1984). The German oak and pine chronologies extend further back to 11.9 ka, and there appears to be considerable potential for extending back into the late-glacial (Friedrich *et al.*, 2001). In the USA, giant sequoia and bristlecone pine allow ages to be obtained for the last 8.7 ka (Ferguson and Graybill, 1983). A comprehensive list of dendrochronological services can be found at http://web.utk.edu/~grissino/links.htm#institutes.

There are numerous texts that provide excellent introductions to dendrochronology. Examples include Baillie (1995, 1999) and Towner (2002). An online searchable bibliographic database for dendrochronology is also freely available (http://www.ltrr.arizona.edu/archive/biblio.html).

Method

Sampling of material is dependent on the nature of the wood. Living material can be sampled using an increment corer that removes a 5 mm diameter sample, non-destructively, from the centre (pith) of the tree. Typically the cores are fragile and should be placed in a protective tube (e.g. a straw). Alternatively, dead trees can be sampled to obtain an entire cross-section, using a hand or chainsaw. Typically, sampling should be taken above the root plate and below the branches to avoid knots and other irregularities in the tree ring pattern. The development of a chronology is dependent on a statistically significant growth pattern, so it is crucial to obtain up to ten tree samples. Single sample records are susceptible to recording changes that were not regional, but sample specific, thus complicating any dataset.

Within the laboratory, cores and sections are attached to mounts, prior to sanding and polishing. Various different measures can be obtained from the rings. The most common is the ring width, allowing the identification of individual years. Tree dynamics mean that growth can be relatively fast in a juvenile, but will slow down as the tree matures. Thus, ring widths will typically be thinner towards the outer part of the trunk, following an almost exponential trend. To compare statistically the individual trees and their dataset that are growing over different periods (and thus reaching maturity at different times), the values must be standardized, using a least-squares growth trend curve to produce a variation around a mean (e.g. Baillie, 1995). Statistical measures (t-test) can then be undertaken to test the degree of correspondence to the master chronology (how closely the pattern of the wiggles and the unknown sample(s) match). Alternative measures include skeleton plotting where individual ring widths are drawn on graph paper (an example of the method can be seen at http://tree.ltrr.arizona.edu/skeletonplot/introcrossdate.htm) or tree ring density using image analysis to extract cell detail (Briffa *et al.*, 1996). The choice of measurement is often dependent on the location of study. For instance, extreme environmental changes recorded in the USA datasets allow the use of skeleton plots, but the changes in European oaks are generally considerably less, necessitating the use of ring widths (Hillam, 1979).

Potential problems

For precise ages to be obtained from dendrochronology, it is crucial that the trees sampled record annually-resolved rings. In some tropical regions, many species can continue growth throughout the year and so should be avoided. Furthermore, trees growing in areas that are susceptible to environmental stresses, and therefore highly sensitive to annual change (producing characteristic rings rather than complacent patterns), are more likely to record regional patterns. Unfortunately, however, variations in one growing season can generate false rings (where growth is interrupted due to poor conditions but then resumes in the same year when conditions are optimum) or missing rings (where conditions are poor throughout the year resulting in no growth). Multiple samples spanning the period of interest can, however, identify such problems. Where the outer sapwood is missing, estimates must be made of the typical number of associated rings and then added on to derive the age of a tree's death. In British oak, for instance, a figure of 30 is typically given (Hughes *et al.*, 1981), but the ages thus obtained must be considered a minimum.

4.3.4.2 Varve dating

Principles

Varve dating is the exploitation of annually-deposited laminae that form as couplets within lake (and in some cases marine) anaerobic sediments. If truly annual, varves provide a powerful chronological tool for placing environmental change within a calendar timescale. The earliest use of varves for dating was in the early twentieth century by De Geer who suggested that the observed couplets of silt and sand could be associated with glacial melt. During spring and summer, glacial outwash flows into proglacial lakes and the coarser material settles relatively rapidly to form a first unit of sand. Subsequent reduction of glacial melt later in the year results in the eventual deposition of finer sediment that had previously been held in suspension, thus forming a second silt horizon. As varves are a function of the degree of glacial melt (Sander *et al.*, 2002), the thickness of individual varves can vary on a yearly basis, reaching up to several centimetres. Measurement of glaciolacustrine varved sequences across Scandinavia allowed the development of floating chronologies, comprising distinctive patterns of varves that could be used to cross-date and derive a master chronology, in principle similar to dendrochronological methods. By measuring these couplets, De Geer was able to reconstruct ice retreat across Scandinavia over the last 12 ka (De Geer, 1912).

Not all varves, however, are formed in glacial environments. In more temperate environments, sedimentation and biomass production can vary seasonally. These variations include organic productivity (including diatom blooms) and iron oxide precipitation (e.g. Simola *et al.*, 1981; Peglar *et al.*, 1984). Although traditionally thinner than glaciolacustrine (typically an order of magnitude smaller), non-glacial varves provide the opportunity for high-precision chronological control in areas distal to past glacial activity. The Swedish chronology has been modified since De Geer (Wohlfarth, 1996), and varve chronologies have since been developed elsewhere in the world, including the USA (Stihler *et al.*, 1992), Japan (Kitagawa and van der Plicht, 1998), Germany (Brauer *et al.*, 1999) and Turkey (Landmann *et al.*, 1996). The results generated have been used to derive measures of changing cosmogenic nuclide productivity (Snowball and Sandgren, 2002), dating of individual tephra horizons (Zillén *et al.*, 2002), comparison to relatively short-lived radionuclides (Stihler *et al.*, 1992; Nijampurkar *et al.*, 1998), radiocarbon dating and calibration (Kitagawa and van der Plicht, 1998) and environmental change (Peglar, 1993).

Excellent overviews of varve dating can be found in O'Sullivan (1983) and Verosub (2000).

Method

If the varves are exposed in the field and easily observed by the naked eye, it is possible to mark the thickness of individual horizons by laying a strip of paper against the exposure. This may prove problematic in dry conditions as the boundaries may not be clear. Conversely, if too wet, the silt horizons may be more easily eroded compared to sand horizons, biasing the record. If these problems are encountered, samples can be taken using monoliths or small blocks and then dried back in the laboratory. If coring is required, it is crucial to obtain samples that have not been disturbed. In many instances, several coring methods may be needed, including freeze drying, gravity and Livingstone corers (Lamoureux, 2001). Samples are returned to the laboratory where they can be cleaned directly or impregnated with an epoxy resin and turned into a thin-section prior to measurement (Tiljander *et al.*, 2002). Analysis of the individual varves can then be undertaken by a range of methods, including scanning electron microscope, photographs and x-rays that can be investigated by image analysis (Cooper, 1997). Crucially, sedimentological and/or biological investigations must be undertaken to determine whether the cores obtained are truly varved (Card, 1997). Errors in the chronologies can be determined by the variance between replicate varve counts over key sections and the entire core (Lamoureux, 2001). This can be reduced by cross-dating to derive a continuous chronology from individual core sections (typically using skeleton plots available to dendrochronologists). Such an approach allows the identification of missing parts of sequences (Zolitschka, 1996; Lamoureux, 1999).

Potential problems

Sections should be investigated for missing or disturbed horizons that, if ignored, can introduce significant errors into the chronology (e.g. Wohlfarth, 1996). Where varved sections are disturbed, deviations between the counts of varves can vary by up to 25 per cent, but in ideal circumstances can be as low as 3 per cent (Ojala and Saarnisto, 1999). Where a varved sequence is considered 'floating' (for instance where the upper part of the sediments are disturbed), comprehensive radiocarbon dating of the laminated sediments can be undertaken to wiggle-match ^{14}C to global atmospheric fluctuations (Kitagawa and van der Plicht, 1998).

4.3.5 Radiometric dating

4.3.5.1 Radioactive decay

Radiometric dating methods are defined here as those techniques that measure changes in the isotopic composition (i.e. isotopic) or cumulative effects on minerals (i.e. radiogenic) of radioactive decay. To fully understand these methods it is crucial to understand the basics of radioactive decay.

An atom is the smallest individual particle that retains all the properties of a given chemical element. It is composed of electrons (negative charged particles) that can be seen as spinning in a defined path around a nucleus. The nucleus is the central portion of an atom containing protons (positive charge particles abbreviated to 'Z') and neutrons (neutral charge particles abbreviated to 'N'). Of relevance to radioactive decay is the mass number (A):

$$A = Z + N$$

An atom is generally neutralized by having the same number of protons as electrons. This is not always the case, however, and deficiencies or surpluses of electrons can occur, creating atoms that are positively (number of protons exceeds electrons) or negatively (number of electrons exceeds protons) charged (cation or anion respectively). Isotopes are atoms with the same number of protons (thus sharing similar chemical properties) but differing number of neutrons (thus having slightly different physical properties). In the scientific literature, the mass number is written as a superscript, preceding the symbol for the element. For instance, in the most simple case, hydrogen (H) has three different isotopes, including deuterium (^{2}H or D) and tritium (^{3}H) (Fig. 4.8). In some instances, the atomic number (number of protons) is given as a superscript, but for the practising environmental archaeologist this is rarely used.

If the combination of protons and neutrons in an isotope is unstable, radioactive decay can take place. A wide range of types of radioactive decay occur in nature that can be used for dating purposes. Of these, the most relevant are:

1 Alpha (α) particles – for nuclides with $A>58$, as well as ^{5}He, ^{5}Li and ^{6}Be, decay can take place by the emission of an alpha particle. In essence, this is an He nucleus (i.e. 2 protons, 2 neutrons, 0 electrons).
2 Beta (β) particles – the emission of an electron expelled from a nucleus as a negative beta particle. Several types of beta decay can take place. Most simply, the loss of the electron converts the neutron of the parent to a proton, increasing the atomic number by 1 but the mass number (A) remains unchanged. In addition, it is also possible to get positron emissions where a proton turns into a neutron and a positron (positively charged electron), with the latter being emitted. Thus, in the daughter, $Z-1$, $N+1$. The final type of beta decay is known as electron capture. Here, the nucleus captures an electron from the inner shell which basically turns a proton into a neutron. Electrons 'fall' from the outer shells producing a cascade of X-ray emissions.
3 Gamma (γ) rays – a powerful form of radiation but with no associated change in mass. In effect, the nucleus falls down to a lower energy state and, in the process, emits a high-energy photon (the gamma particle).

An atom that undergoes atomic transformation is termed the parent/mother nuclide while the product is referred to as the daughter nuclide. An important term in decay is the half-life ($t_{0.5}$) which is the period of time required to reduce a given quantity of parent nuclide by one half:

$$t_{0.5} = \frac{\ln 2}{\lambda}$$

where λ is the decay constant of the parent nuclide. As a result, radioactive decay is exponential (Fig. 4.9).

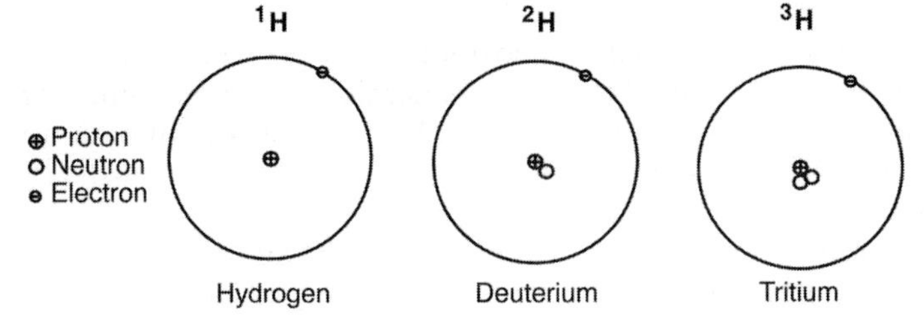

FIGURE 4.8 Isotopes of hydrogen.

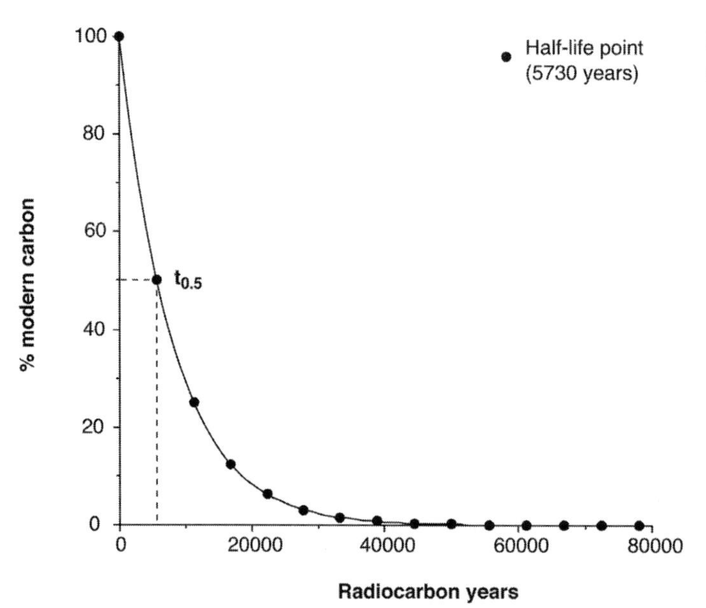

FIGURE 4.9 Exponential radioactive decay of ^{14}C.

Typically after around eight half-lives, background levels of the radioisotope are reached and it generally cannot be used for dating purposes. This, however, is not always the case, with increasing technological developments allowing ever more sophisticated detection limits and thus allowing greater dating ranges using isotopic dating methods.

4.3.5.2 Radiocarbon

Principles

Radiocarbon dating has almost single-handedly transformed our understanding of the timing of events and rates of change in the archaeological and environmental records. In the mid-1940s, an American chemist Willard Libby proposed that a radioactive form of carbon (radiocarbon or ^{14}C) should be produced by the interaction of cosmic radiation (formed deep in space) with nitrogen gas in the upper atmosphere. He argued that as radiocarbon would rapidly be converted to carbon dioxide (CO_2) and taken up by plants during photosynthesis, it should exist in all organic matter by moving up the food chain (Fig. 4.10). By the end of the 1940s, Libby (in collaboration with other workers) demonstrated that the radiocarbon content of the air was uniform throughout the world (and therefore not restricted to specific regions) (Anderson *et al.*, 1947; Libby *et al.*, 1949) and made the first independent age estimates of key archaeological sites and samples (e.g. Arnold and Libby, 1951).

In contrast to hydrogen, carbon (C) has six protons and in its most common form (99 per cent of all carbon) has six neutrons, thus, ^{12}C. A less common isotope (but still stable) is ^{13}C, which forms approximately 1 per cent of all carbon, and has seven neutrons. The naturally occurring radioactive isotope of carbon, ^{14}C (10^{12} of all carbon), however, has six protons and eight neutrons. Radiocarbon (^{14}C) decays via the emission of a β-particle to form ^{14}N. When originally measured by Libby, the half-life of radiocarbon was calculated to be just over 5,720 ± 47 years (Arnold and Libby, 1949). Several other workers, however, reported this to be wrong and suggested a slightly lower value of around 5,568 ± 30 years. The latter half-life was subsequently adopted and by the time it was realized to be wrong (the correct value of 5730 ± 40 years is virtually the same as Libby first maintained) it was too late to change the value, as the earlier age calculations would no longer be comparable to future determinations. We therefore continue to use the 'incorrect' value of 5,568 years and refer to it as the Libby half-life to differentiate it from the real one of 5,730 years. In practice, it makes little difference

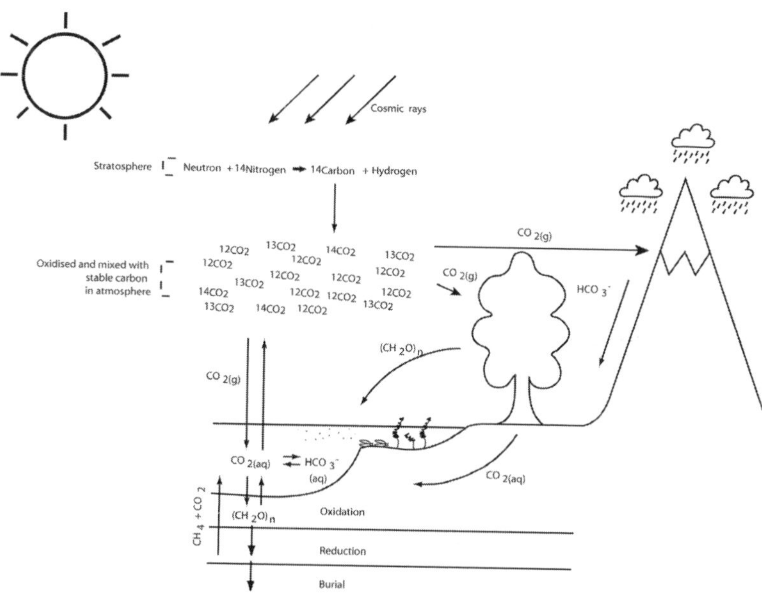

FIGURE 4.10 Schematic showing ^{14}C production and cycling. (Turney, 1999)

as all radiocarbon ages are calculated using the Libby half-life and when calibrated to calendar years (see below), a suitable correction is made. As a result of exponential decay, after about eight to nine half-lives, the amount of radiocarbon in a material is so low as to be virtually non-existent, limiting the dating range to somewhere between 60 and 40 ^{14}C ka BP, depending on the samples being submitted and the laboratory used.

Method

When sampling, the radiocarbon community strongly encourages users to contact a laboratory and take advice on sampling strategies. A typical radiocarbon age will be of the order of several hundred pounds or US dollars and it is in the users' interest to maximize the investment for scientific gain. A list of all operating radiocarbon laboratories is available at http://www.radiocarbon.org/Info/ with direct internet links to virtually all of the laboratories. Wherever possible, materials should be sampled under sterile, laboratory conditions, thereby minimizing any modern contamination.

During most of the past 50 years, radiocarbon dating has been dominated by 'conventional' methods, such as Gas Proportional Counting (GPC) or Liquid Scintillation Counting (LSC). For LSC, the sample is converted to benzene via CO_2, whereby beta emissions are detected by the addition of a scintillant, phosphor (butyl-PBD). The latter emits a photon of light when decay takes place which is detected by a photomultiplier tube in the LSC. In contrast, GPC uses just CO_2 that is introduced into a counting chamber where beta decay results in a current pulse that can be detected.

The basic premise of both these methods is the measurement of individual ^{14}C atom decay. The error term reported with ^{14}C ages is a measure of the extent of the scatter in decay (counts) over the duration of measurement. Statistically we know that given an infinite number of measurements of ^{14}C decay, the curve will assume a normal distribution. Thus, there is a 68 per cent chance that the true age will fall within the 1σ limits. For radioactive decay, the standard deviation is approximately given by the square root of the number of counts (decay) measured ($\sqrt{x}$) i.e. given a sufficient period of counting, a relatively high precision can be attained. Although such methods can allow high-precision ages to be obtained (typically ± 20 ^{14}C years at 1σ for samples of mid-Holocene age), relatively large amounts of material are required (c. 100 g of wood is required per sample). For the environmental archaeologist, such quantities of material are rarely available. As a result, considerably smaller (though still relatively large) samples can be dated by LSC (conventional counting), though the level of precision is inherently lower than that possible with larger samples. The size of the sample varies depending on the carbon content and degree of preservation, but a typical amount for conventional dating of wood is 20–30 g, charcoal is 5–20 g, peat is 50–60 g and shell is 35–40 g. Early concerns about different age reports of identically aged samples indicates there is some variability in standards between laboratories (e.g. Scott *et al.*, 1998) and users should check whether a laboratory meets international standards.

Since the late-1970s, however, it has been demonstrated that ^{14}C can be detected directly in organic samples of milligram size (in the form of graphite) using accelerators developed for nuclear physics research. This allows individual radiocarbon atoms to be counted prior to decay (Bennett *et al.*, 1977; Nelson *et al.*, 1977), thus providing the potential for high-resolution analysis. Recent technological advances have reduced the size of AMS systems and it is now possible to use 0.25 MeV systems for dating purposes. These systems require considerably smaller housing space than their higher energy counterparts and have proved extremely useful for the analysis of relatively small samples. Much like conventional methods, the size of the sample varies depending on the carbon content and degree of preservation, but a typical quantity is at least an order of magnitude smaller than that required by conventional methods.

When ages are reported, they are quoted as an age from year zero which is taken as AD 1950 (BP or Before Present), when the ^{14}C content of the atmosphere was approximately in equilibrium, prior to nuclear bomb testing. Ages are quoted with the measurement error only, and this is typically given at 1 standard deviation (σ) i.e. 68 per cent confidence limits. It must be remembered that this does not include uncertainties regarding the sample integrity, contamination, calibration etc. and should not be taken as a total error value (see below).

Potential problems

In the method, several assumptions are made. These include: (a) the atmosphere has had the same ^{14}C concentration in the past as 'now'; (b) the biosphere has the same overall concentration as the atmosphere, i.e. rapid mixing between reservoirs; and (c) the death of a plant or animal is the point at which ^{14}C ceases to exchange with the environment. We now know that these assumptions are not always true. Several factors must be considered when interpreting radiocarbon age and at no time should ages be uncritically compared directly to calendar (sidereal) dating methods.

When carbon is fixed by living material, the lighter isotopes are preferentially taken up. As a point of reference, the stable isotopic ratio of ^{13}C to ^{12}C ($\delta^{13}C$) is routinely measured. The degree of discrimination (or fractionation) of ^{12}C over ^{14}C is approximately twice that over ^{13}C and hence can be easily corrected for by using $\delta^{13}C$. In radiocarbon dating, an arbitrary standard of −25‰ is taken (typical of C3 plants) and ages are corrected to this value. This is crucial, for instance in marine and lacustrine environments, where carbonates will typically have a $\delta^{13}C$ of approximately 0‰ creating an apparent age difference of 400 ^{14}C years to a plant living at the same time (the carbonate sample will appear 400 years too young). A difference of 1‰ in $\delta^{13}C$ is therefore equivalent to 16 ^{14}C years. The $\delta^{13}C$ value also provides a handle on whether the material was truly in equilibrium with atmospheric ^{14}C values. For instance, if geologically-old carbon (e.g. coal and limestone) had been used by living matter as a carbon source (especially in aquatic environments) or formed part of a bulk sample submitted, erroneously old ages may be reported, despite being corrected to −25‰.

In marine environments, things are complicated further by the global thermohaline circulation. In areas of intense ocean upwelling at the extreme ends of the conveyor belt, such as in tropical Pacific areas, old waters release relatively old CO_2 that has experienced ^{14}C decay during the course of its travel from deepwater formation in the North Atlantic and Southern Ocean. Such upwelling can produce marine reservoir ages of ocean samples of the order of up to 1600 ^{14}C years too old (Bard *et al.*, 1990). In areas of less extreme upwelling, marine reservoir ages of ~400 ^{14}C years are typically reported, though there appears to be considerable spatial (Sikes *et al.*, 2000; Reimer and Reimer, 2001; Siani *et al.*, 2001) and temporal (Reimer *et al.*, 2002) variability. In some instances, the upwelling of old waters has even been shown to have a significant effect on terrestrial vegetation growing in close proximity (Hua *et al.*, 2000).

In terrestrial environments, percolating humic acids (Tornqvist *et al.*, 1990; Moar and Suggate, 1996), rootlets and bacterial deposits (Gove *et al.*, 1997; Turney *et al.*, 2000) have all been identified as potential causes of young contamination and must be avoided as much as possible in advance by careful sampling and/or screening. The converse is true of older material reworking into deposits, artificially ageing the sample (McGlone and Wilmshurst, 1999; Turney *et al.*, 2000), though the identification of such material is problematic.

Radiocarbon years, unfortunately, do not directly equate to calendar years because of the changes in the cosmic ray flux, solar intensity and changes in the carbon cycle (including thermohaline circulation strength) (Stuiver *et al.*, 1991; Goslar *et al.*, 2000; Bond *et al.*, 2001). Thus, a large proportion of the calibration curve with which the international scientific community converts radiocarbon ages to calendar years was constructed from measurements on absolutely-dated tree rings (Stuiver *et al.*, 1998). The present state of the calibration curve has reached 11,857 dendro years B.P, using a combination of Irish and German oak and pine chronologies measured over the years at Queen's University Belfast and University of Washington (Seattle) (Stuiver *et al.*, 1998). In this period, industrial (diluting the ^{14}C content of the atmosphere), atomic (enriching the atmosphere in ^{14}C) and long-term drift can be detected in the difference between radiocarbon and calendar years. The latter is as much as ~1500 years by 12,000 dendro years, but also includes periods of constant radiocarbon age ('plateaux') that may represent several hundred calendar years.

Unfortunately, prior to 12,000 dendro years, material is scarce in Europe due to limited forest development during the Last Glacial Maximum, though a near complete calibration curve has been reported from Germany (Friedrich *et al.*, 2001). As a result, alternative methods have been applied to attempt to identify longer-term ^{14}C fluctuations, all with some inherent problems. These methods include U-series dated corals (Bard *et al.*, 1990) and carbonates (Schramm *et al.*, 2000), lacustrine (Kitagawa and van der Plicht, 1998) and marine (Hughen *et al.*, 2000, 2004) varved sequences, geomagnetic records (Laj *et al.*, 2002), and marine planktonic foraminifera linked to the $\delta^{18}O$ Greenland ice-core records (Voelker *et al.*, 2000). In addition, there has been an intriguing recent discovery of ^{14}C excursions; the most intense of these is recorded in a speleothem record from the Bahamas (Beck *et al.*, 2001) that occurs between 44,300 and 43,300 cal BP and appears to cause a 10,000 year offset with radiocarbon ages. The calendar ages in this record are provided by paired TIMS U and Th analyses and although there are uncertainties regarding the degree of dead carbon contribution to the speleothem, similar excursions (albeit smaller) are also recorded in paired U-series and ^{14}C dating of corals from Papua New Guinea (Yokoyama *et al.*, 2000). If corroborated by other studies, not only will age inversions be expected in both marine and terrestrial sequences, but these might appear close to the limits of radiocarbon dating and therefore be incorrectly interpreted as not reliable. Nevertheless, there is no universally accepted radiocarbon calibration curve, and the development of a continuous dendro-dated calibration curve will be the long-term goal for the radiocarbon community. In the meantime it is advised that, if ages are calibrated beyond 12,000 years, the curve used should be reported and the original ages provided for future users.

For those samples that fall within the accepted radiocarbon calibration curve, numerous online or freely downloadable programs are available to the user. These include CALIB 4 (http://depts.washington.edu/qil/calib/), OxCal (http://www.rlaha.ox.ac.uk/orau/06_ind.htm) and Groningen (http://www.cio.phys.rug.nl/HTML-docs/carbon14/cal25.html). These programs provide calibrated ages for individual and, in some cases,

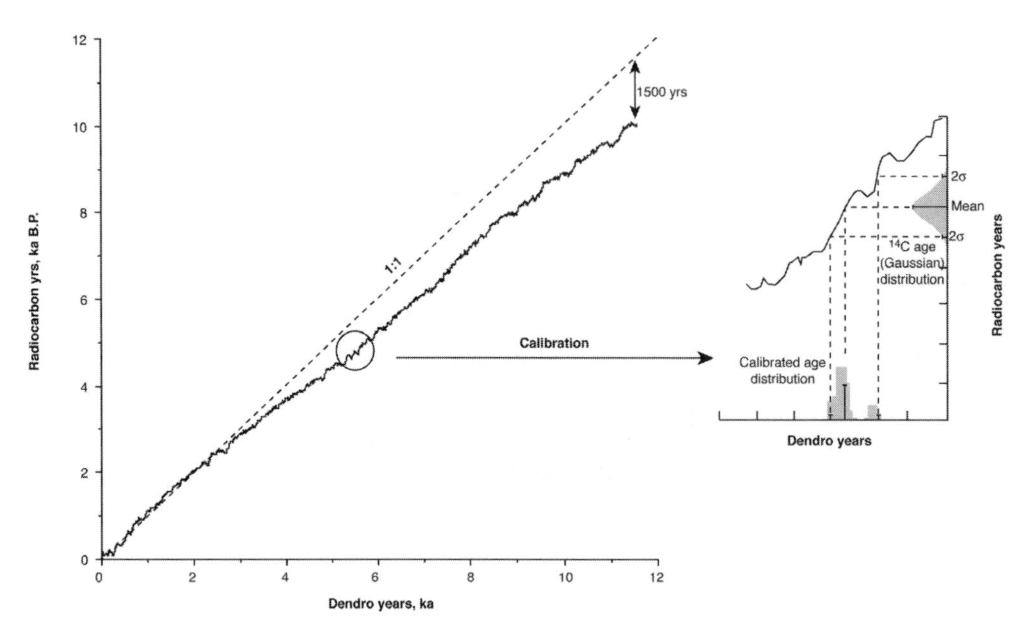

FIGURE 4.11 Holocene radiocarbon calibration and schematic of age calibration.

multiple ages, as well as errors. As can be seen in Fig. 4.11, the errors of calibrated ages, unlike radiocarbon ages, are not symmetrical due to the complexities of the calibration curve, often resulting in significantly larger errors than the raw radiocarbon ages. One of the statistical developments to attempt to limit this effect is the application of Bayesian statistics to a comprehensive suite of ^{14}C measurements (Buck *et al.*, 1991). This allows the incorporation of other ages and stratigraphic information to identify statistical outliers and constrain the precision of individual radiocarbon ages for calibrating. Ages can be evaluated using a Bayesian form of Markov Chain Monte Carlo analysis available on the OxCal Calibration program (Ramsey, 1998) that allows the analysis to be constrained by recognizing that the radiocarbon ages were obtained in stratigraphic order. Thus, a set of prior (unconstrained) and posterior (constrained) probability distributions can be generated, with the difference expressed as an agreement index. Strong agreement would give a value of 100 per cent or more, with rejection of individual ages and series of ages set at 60 per cent (Ramsey, 1998). Incorporation of independently derived ages (e.g. luminescence) allows further precision.

As a result of technological developments, there has been an increase in the number of AMS laboratories and the resultant radiocarbon measurements by this method (though with an associated closure of approximately 40 per cent of conventional laboratories). A key problem is the time taken to prepare and graphitize samples that can then be fed into the AMS system, though gas systems are now available. In addition, there are sources of errors that limit the practical levels of precision attainable with AMS (compared to LSC), including target size and positioning, chemical blank variation, variable sample contamination and the measurement of interdependency (memory effects) (Jull *et al.*, 1997).

The degree of sample contamination and verification of the reliability of the ages becomes more difficult to quantify as the age of the sample increases. In practice, this means that many laboratories will only quote ^{14}C ages to an upper limit of 40 ka ^{14}C BP with ages greater than this generally considered to be indistinguishable from background levels. The so-called 'radiocarbon barrier' and the difficulty of making reliable age estimates at <1 per cent modern carbon levels has hindered progress in a range of studies from many disciplines. Recently, a chemical pretreatment and stepped combustion vacuum line has been developed (Bird, *et al.*, 1999, 2003; Turney *et al.*, 2001a). Samples are given a sequential pretreatment with HCl, HF and NaOH followed by a $K_2Cr_2O_7/H_2SO_4$ oxidation at 60°C for 14 hours (Bird and Gröcke, 1997). Combustions are performed in a vacuum line that is insulated from the atmosphere by a second backing. Three temperatures are used (330°, 630° and 850°C) with a graphite target produced from the CO_2 evolved during each combustion step for ^{14}C AMS analysis (acid-base-wet oxidation with stepped combustion or ABOX-SC). In this way, the progressive removal of any contamination can be monitored, and a high degree of confidence can be placed on the final age assessment. The total pretreatment, combustion, graphitization and measurement blank for the procedure is equivalent to $0.04 \pm 0.02\%M$ (1σ, n = 14), or an 'age' of approximately 60,000 years for a 1 mg graphite target, significantly greater than 40 ka ^{14}C BP.

4.3.5.3 Luminescence and electron spin resonance

Principles

Luminescence and electron spin resonance are often collectively referred to as trapped charge dating (Grün *et al.*, forthcoming) and are based on the measurement of the accumulation of electrons within minerals following burial. The term luminescence was first suggested to differentiate light emitted by minerals following heating (incandescence) to that caused by an alternative source (luminescence). Daniels

Box 4.2 Radiocarbon dating early agriculture

Prior to the 1980s, the conventional view of agricultural origins indicated that farming was developed by hunter-gatherers who settled in the Fertile Crescent within the Near East sometime around 10 ka BP to sow cereal grains. This timing dovetailed with the termination of the Pleistocene (and the onset of ameliorating climatic conditions) and fell in nicely with the development of agriculture being a prerequisitite for civilization (Pringle, 1998). Much of this evidence relied on the dating of charred plant seeds, buried bones, and starch grains on stone tools. More recently, however, a large amount of new data has become available, overturning many of our earlier ideas concerning agricultural development, much of which has been undertaken using AMS dating associated with microfossils.

In the Near East, radiocarbon ages of charred rye grains indicate cultivation commenced by at least 11.1 ka BP, probably in response to a decline in wild plants due to deteriorating climatic conditions associated with the Younger Dryas stadial (Hillman *et al.*, 2001). Studies now indicate that in the New World, the earliest cultivation was not in one specific centre. Previous work using visible remains appeared to suggest that agriculture developed around 9 ka BP in Mexico with the exploitation of squash several thousand years before maize (Smith, 1997). More recently, however, by extracting phytoliths of domesticated squash and gourd within Ecuadorian archaeological sites, direct radiocarbon dating of the microfossils has been obtained by measuring the ^{14}C content of the occluded carbon (Kelly *et al.*, 1991). Ages as early as 10.1 ka BP have now been obtained using this approach (Piperno and Stothert, 2003), indicating domestication in lowland South America was independent and may have been slightly earlier than Mesoamerica. In contrast, in the middle reaches of the Yangtze river, the identification of cultivated rice has been dated to 14.0 ka BP (Yasuda, 2002), significantly earlier than the agricultural developments in the Near East.

(1953) was the first to suggest that luminescence emissions might be used a geochronological tool.

Luminescence dating is an established method for determining the time elapsed since quartz and feldspar grains were last subjected to heat or light. Two principal methods exist. By heating the sample in a laboratory to induce light emissions, thermoluminescence dating (TL) has been used extensively to date pottery (Kennedy and Knopff, 1960; Aitken *et al.*, 1964), burnt artefacts (Göksu and Fremlin, 1972; Valladas *et al.*, 1987, 1988; Mercier *et al.*, 1991; Henshilwood *et al.*, 2002) and unheated sediments (Shelkoplyas and Morozov, 1965; Li and Sun, 1982; Roberts *et al.*, 1990; Rendell, 1992). Alternatively, light can also induce subsequent luminescence from minerals (Huntley *et al.*, 1985). This approach is referred to as photoluminescence, or more commonly, optically stimulated luminescence (OSL), and has been extensively used to date sediments in a range of archaeological (e.g. Roberts *et al.*, 1994; Roberts, 1997; Henshilwood *et al.*, 2002) and environmental settings (Lian and Shane, 2000; Radtke *et al.*, 2001). This work has been considerably expanded following the development of single aliquot and grain methods allowing precise measurements of OSL to be made (Duller, 1994). Similar to luminescence, electron spin resonance (ESR) allows the measurement of free electrons but here it is achieved by utilizing a magnetic field to measure the trapped signal. This approach was first suggested as a dating method by Zeller *et al.* (1967) and subsequent work has largely focused on tooth enamel which has had a considerable impact on our understanding of modern human evolution (Schwarcz *et al.*, 1988, 1989; Stringer *et al.*, 1989; Grün *et al.*, 1990, 1997, 1998; Grün and Beaumont, 2001).

All trapped charge dating methods are based on the fact that natural crystals are not perfectly formed. To best understand the principles, the energy levels of many minerals (e.g. quartz, feldspar) can be shown as an energy band model (Fig. 4.12). Here, electrons can be held at two energy levels, a valence band that has a ground state, and a conduction band that has a higher energy level (see Fig. 4.12). Minerals, when buried, are exposed to the radiation flux

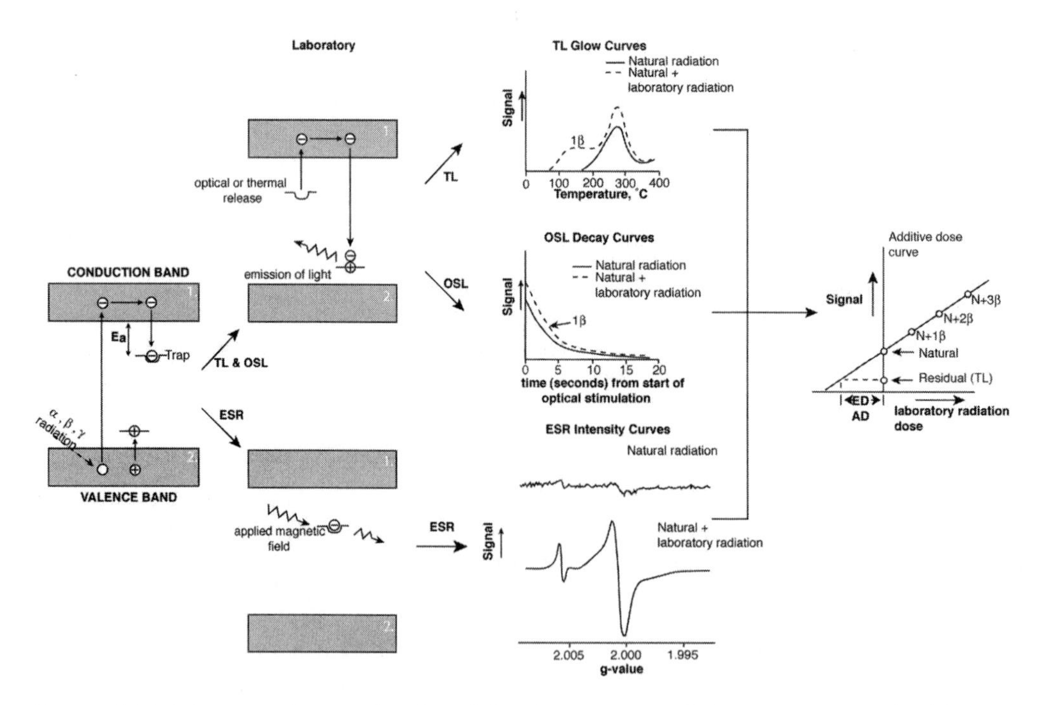

FIGURE 4.12 Schematic of energy band model and summary of different stages involved in age estimation using thermoluminescence, optically stimulated luminescence and electron spin resonance.

(the dose rate) from the radioactive decay of ^{238}U, ^{235}U, ^{232}Th (and their daughter products) and ^{40}K in the deposit, and from cosmic rays. When ionized, electrons are sufficiently excited to move out of the valence band and transfer to the conduction band. Typically, the electrons immediately return to the valence band. In some instances, however, where defects ('traps') exist within the crystal lattice, the electrons are trapped in the conduction band, preventing them from returning to the ground state. Thus, minerals accumulate electrons due to continual bombardment of alpha and beta particles and gamma rays. To exploit this phenomena for dating purposes, it is crucial that the traps are initially empty immediately prior to deposition ('zeroed'). If a mineral is heated to a high enough temperature (ca. 400°C) or exposed to sufficient sunlight, some or all of the traps will be emptied of electrons. The optimal depositional environment for OSL is that which permits lengthy exposure to sunlight. In the case of faunal remains, very few or no electrons are trapped when alive, so the zeroing event for ESR can be assumed to be the time of death. By measuring this dose rate and the amount of electrons that have become stored over time – by heating in the laboratory (TL), exposing to a known wavelength (OSL) or exposing to a fluctuating magnetic field (ESR), we can determine the time since the sample was last zeroed.

Extensive, detailed reviews of methods and applications of luminescence and ESR dating are given by Aitken (1985), Grün (1989), Wintle (1993), Duller (1996), Roberts (1997), Forman *et al.* (2000), Schwarcz and Lee (2000), Blackwell (2001), Lian and Huntley (2001) and Grün *et al.* (forthcoming).

Methods

To determine an age using luminescence or ESR, certain criteria must be met (Grün *et al.*, forthcoming). These include:

- the traps are fully zeroed or have an inherent level that can be experimentally determined;
- the signal intensity increases in proportion to the dose received;
- the number of traps is constant or changes in a predictable manner;
- the signal is not influenced by sample preparation.

To sample material for luminescence dating, most investigations are undertaken on stratigraphic exposures (Lian and Huntley, 2001). Typically the face is cleaned and a lightproof container gently hammered in to sample up to 1kg of material, though in extreme cases, several kilograms may be necessary (Lian and Huntley, 2001). If the sediment is sufficiently cohesive, the sample can be collected as a block. If loose, the material must be collected at night after clearing the face of material that may be zeroed by contemporary sunlight. The sample must be protected at all times from sunlight or moisture loss during transportation to the laboratory. All samples are prepared in the laboratory under subdued light (normally red or orange), where organic matter and carbonates are removed, prior to sieving to extract a specific size range (typically 4–11 μm, 90–125 μm or 250–350 μm) for analysis. Detailed sample preparation methods in the laboratory are given by Wintle (1997).

The rate at which free electrons are produced and traps filled is proportional to the concentration of radioactive elements in minerals and surroundings. Thus, to make a luminescence age estimate, the following equation is used:

$$\text{Luminescence age} = \frac{\text{Equivalent dose (ED)}}{\text{Dose rate (D)}}$$

where the ED is the radiation dose received by the sample since zeroing and the D is annual exposure to radiation through the combined 'internal' radioactive content of the sample (i.e. concentration of U, Th and K) and 'external' radioactive content. The radiation dose is measured using the SI unit Gray (Gy).

If the stratigraphy of the site is complex, it is best to use an *in situ* gamma ray spectrometer to determine the external dose rate for D, though it is possible to sample all different sediment types within 50 cm in conjunction with detailed field notes. Alternatively, *in situ* dosimetry can be undertaken for the external rate by the use of capsules that can be left to determine a longer-term record of the dose rate (Bøtter-Jensen *et al.*, 1997). If the sample is within a few metres of the surface, details of surface features including past geomorphological processes (e.g. former ice cover) should be noted to evaluate the dose rate associated with cosmic rays, the latter being dependent on altitude and latitude (Prescott and Hutton, 1994). Typically, a subsample of the sediments or artefact is used to determine the internal radioisotopic concentration.

Numerous methods exist for the determination of the ED, including total bleach, partial bleach, regeneration and additive-dose methods (Lian and Huntley, 2001). In samples that have been exposed to heat and are to be dated using the TL method, the ED is typically calculated using the additive-dose method (Fig. 4.12). Here, several aliquots are deposited on small steel discs and irradiated with known doses (using beta or gamma rays). The TL signal is recorded as a function of temperature (generating a series of glow curves). Each glow curve produces multiple peaks that correspond to overlapping traps. The higher the temperature, the greater the energy that is required to release the electrons from their respective traps. In practice, the lower temperature traps are inherently unstable as the amount of energy required for electron release is relatively small. Thus, the most stable temperature signal associated with one trap (usually in the range of 200–400°C) is used to plot the luminescence signal associated with the natural and known irradiation values. These values are then extrapolated back to where the luminescence signal would be zero, thus determining the equivalent dose, i.e. the dose received by the sample from time of deposition to collection.

For TL dating where the sample was exposed to sunlight, a proportion of traps are light-insensitive, resulting in an inherent signal that was not removed prior to deposition. Such samples typically require a total (Singhvi *et al.*, 1982) or partial bleach (Wintle and Huntley, 1980) approach. In the former, the sample is bleached by being exposed to laboratory light comparable to natural sunlight, to determine the inherent signal carried over in the sample during deposition. In some instances, however, traps may have a range of light sensitivities and not all of them will have been emptied by this method. A partial bleach may therefore be required, where the sample is exposed to only a brief, limited wavelength of light. Finally, the regenerative method can be used. Here, the natural luminescence of a sample is first measured and then set to zero prior to irradiating and subsequent measurement until the signal is greater than the original (natural) value at the time of sampling, but this does appear consistently to underestimate the sample age.

OSL exploits the fact that by using laboratory light of a specific wavelength, the most light-sensitive traps can be preferentially sampled, thereby allowing the measurement of sediments that have been exposed to just a few seconds of light prior to burial. Originally lasers were used to generate green light (Ugumori and Ikeya, 1980) but now relatively cheap diodes can be used to produce the desired wavelength (Lian and Huntley, 2001). Multi-aliquot samples are typically irradiated with known doses as in TL but then preheated to remove electrons that have since transferred to shallow, short-lived traps. Unfortunately, however, preheating can also transfer samples to more stable traps. Hence, prior to preheating, a second set of aliquots are exposed to laboratory light following irradiation to empty the light-sensitive traps. These aliquots are then run alongside the natural samples to quantify and correct for the extent of the transfer. Samples are then exposed to a specific wavelength (e.g. green, red or blue) for typically 0.1 seconds or less and the degree of luminescence is measured over time, producing a decay curve (Fig. 4.12). For samples where the traps are approaching dose saturation (i.e. the signal is starting to level out at high laboratory doses), a combination of additive and regenerative methods may be required ('the slide method'; Prescott *et al.*, 1993). More recently, the development of single aliquot (Duller, 1994) and single grain (Liritzis *et al.*, 1994; Murray and Roberts, 1997) OSL allow the analysis of just a few milligrams and less of material that might be considered unsuitable for multi-aliquot methods (Roberts, 1997). Luminescence ages can typically range from 1 to 150 ka, though in rare cases ages in excess of 800 ka have been reported (Lian and Huntley, 2001).

For ESR, faunal samples should be collected in a watertight bag with associated sediment. In the laboratory, the outer 50 μm surface of the sample is physically removed to isolate any influence of external alpha irradiated material (Grün *et al.*, forthcoming). The remaining material is typically ground to a diameter of 200–400 μm (Grün, 1989). To determine an ESR age, the following equation is used:

$$\text{ESR age} = \frac{\text{Accumulated dose (AD)}}{\text{Dose rate (D)}}$$

where the accumulated dose is the equivalent of the ED when determining a luminescence age. For D, similar methods are used as for luminescence. In teeth, however, virtually no K and Th are present, thus only the U concentration has to be determined for the internal dose (Grün and Stringer, 1991). For ESR, a further complication exists in that teeth and bones exhibit post-depositional uranium uptake and cannot be easily determined. Until recently, two models were used to calculate D: early U-uptake (EU) where uranium accumulated shortly after burial; and linear U-uptake (LU) for continuous uranium accumulation. Where the uranium content is relatively low, typically the differences between the two age models are relatively small. In high uranium content samples, however, there can be considerable differences (in extreme instances the LU age may be twice that of the EU), especially where the external dose rate is lower. More accurate uranium modelling may be achieved by combined U-series and ESR dating (Grün and McDermott, 1994), with considerable potential through the use of LA-ICP-MS (Eggins *et al.*, 2003). U-series ages are calculated on the assumption that uranium uptake is early and the system remains geochemically closed afterwards. Thus, U-series ages should agree with ESR-EU ages, while any leaching of uranium should lead to younger apparent ages.

To determine the AD, samples are placed in a cavity within an ESR spectrometer and exposed to microwaves within a strong magnetic field. Unpaired electrons (paramagnetic centres) exhibit an induced magnetic current that is created by their own spin. Thus, the trapped electrons have a magnetic moment which is oriented parallel to the magnetic field (ground state). When the magnetic field is changed, the orientation can flip into an excited state that is in the opposite direction (resonate). Resonation takes place when a magnetic field exerts the same energy as the difference between the two spin states. When the electrons resonate, microwave energy is absorbed: the greater the absorption, the greater the number of electrons in a trap. To allow comparability between different ESR units, the spectrum is described by a g-value (Fig. 4.12), which is a measure of the ratio of microwave frequency to magnetic field strength (Grün and Stringer, 1991). Although traditionally less precise than luminescence (<20 per cent at 1σ), ages from as young as 0.5 ka through the entire Quaternary have been reported (Blackwell, 2001), making it an important geochronological tool, particularly within palaeoanthropological contexts.

Potential problems

Many of the errors are common to all trapped charged dating techniques. Of particular note is the assumption that each radioisotope is in secular equilibrium (i.e. the parent and daughter isotopes have not migrated during the period of the deposition). The luminescence samples themselves are in radioactive equilibrium but a common problem for all trapped charge methods is differing moisture and organic levels (spatially and temporally) in the deposit. Organics can absorb uranium (the basis of U-series dating of organic sediments) and water attenuates the dose rate, thus higher water contents increase the final age estimates. This is an important consideration in areas that are known to have experienced significant climate change during and since sample deposition. Furthermore, recent applications of single grain analyses using OSL have allowed the identification of contaminated deposits through weathered rubble and/or bedrock (with filled traps) (Roberts *et al.*, 1998a). Thus, the incorporation of such grains will bias the sample to artificially older ages when using multi-aliquot methods and must be carefully considered when sampling.

Traditionally luminescence has had a reputation of relatively poor precision. Recent developments, however, have made important advances in producing high-precision, accurate luminescence age determinations. On site gamma spectrometry, *in situ* dosimetry capsules, single aliquot regeneration protocol, the use of blue light diodes for analysis, high-precision calibration and duplicate analyses reduce the errors for OSL analyses to as low as 3 per cent at 1σ.

For ESR, one of the greatest uncertainties is the open system modelling required to derive an age. In some instances, EU and LU ages are significantly different, making it problematic to assess the accuracy of each without independent age control. The recent developments in LA-ICP-MS (Eggins *et al.*, 2003), however, will hopefully greatly improve the situation in the near future. Also, it must be noted that although ESR can be used on carbonates, the accuracy is far greater using direct isotopic methods (e.g. U-series) (Turney *et al.*, 2001a).

4.4 Multi-dating examples

Although much of this chapter has focused on individual dating techniques, it is crucial, particularly in environmental archaeology, that multi-dating of sequences is undertaken. Not only does this provide greater confidence that the chronological models are robust, but it also allows a greater confidence in objectively testing hypotheses of change. The following cases are examples of situations where multi-dating has been undertaken at the micro-, meso-, macro- and megascale levels.

4.4.1 Microscale – early hominin settlement at Devil's Lair, Australia

Devil's Lair (30°9′S, 115°4′E) is a single chamber cave (floor area 200 m^2) formed in the Quaternary dune limestone of the Leeuwin-Naturaliste Ridge, 5 km from the modern coastline. The stratigraphic sequence of the floor deposit consists of 660 cm of sandy sediments, with >100 distinct layers, intercalated with flowstone and other indurated deposits (Fig. 4.13). The sandy sediments have been interpreted as being washed into the cave from the brown sandy soils and dunes outside the entrance, probably during periods of exceptionally heavy rainfall (Dortch and Merrilees, 1973), while the flowstone layers probably represent periods of negligible sedimentation (Dortch, 1984).

Archaeological evidence for intermittent human occupation extends down to layer 30 (~350 cm depth), with hearths, bone, and stone artefacts found throughout. The lower part of layer 30 represents a fan of redeposited topsoil that accumulated rapidly after widening of the cave mouth, and contains the earliest evidence for occupation of the cave itself. Below layer 30, only a half-dozen stone artefacts have been identified, including a single specimen each from layers 32–35, 37 and 38. No artefacts have been found below layer 38.

The original chronology (Dortch, 1979a, 1979b, 1984; Dortch and Dortch, 1996) relied on liquid scintillation ^{14}C analysis of acid-washed or ABA (acid-base-acid) pretreated charcoal fragments. The redating program here included AMS radiocarbon dating of hand-picked charcoal fragments using ABOX-SC and ABA, emu eggshell ^{14}C, OSL, ESR and U-series.

The original radiometric ^{14}C results on charcoal (Dortch, 1979a, 1979b, 1984) provide a coherent chronology for the upper part of the Devil's Lair sequence. Many of them are consistent with the new ages for the upper part of the sequence based on AMS-^{14}C, OSL and ESR (Fig. 4.13). The early occupation of Devil's Lair is bracketed by the ages obtained from layer 28 (345 cm) and layer 39 (496 cm). Below layer 28 the original conventional ages do not increase systematically with depth and only one age is older than 35 ka BP. By contrast, both the ABA and ABOX-SC AMS-^{14}C ages from units 28 and below are all older than 35 ka BP. Whereas the ABOX-SC ^{14}C ages continue to increase systematically to the base of the sequence, the ABA ^{14}C ages plateau at 42–40 ka BP, suggesting that the latter samples have reached a 'radiocarbon barrier'.

The ABOX-SC radiocarbon results and the optical ages consistently overlap at 1σ. In particular, beyond 40 ka BP, all results are in correct

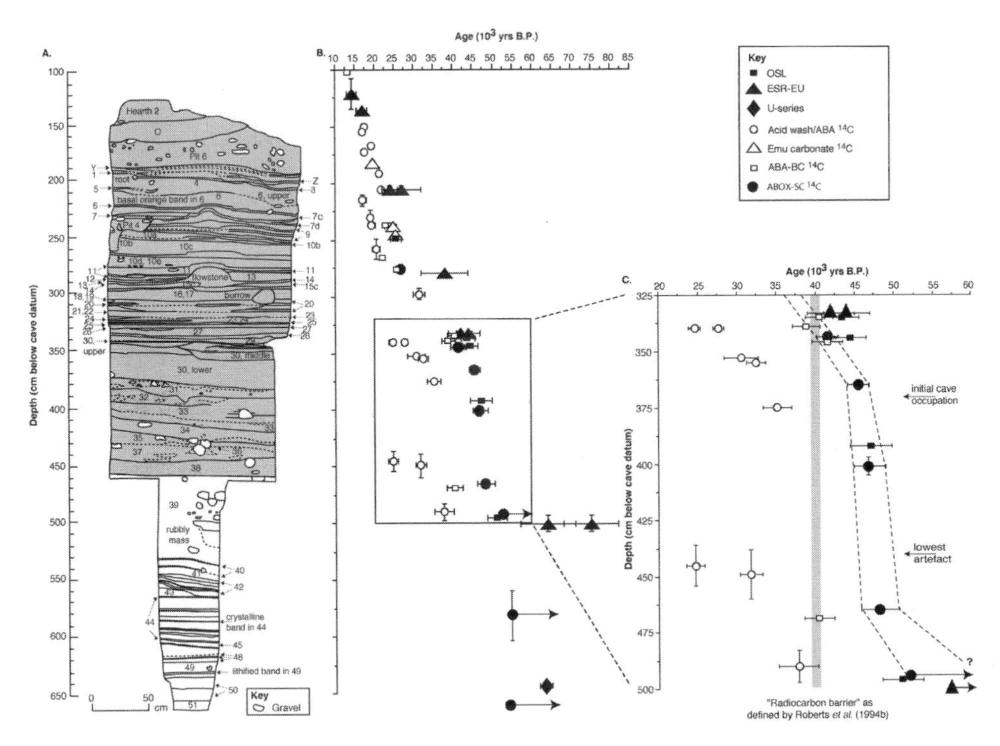

FIGURE 4.13 Summary lithostratigraphy and multi-dating obtained from the Devil's Lair sequence. The grey fill represents stratigraphic units containing cultural remains. (Reprinted from *Quaternary Research* **55**, Turney, C.S.M., Bird, M.I., Fifield, L.K., Roberts, R.G., Smith, M.A., Dortch, C.E., Grün, R., Lawson, E., Ayliffe, L.K., Miller, G.H., Dortch, J. and Cresswell, R.G., Early human occupation at Devil's Lair, south-western Australia 50,000 years ago, 3–13) Copyright (2001), with permission from Elsevier

stratigraphic order. Moreover, they suggest a rapid rate of sedimentation between layers 30 and 38, consistent with the presence of cut-and-fill structures and multiple, partly superimposed, channels that have convoluted or angular sections (Dortch, 1979a). The discord between the other charcoal ^{14}C datasets and the new ABOX-SC ^{14}C chronology provides strong evidence that conventional charcoal pretreatment strategies are inadequate for removing younger carbon contaminants from these samples.

The ESR-EU ages for layer 39 (75.0 $\pm$ 7 and 64.0 $\pm$ 7 ka) are older than the optical age for this layer (51,100 $\pm$ 2600 yrs). Additional U-series analysis of the dental material would be required to further refine the ESR age estimates for this layer (Grün *et al.*, 1999). The ABOX-SC ^{14}C ages for this and deeper levels (~48 ka BP and older) overlap with background values, but a maximum age for the cave fill is given by the detritally-corrected uranium series age for the basal flowstone at 660 cm (63.2 $\pm$ 0.8 ka), which is ~1.6 m below layer 39. The new chronology indicates that the base of the stratigraphic sequence at Devil's Lair is in excess of 55 ka BP. The earliest evidence for people in the vicinity of the cave, represented by the artefacts between layers 32–38, dates to ~48–47 ka BP while the earliest evidence for actual occupation of the cave dates to 45.5 ka BP.

4.4.2 Meso- and macroscale – early hominin settlement across Australia

The conventional radiocarbon chronology for Australian human colonization suggests an arrival time of approximately 40 ka ^{14}C BP (O'Connell and Allen, 2004), suggested from archaeological sites such as Upper Swan

(Pearce and Barbetti, 1981), Carpenter's Gap 1 (O'Connor, 1995), Puritjarra (Smith, 1987), Wareen Cave (Allen, 1996), Ngarrabullgan (David *et al.*, 1997) and Mimbi Caves (Balme, 2001) (Fig. 4.14). In contrast, alternative dating techniques (such as TL, OSL (multi-aliquot and single-grain) and electron spin resonance) and recent developments in the pretreatment and graphitization of samples for radiocarbon dating, at Malakunanja II, Nauwalabilia I and Devil's Lair, have implied human arrival by at least 45–55 ka cal BP (Fig. 4.14) (Roberts *et al.*, 1990, 1994, 1998b; Turney *et al.*, 2001a). The modern hominid skeleton Mungo III burial has been reportedly dated by ESR, OSL and U-series at 62 ± 6 ka (Thorne *et al.*, 1999) while more recently, extensive OSL dating of the stratigraphy suggests a date of burial between 50.1 ± 2.4 and 45.7 ± 2.3 ka (Bowler *et al.*, 2003), immediately prior to long-term increased aridity suggested by stratigraphic evidence for low-lake levels.

In addition to the above results, dating of environmental change inferred to be as a result of human arrival supports the older ages. These events include the extinction of the large flightless bird *Genyornis newtoni* and disruption of the vegetation cover in the Lake Eyre basin at approximately 50 ka cal BP (based on AAR, ^{14}C, OSL and U-series; Johnson *et al.*, 1999; Miller *et al.*, 1999b), continental-wide extinction of megafauna centred on 46.4 ka (based on OSL and U-series; Roberts *et al.*, 2001) and increases in burning recorded at Lynch's Crater and ODP-820 at around 45 ka BP (based on ^{14}C and oxygen isotope stratigraphy respectively; Moss and Kershaw, 2000; Turney *et al.*, 2001b). The ages obtained by these new methods are significantly beyond the age of human arrival suggested by traditional radiocarbon pretreatment methods and are considered unlikely to result from significant post-depositional disturbance of artefacts (Turney *et al.*, 2001c; Bird *et al.*, 2002).

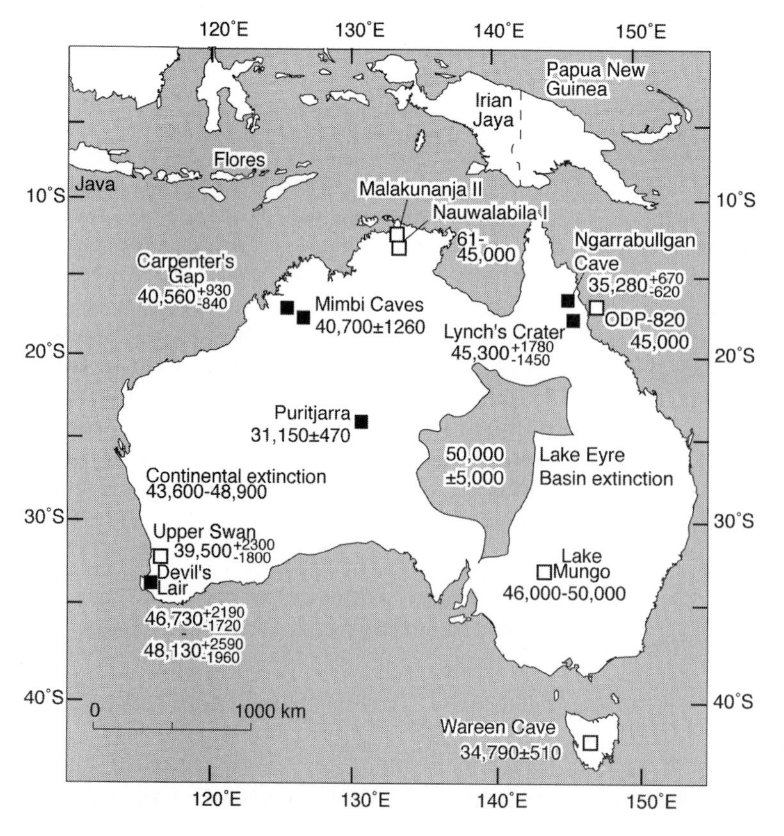

FIGURE 4.14 Principal sites associated with early human colonization of Australia. (Turney and Bird, 2002)

4.4.3 Megascale – human evolution and migration during the late Quaternary

The 'Multi-Regional' hypothesis of human evolution argues that modern humans arose by evolutionary changes in pre-existing (i.e. 1 Ma) hominin populations in many parts of the Old World, while the 'Out of Africa' hypothesis holds that modern humans first appeared in Africa less than 200,000 years ago and then dispersed across the world, eclipsing all earlier hominin populations (Stringer, 2002). Variants on these two models have also been proposed (Stringer, 2002), and one of the reasons that multiple views persist is the lack of adequate age control.

The earliest evidence of human evolution and behaviour is found in southern Africa. Early art, tools and skeletal material associated with modern humans have been dated by ESR, TL and OSL to between 150 and 70 ka (Brauer *et al.*, 1997; Grün and Beaumont, 2001; Henshilwood *et al.*, 2002), while modern human movement out of Africa is evidenced by a small child's burial in Eygpt, dated by OSL to between 80.4 and 49.8 ka (Vermeersch *et al.*, 1998). Dating using TL and ESR suggests significant migrations of *Homo neanderthalensis* (Neanderthal) and modern humans across the Middle East during the late-Quaternary with occupation of sites by the former between 120 and 40 ka (Grün and Stringer, 1991), while the latter appears to have been in similar locations, 90 and 50 ka ago, suggesting overlap between the species over a limited spatial area.

The settlement of Australia is an excellent ongoing test of these hypotheses. Despite estimated sea level changes of 130 m associated with glacial-interglacial cycles (Chappell *et al.*, 1996), Australia has remained an island, making any age associated with early settlement an indication of modern human behaviour (Davidson and Noble, 1992; O'Connell and Allen, 2004). An obvious route of entry to Australia is through Indonesia, where U-series and ESR ages for animal teeth associated with skulls of *Homo erectus* suggest that this species may have survived as a relict population until the latter part of Oxygen Isotope Stage 3. LU and EU ESR ages of 46.4 ka and 27.3 ka respectively (Swisher *et al.*, 1996), imply that the more archaic species co-existed for some tens of millennia with *H. sapiens*. These ages have met with some controversy, however, as U-series for the same material reported an age of 31.0 ± 0.2 ka, i.e. older than the EU ESR ages (Grün and Thorne, 1997). An excellent overview of the dating of early humans in Southeast Asia and Australia is given in Duller (2001).

In addition to the pronounced climatic changes identified during the late-Quaternary, radiocarbon, luminescence and electron spin resonance (ESR) dating of artefacts and associated skeletal remains indicate that *Homo sapiens* migrated into Europe from Africa during Oxygen Isotope Stage 3, with the subsequent demise of the more archaic Neanderthals (Stringer and Grün, 1991; d'Errico and Goñi, 2003). ^{14}C dating of Neanderthal remains suggest they survived in localized areas until at least 28 ka BP (Smith *et al.*, 1999), though the recent find of a proposed hybridized modern human-Neanderthal in Iberia ^{14}C dated to 20 ka BP (Duarte *et al.*, 1999) suggests relict populations may have survived until at least this time.

In the Americas, considerable controversy surrounds human arrival. The traditional view is that humans arrived from Asia across the Bering Strait at relatively low sea levels following the collapse of the Laurentide Ice Sheet at the end of the last ice age (Dixon, 2001). Earlier studies suggested an arrival date in northwestern America sometime around 12 ka BP. More recently, however, a considerable body of evidence has been built up to challenge this view. For instance, TL ages obtained from burnt artefacts at Caverna da Pedra Pintada (Brazil) suggest two periods of occupation, centred on 17.5–14.8 ka and 11.9–9.5 ka (Roosevelt *et al.*, 1996; Michab *et al.*, 1998). Furthermore, Pedra Furada (Brazil) is a 5 m deposit suggesting human occupation >45 ka (Bahn, 1993). Radiocarbon ages are reported in excess of 48 ka BP (Santos *et al.*, 2003), while within the 5 m sandy deposits, around 160 hearths have been recorded. Luminescence on the hearth stones indicates temperatures >450°C, significantly above natural fires that are typically <300°C (Parenti *et al.*, 1996). In

Chile, there is strong evidence from Monte Verde (Adovasio and Pedler, 1997; Dillehay, 1997; Meltzer, 1997) of 13–12.5 ka ^{14}C BP for human occupation, while the Peruvian coastal site of Quebrada Jaguay has relatively early ^{14}C ages of 11 ka BP (Sandweiss *et al.*, 1998) suggesting rapid migration down the continent was achieved by boat. Further discussion of human arrival in the Americas can be found in Section 3.4.4.

Thus, despite considerable resources being invested in the global issue of recent human evolution and migration, no consensus has been reached. Indeed, the picture is significantly more complicated than previously suspected only a few years ago. What can be said with certainty, however, is that significant discoveries can be anticipated in the near future and that dating these finds will continue to play a crucial role in our understanding of human evolution and migration.

REFERENCES

Abelson, P.H. (1954) Amino acids in fossils. *Science* **119**, 576.

Abelson, P.H. (1956) Paleobiochemistry. *Scientific American* **195**, 83–92.

Adovasio, J.M. and Pedler, D.R. (1997) Monte Verde and the antiquity of humankind in the Americas. *Antiquity* **71**, 573–580.

Aitken, M.J. (1974) *Physics and Archaeology*. Oxford: Clarendon Press.

Aitken, M.J. (1985) *Thermoluminescence Dating*. New York: Academic Press.

Aitken, M.J. (1990) *Science-based dating in Archaeology*. London: Longman.

Aitken, M.J., Tite, M.S. and Reid, J. (1964) Thermoluminescent dating of ancient ceramics. *Nature* **202**, 1032–1033.

Allen, J. (1996) *Report of the Southern Forests Archaeological Project. Volume 1: Site Descriptions, Stratigraphies and Chronologies*. La Trobe University: School of Archaeology.

Anderson, E.C., Libby, W.F., Weinhouse, S., Reid, A.F., Kirshenbaum, A.D. and Grosse, A.V. (1947) Radiocarbon from cosmic radiation. *Science* **105**, 576.

Arnold, J.R. and Libby, W.F. (1949) Age determinations by radiocarbon content: checks with samples of known age. *Science* **110**, 678–680.

Arnold, J.R. and Libby, W.F. (1951) Radiocarbon dates. *Science* **113**, 111–120.

Bada, J.L., Wang, X.S. and Hamilton, H. (1999) Preservation of key biomolecules in the fossil record: Current knowledge and future challenges. *Philosophical Transactions of the Royal Society of London* **354**, 77–87.

Bahn, P.G. (1993) 50,000-year-old Americans of Pedra Furada. *Nature* **362**, 114–115.

Baillie, M. (2000) *Exodus to Arthur*. London: Batsford.

Baillie, M.G.L. (1995) *A slice through time: dendrochronology and precision dating*. London: Routledge.

Baillie, M.G.L. (1999) A view from outside: recognising the big picture. In (K.J. Edwards and J.P. Sadler, eds.) Holocene Environments of Prehistoric Britain. *Quaternary Proceedings* **7**. Chichester: John Wiley and Sons Ltd.

Baillie, M.G.L., Hillam, J., Briffa, K.R. and Brown, D.M. (1985) Re-dating the English art-historical tree-ring chronologies. *Nature* **315**, 317–319.

Balme, J. (2001) Excavations revealing 40,000 years of occupation at Mimbi Caves in south central Kimberley of Western Australia. *Australian Archaeology* **50**, 1–5.

Bard, E., Hamelin, B., Fairbanks, R.G. and Zindler, A. (1990) Calibration of the ^{14}C timescale over the past 30,000 years using mass spectrometric U-Th ages from Barbados corals. *Nature* **345**, 405–410.

Beck, J.W., Richards, D.A., Edwards, R.L., Silverman, B.W., Smart, P.L., Donahue, D.J., Hererra-Osterheld, S., Burr, G.S., Calsoyas, L., Jull, A.J.T. and Biddulph, D. (2001) Extremely large variations of atmospheric ^{14}C concentration during the last glacial period. *Science* **292**, 2453–2458.

Bennett, C.L., Beukens, R.P., Clover, M.R., Gove, H.E., Liebert, R.B., Litherland, A.E., Purser, K.H. and Sonderheim, W.E. (1977) Radiocarbon dating using electrostatic accelerators: negative ions provide the key. *Science* **198**, 508–510.

Berry, A.J. (1928) The volcanic ash deposits of Scinde Island with special reference to the pumice bodies called chalazoidites. *Transactions of the New Zealand Institute* **59**, 571–608.

Bettinger, R.L. and Eerkens, J. (1999) Point typologies, cultural transmission, and the spread of bow-and-arrow technology in the prehistoric Great Basin. *American Antiquity* **64**, 231–242.

Bicknell, P. (2000) Late Minoan IB marine ware, the marine environment of the Aegean, and the Bronze Age eruption of the Thera volcano. In (W.G. McGuire, D.R. Griffiths, P.L. Hancock and I.S. Stewart, eds.) *The Archaeology of Geological Catastrophes*, 95–103. London: Geological Society Special Publications 171.

Bindon, P. and Raynal, J.-P. (1998) Humans and volcanoes in Australia and New Guinea. *Quaternaire* **9**, 71–75.

Bird, M.I. and Gröcke, D.R. (1997) Determination of the abundance and carbon-isotope composition of elemental carbon in sediments. *Geochimica et Cosmochimica Acta* **61**, 3413–3423.

Bird, M.I., Ayliffe, L.K., Fifield, L.K., Turney, C.S.M., Cresswell, R.G., Barrows, T.T. and David, B. (1999) Radiocarbon dating of 'old' charcoal using a wet oxidation-stepped combustion procedure. *Radiocarbon* **41**, 127–140.

Bird, M.I., Turney, C.S.M., Fifield, L.K., Jones, R., Ayliffe, L.K., Palmer, A., Cresswell, R. and Robertson, S. (2002) Radiocarbon analysis of the early archaeological site of Nauwalabila I, Arnhem Land, Australia: implications for sample suitability and stratigraphic integrity. *Quaternary Science Reviews* **21**, 1061–1075.

Bird, M.I., Turney, C.S.M., Fifield, L.K., Smith, M.A., Miller, G.H. and Magee, J. (2003) Radiocarbon dating of organic- and carbonate-carbon in Genyornis and Dromaius eggshell using stepped combustion and stepped acidification. *Quaternary Science Reviews* **22**, 1805–1812.

Birks, H.J.B. and Birks, H.H. (1980) *Quaternary Palaeoecology*. Cambridge: Cambridge University Press.

Biswas, D.K., Hyodo, M., Taniguchi, Y., Kaneko, M., Katoh, S., Sato, H., Kinugasa, Y. and Mizuno, K. (1999) Magnetostratigraphy of Plio-Pleistocene sediments in a 1700-m core from Osaka Bay, southwestern Japan and short geomagnetic events in the middle Matuyama and early Brunhes chrons. *Palaeogeography, Palaeoclimatology, Palaeoecology* **148**, 233–248.

Blackwell, B.A. (2001) Electron Spin Resonance (ESR) dating in lacustrine environments. In (W.M. Last and J.P. Smol, eds.) *Tracking Environmental Changes in Lake Sediments: Physical and Chemical Techniques*, 283–369. Kluwer Academic Publishers, Dordrecht, The Netherlands.

Bond, G., Kromer, B., Beer, J., Muscheler, R., Evans, M.N., Showers, W., Hoffman, S., Lotti-Bond, R., Hajdas, I. and Bonani, G. (2001) Persistent solar influence on North Atlantic climate during the Holocene. *Science* **294**, 2130–2136.

Borchardt, G.A., Aruscavage, P.J. and Millard Jr., H.T. (1972) Correlation of the Bishop Ash, a Pleistocene marker bed, using instrumental neutron activation analysis. *Journal of Sedimentary Petrology* **42**, 301–306.

Bøtter-Jensen, L., Agersnap Larsen, N., Markey, B.G. and McKeever, S.W.S. (1997) Al_2O_3:C as a sensitive OSL dosemeter for rapid assessment of environmental photon dose rates. *Radiation Measurements* **27**, 295–298.

Boujot, C. and Cassen, S. (1993) A pattern of evolution for the Neolithic funerary structures of the west of France. *Antiquity* **67**, 477–492.

Bowen, D.Q., Hughes, S., Sykes, G.A. and Miller, G.H. (1989) Land-sea correlations in the Pleistocene based on isoleucine epimerization in non-marine molluscs. *Nature* **340**, 49–51.

Bowler, J.M., Johnston, H., Olley, J.M., Prescott, J.R., Roberts, R.G., Shawcross, W. and Spooner, N.A. (2003) New ages for human occupation and climatic change at Lake Mungo, Australia. *Nature* **421**, 837–840.

Bowles, J., Gee, J., Hildebrand, J. and Tauxe, L. (2002) Archaeomagnetic intensity results from California and Ecuador: Evaluation of regional data. *Earth and Planetary Science Letters* **203**, 967–981.

Box, G.E.P. and Jenkins, G.M. (1976) *Time Series Analysis: Forecasting and Control*. Holden-Day, San Francisco.

Boygle, J. (1999) Variability of tephra in lake and catchment sediments, Svínavatn, Iceland. *Global and Planetary Change* **21**, 129–149.

Brainerd, G.W. (1951) The place of chronological ordering in archaeological analysis. *American Antiquity* **16**, 301–313.

Brauer, A., Endres, C., Günter, C., Litt, T., Stebich, M. and Negendank, J.F.W. (1999) High resolution sediment and vegetation responses to Younger Dryas climate change in varved lake sediments from Meerfelder Maar, Germany. *Quaternary Science Reviews* **18**, 321–329.

Bräuer, G., Yokoyama, Y., Falguéres, C. and Mbua, E. (1997) Modern human origins backdated. *Nature* **386**, 337–338.

Briffa, K.R., Jones, P.D., Schweingruber, F.H., Shiyatov, S.G. and Vaganov, E.A. (1996) Development of a North Eurasian chronology network: Rationale and preliminary results of comparative ring-width and densitometric analyses in northern Russia. In (J.S. Dean, D.M. Meko and T.W. Swetnam, eds.) *Tree-Rings, Environment and Humanity*. Proceedings of the International Conference, Tucson, Arizona, 17–21 May 1994. *Radiocarbon* **(1996)**, 25–42.

Buck, C.E., Kenworthy, J.B., Litton, C.D. and Smith, A.F.M. (1991) Combining archaeological and radiocarbon information: a Bayesian approach to calibration. *Antiquity* **65**, 808–821.

Cahen, D. and Moeyersons, J. (1977) Subsurface movements of stone artefacts and their implications for the prehistory of Central Africa. *Nature* **266**, 812–815.

Card, V.M. (1997) Varve-counting by the annual pattern of diatoms accumulated in the sediment of Big Watab Lake, Minnesota, AD 1837–1990. *Boreas* **26**, 103–112.

Carey, S.N. and Sigurdsson, H. (1982) Influence of particle aggregation on deposition of distal tephra from the May 18, 1980, eruption of Mount St. Helens volcano. *Journal of Geophysical Research* **87**, 7061–7072.

Caseldine, C., Baker, A. and Barnes, W.L. (1999) A rapid, non-destructive scanning method for detecting distal tephra layers in peats. *The Holocene* **9**, 635–638.

Cattell, R.B. (1966) The scree test for the number of factors. *Multivariate Behavorial Research* **1**, 245–276.

Chappell, J., Omura, A., Esat, T., McCulloch, M., Pandolfi, J., Ota, Y. and Pillans, B. (1996) Reconciliation of late Quaternary sea levels derived from coral terraces at Huon Peninsula with deep sea oxygen isotope records. *Earth and Planetary Science Letters* **141**, 227–236.

Charman, D.J. and Grattan, J. (1999) An assessment of discriminant function analysis in the identification and correlation of distal Icelandic tephras in the British Isles. In (C.R. Firth and W.J. McGuire, eds.) *Volcanoes in the Quaternary*, 147–160. London: Geological Society Special Publications 161.

Cioni, R., Levi, S. and Sulpizio, R. (2000) Apulian Bronze Age pottery as a long-distance indicator of the Avellino Pumice eruption (Vesuvius, Italy). In (W.G. McGuire, D.R. Griffiths, P.L. Hancock and I.S. Stewart, eds.) *The Archaeology of Geological Catastrophes*, 159–177. London: Geological Society Special Publications 171.

Colman, S.M., Pierce, K.L. and Birkeland, P.W. (1987) Suggested terminology for Quaternary dating methods. *Quaternary Research* **28**, 314–319.

Cooper, M.C. (1997) The use of digital image analysis in the study of laminated sediments. *Journal of Paleolimnology* **19**, 33–40.

d'Errico, F. and Goñi, M.F.S. (2003) Neanderthal extinction and the millennial scale climatic variability of OIS 3. *Quaternary Science Reviews (Quaternary Geochronology)* **22**, 769–788.

Daniels, F., Boyd, C.A. and Saunders, D.F. (1953) Thermoluminescence as a research tool. *Science* **117**, 343–349.

David, B., Roberts, R., Tuniz, C., Jones, R. and Head, J. (1997) New optical and radiocarbon dates from Ngarrabullgan Cave, a Pleistocene archaeological site in Australia: implications for the comparability of time clocks and for the human colonisation of Australia. *Antiquity* **71**, 183–188.

Davidson, I. and Noble, W. (1992) Why the first colonisation of the Australian region is the earliest evidence of modern human behaviour. *Archaeology in Oceania* **27**, 113–119.

De Geer, G. (1912) A geochronology of the last 12,000 years. *Report of the 11th International Geological Congress* **1**, 241–253.

Dethlefson, E.S. and Deetz, J. (1966) Death's heads, cherubs and willow trees: Experimental archaeology in colonial cemeteries. *American Antiquity* **31**, 502–510.

Dillehay, T.D. (1997) *Monte Verde: A late Pleistocene settlement in Chile 2: the archaeological context*. Washington (D.C.): Smithsonian Institution Press.

Dimbleby, G.W. (1957) Pollen preservation. *New Phytologist* **56**, 12–16.

Dixon, E.J. (2001) Human colonization of the Americas: timing, technology and process. *Quaternary Science Reviews* **20**, 277–299.

Dortch, C. (1979a) Devil's Lair, an example of prolonged cave use in south-western Australia. *World Archaeology* **10**, 258–279.

Dortch, C.E. (1979b) 33,000 year old stone and bone artefacts from Devil's Lair, Western Australia. *Records of the Western Australian Museum* **7**, 329–367.

Dortch, C. (1984) *Devil's Lair, A Study In Prehistory*. Perth: Western Australia Museum.

Dortch, C.E. and Dortch, J. (1996) Review of Devil's Lair artefact classification and radiocarbon chronology. *Australian Archaeology* **43**, 28–31.

Dortch, C.E. and Merrilees, D. (1973) A salvage excavation in Devil's Lair, Western Australia. *Journal of the Royal Society of Western Australia* **54**, 103–113.

Douglass, A.E. (1919) *Climatic Cycles and Tree Growth I*. Washington: Carnegie Institute.

Downey, W.S. and Tarling, D.H. (1984) Archaeomagnetic dating of Santorini volcanic eruptions and fired destructive levels of late Minoan civilization. *Nature* **309**, 519–523.

Duarte, C., Mauríco, J., Pettitt, P.B., Souto, P., Trinkaus, E., van der Plicht, H. and Zilhão, J. (1999) The early Upper Paleolithic human skeleton from the Abrigo do Lagar Velho (Portugal) and modern human emergence in Iberia. *Proceedings of the National Academy of Sciences* **96**, 7604–7609.

Dugmore, A. (1989) Icelandic volcanic ash in Scotland. *Scottish Geographical Magazine* **105**, 168–172.

Dugmore, A.J. and Newton, A.J. (1992) Thin tephra layers in peat revealed by X-radiography. *Journal of Archaeological Science* **19**, 163–170.

Dugmore, A.J., Newton, A.J., Sugden, D.E. and Larsen, G. (1992) Geochemical stability of fine-grained silicic Holocene tephra in Iceland and Scotland. *Journal of Quaternary Science* **7**, 173–183.

Dugmore, A., Larsen, G. and Newton, A.J. (1995) Seven tephra isochrones in Scotland. *The Holocene* **5**, 257–266.

Duller, G.A.T. (1994) Luminescence dating of sediments using single aliquots: New procedures. *Quaternary Science Reviews* **13**, 149–156.

Duller, G.A.T. (1996) Recent developments in luminescence dating of Quaternary sediments. *Progress in Physical Geography* **20**, 127–145.

Duller, G.A.T. (2001) Dating methods: the role of geochronology in studies of human evolution and migration in southeast Asia and Australasia. *Progress in Physical Geography* **25**, 267–276.

Eastwood, W.J., Pearce, N.J.G., Westgate, J.A. and Perkins, W.T. (1998) Recognition of Santorini (Minoan)

Tephra in lake sediments from Gölhisar Gölü, southwest Turkey by laser ablation ICP-MS. *Journal of Archaeological Science* **25**, 677–687.

Eggins, S., Grün, R., Pike, A., Shelley, A. and Taylor, L. (2003) ^{238}U, ^{232}Th profiling and U-series isotope analysis of fossil teeth by laser ablation ICPMS. *Quaternary Science Reviews* **22**, 1373–1382.

Everitt, B.S. (1978) *Graphical techniques for multivariate data*, 5–41. London: Heinemann Educational Books Ltd..

Ferguson, C.W. and Graybill, D.A. (1983) Dendrochronology of bristlecone pine: A progress report. *Radiocarbon* **25**, 287–288.

Forman, S.L., Pierson, J. and Lepper, K. (2000) Luminescence geochronology. In (J.S. Noller, J.M. Sowers and W.R. Lettis, eds.) *Quaternary Geochronology: Methods and Applications*, 157–176. AGU Reference Shelf 4.

Friedrich, M., Kromer, B., Kaiser, K.F., Spurk, M., Hughen, K.A. and Johnsen, S.J. (2001) High-resolution climate signals in the Bølling-Allerød Interstadial (Greenland Interstadial 1) as reflected in European tree-ring chronologies compared to marine varves and ice-core records. *Quaternary Science Reviews* **20**, 1223–1232.

Froggatt, P.C. (1992) Standardization of the chemical analysis of tephra deposits. Report of the ICCT Working Group. *Quaternary International* **13/14**, 93–96.

Göksu, H.Y. and Fremlin, J.H. (1972) Thermoluminescence from unirradiated flints: Regeneration thermoluminescence. *Archaeometry* **14**, 127–132.

Goodfriend, G.A., Macko, S.A., Prewitt, C.T. and Wehmiller, J.F. (1998) Perspectives in amino acid and protein geochemistry. *Amino Acids* **15**, 271–290.

Goslar, T., Arnold, M., Tisnert-Laborde, N., Czernik, J. and Wieckowski, K. (2000) Variations of Younger Dryas atmospheric radiocarbon explicable without ocean circulation changes. *Nature* **403**, 877–880.

Gove, H.E., Mattingly, S.J., David, A.R. and Garza-Valdes, L.A. (1997) A problematic source of organic contamination of linen. *Nuclear Instruments and Methods in Physics Research* **B123**, 504–507.

Grange, L.I. (1931) Volcanic ash showers: a geological reconnaissance of volcanic ash showers of the central part of the North Island. *New Zealand Journal of Science and Technology* **12**, 228–240.

Grün, R. (1989) Electron spin resonance (ESR) dating. *Quaternary International* **1**, 65–109.

Grün, R. and Beaumont, P. (2001) Border Cave revisited: a revised ESR chronology. *Journal of Human Evolution* **40**, 467–482.

Grün, R. and McDermott, F. (1994) Open system modelling for U-series and ESR dating of teeth. *Quaternary Science Reviews* **13**, 121–125.

Grün, R. and Stringer, C.B. (1991) Electron spin resonance dating and the evolution of modern humans. *Archaeometry* **33**, 153–199.

Grün, R. and Thorne, A. (1997) Dating the Ngandong humans. *Science* **38**, 733–741.

Grün, R., Beaumont, P.B. and Stringer, C.B. (1990) ESR dating evidence for early modern humans at Border Cave in South Africa. *Nature* **344**, 537–539.

Grün, R., Davies, S.W., Richards, M. and Stringer, C.B. (forthcoming) The chronology of modern human evolution, Part 1: Methods. *Quaternary Science Reviews.*

Grün, R., Yan, G., McCulloch, M. and Mortimer, G. (1999) Detailed mass spectrometric U-series analyses of two teeth from the archaeological site of Pech de l'Aze II: Implications for uranium migration and dating. *Journal of Archaeological Science* **26**, 1301–1310.

Grün, R., Huang, P.-H., Huang, W., McDermott, F., Thorne, A., Stringer, C.B. and Yan, G. (1998) ESR and U-series analyses of teeth from the palaeoanthropological site of Hexian, Anhui Province, China. *Journal of Human Evolution* **34**, 555–564.

Grün, R., Huang, P.-H., Wu, X., Stringer, C.B., Thorne, A.G. and McCulloch, M. (1997) ESR analysis of teeth from the palaeoanthropological site of Zhoukoudian, China. *Journal of Human Evolution* **32**, 83–91.

Guerrero, B.O., Thompson, R. and Fucugauchi, J.U. (2000) Magnetic properties of lake sediments from Lake Chalco, central Mexico, and their palaeoenvironmental implications. *Journal of Quaternary Science* **15**, 127–140.

Guiot, J., Pons, A., de Beaulieu, J.L. and Reille, M. (1989) A 140,000-year continental climate reconstruction from two European pollen records. *Nature* **338**, 309–313.

Hare, P.E. (1963) Amino acids in the proteins from aragonite and calcite in shells of Mytilus californianus. *Science* **139**, 216–217.

Hare, P.E., von Endt, D.W. and Kokis, J.E. (1997) Protein and amino acid diagenesis dating. In (R.E. Taylor and M.J. Aitken, eds.) *Chronometric Dating in Archaeology*, 261–296. New York: Plenum.

Harrington, J.C. (1954) Dating stem fragments of seventeenth and eighteenth century clay tobacco pipes. *Quarterly Bulletin Archaeological Society of Virginia* **9**, 6–8.

Haynes, R. (1982) *Environmental Science Methods.* London: Chapman and Hall.

Henshilwood, C.S., d'Errico, F., Yates, R., Jacobs, Z., Tribolo, C., Duller, G.A.T., Mercier, N., Sealy, J.C., Valladas, H., Watts, I. and Wintle, A.G. (2002) Emergence of modern human behaviour: Middle Stone Age engravings from South Africa. *Science* **295**, 1278–1280.

Hill, M.O. and Gauch, H.G. (1980) Detrended correspondence analysis, an improved ordination technique. *Vegetatio* **42**, 47–58.

Hillam, J. (1979) Tree-rings and archaeology: Some

problems explained. *Journal of Archaeological Science* **6**, 271–278.

Hillman, G., Hedges, R., Moore, A., Colledge, S. and Pettitt, P. (2001) New evidence of Lateglacial cereal cultivation at Abu Hureyra on the Euphrates. *The Holocene* **11**, 383–393.

Hiscock, P. and Kershaw, A.P. (1992) Palaeoenvironments and prehistory of Australia's tropical Top End. In (J. Dodson, ed.) *The Native Lands: Prehistory and Environmental Change in Australia and the Southwest Pacific.* 43–75. Longman Cheshire.

Hodder, A.P.W., de Lange, P.J. and Lowe, D.J. (1991) Dissolution and depletion of ferromagnesian minerals from Holocene tephra layers in an acid bog, New Zealand, and implications for tephra correlation. *The Holocene* **6**, 195–208.

Hua, Q., Barbetti, M., Jacobsen, G.E., Zoppi, U. and Lawson, E.M. (2000) Bomb radiocarbon in annual tree rings from Thailand and Australia. *Nuclear Instruments and Methods in Physics Research* **B172**, 359–365.

Hughen, K.A., Southon, J.R., Lehman, S.J. and Overpeck, J.T. (2000) Synchronous radiocarbon and climate shifts during the last deglaciation. *Science* **290**, 1951–1954.

Hughen, K., Lehman, S., Southon, J., Overpeck, J., Marchal, O., Herring, C. and Turnbull, J. (2004) 14C activity and global carbon cycle changes over the past 50,000 years. *Science* **303**, 202–207.

Hughes, M.K., Milsom, S.J. and Leggett, P.A. (1981) Sapwood estimates in the interpretation of tree-ring dates. *Journal of Archaeological Science* **8**, 381–390.

Hunt, J.B. and Hill, P.G. (1993) Tephra geochemistry: a discussion of some persistent analytical problems. *The Holocene* **3**, 271–278.

Hunt, J.B. and Hill, P.G. (1996) An inter-laboratory comparison of the electron probe microanalysis of glass geochemistry. *Quaternary International* **34–36**, 229–241.

Huntley, D.J., Godfrey-Smith, D.I. and Thewalt, M.L.W. (1985) Optical dating of sediments. *Nature* **313**, 105–107.

Jacoby, G.C., Workman, K.W. and D'Arrigo, R.D.D. (1999) Laki eruption of 1783, tree rings, and disaster for northwest Alaska Inuit. *Quaternary Science Reviews* **18**, 1365–1371.

Johnson, B.J., Miller, G.H., Fogel, M.L., Magee, J.W., Gagan, M.K. and Chivas, A.R. (1999) 65,000 years of vegetation change in central Australia and the Australian Summer monsoon. *Science* **284**, 1150–1152.

Jull, A.J.T., Beck, J.W. and Burr, G.S. (1997) *Accelerator mass spectrometry.* Proceedings of the Seventh International Conference on Accelerator Mass Spectrometry. Tucson, Arizona, USA, 20–24 May. North Holland.

Katari, K. and Tauxe, L. (2000) Effects of pH and salinity on the intensity of magnetization in redeposited sediments. *Earth and Planetary Science Letters* **181**, 489–496.

Kaufman, D.S. and Manley, W.F. (1998) A new procedure for determining DL amino acid ratios in fossils using reverse phase liquid chromatography. *Quaternary Science Reviews (Quaternary Geochronology)* **17**, 987–1000.

Kelly, E.F., Amundson, R.G., Marino, B.D. and DeNiro, M.J. (1991) Stable isotope ratios of carbon in phytoliths as a quantitative method of monitoring vegetation and climate change. *Quaternary Research* **35**, 222–233.

Kendall, M. and Ord, J.K. (1990) *Time Series*. Sevenoaks: Edward Arnold.

Kennedy, G.C. and Knopff, L. (1960) Dating by thermoluminescence. *Archaeology* **13**, 147–148.

Kershaw, A.P. (1983) A Holocene pollen diagram from Lynch's Crater, north-eastern Queensland, Australia. *New Phytologist* **94**, 669–682.

Kershaw, A.P. (1986) Climatic change and Aboriginal burning in north-east Australia during the last two glacial/interglacial cycles. *Nature* **322**, 47–49.

Kitagawa, H. and van der Plicht, J. (1998) Atmospheric radiocarbon calibration to 45,000 yr B.P: Late Glacial fluctuations and cosmogenic isotopic production. *Science* **279**, 1187–1190.

Kroeber, A.L. (1916) Zuñi culture sequences. *Proceedings of the National Academy of Sciences* **2**, 42–45.

Lageard, J.G.A., Chambers, F.M. and Thomas, P.A. (1999) Climatic significance of the marginalization of Scots pine (*Pinus sylvestris* L.) c. 2500 BC at White Moss, south Chesire, UK. *The Holocene* **9**, 321–331.

Laj, C., Kissel, C., Mazaud, A., Channell, J.E.T. and Beer, J. (2000) North Atlantic palaeointensity stack since 75 ka (NAPIS-75) and the duration of the Laschamp event. *Philosophical Transactions of the Royal Society of London* **A358**, 1009–1025.

Laj, C., Kissel, C., Mazaud, A., Michel, E., Muscheler, R. and Beer, J. (2002) Geomagnetic field intensity, North Atlantic Deep Water circulation and atmospheric $\Delta^{14}C$ during the last 50 kyr. *Earth and Planetary Science Letters* **200**, 177–190.

Lamoureux, S. (2001) Varve chronology techniques. In (W.M. Last and J.P. Smol, eds.) *Tracking Environmental Changes in Lake Sediments: Physical and Chemical Techniques*, 247–260. Dordrecht, The Netherlands: Kluwer Academic Publishers.

Lamoureux, S.F. (1999) Spatial and interannual variations in sedimentation patterns recorded in nonglacial varved sediments from the Canadian High Arctic. *Journal of Paleolimnology* **21**, 73–84.

Landmann, G., Reimer, A., Lemcke, G. and Kempe, S. (1996) Dating Late Glacial abrupt climate changes in the 14,570 yr long continuous varve record of Lake Van, Turkey.

Palaeogeography, Palaeoclimatology, Palaeoecology **122**, 107–118.

Lane-Fox, A. (1875) On the evolution of culture. *Notices of the Proceedings of the Royal Institute of Great Britain* **7**, 496–520.

Li, H.-H., and Sun, J.-Z. (1982) Age of loess determined by thermoluminescence TL dating of quartz. In (T.L. Liu, ed.) *Quaternary Geology and Environment of China*, 39–41. Beijing: China Ocean Press.

Lian, O.B. and Huntley, D.J. (2001) Luminescence dating. In (W.M. Last and J.P. Smol, eds.) *Tracking Environmental Changes in Lake Sediments: Physical and Chemical Techniques*, 261–282. Dordrecht, The Netherlands: Kluwer Academic Publishers.

Lian, O.B. and Shane, P.A. (2000) Optical dating of paleosols bracketing the widespread Rotoehu tephra, North Island, New Zealand. *Quaternary Science Reviews* **19**, 1649–1662.

Libby, W.F., Anderson, E.C. and Arnold, J.R. (1949) Age determination by radiocarbon content: world-wide assay of natural radiocarbon. *Science* **109**, 227–228.

Liritzis, I., Galloway, R.B. and Theocaris, P.S. (1994) Thermoluminescence dating of ceramics revisited: Optically stimulated luminescence of quartz single aliquot with green light-emitting diodes. *Journal of Radioanalytical and Nuclear Chemistry* **188**, 189–198.

Lowe, D.J. and Hunt, J.B. (2001) A summary of terminology used in tephra-related studies. In (E. Juvigne and J.-P. Raynal, eds.) *Tephras: Chronology and Archeology*, 18–22. Les dossiers de l'Archéo-Logis No. 1, Goudet.

Lowe, J.J. and Turney, C.S.M. (1997) Vedde Ash layer discovered in small lake basin on Scottish mainland. *Journal of the Geological Society of London* **154**, 605–612.

Lowe, J.J. and Walker, M.J.C. (1997) *Reconstructing Quaternary Environments*. Harlow: Longman.

Mackie, E.A.V., Davies, S.M., Turney, C.S.M., Dobbyn, K., Lowe, J.J. and Hill, P.J. (2002) The use of magnetic separation techniques to detect basaltic microtephra in last glacial-interglacial transition (L.G.I.T.; 15–10 ka cal. BP) sediment sequences in Scotland. *Scottish Journal of Geology* **38**, 21–30.

Makridakis, S.G., Wheelwright, S.C. and Hyndman, R.G. (1998) *Forecasting: Methods and Applications 3rd edition*. New York: John Wiley and Sons, Inc.

Mann, P.S. (1998) *Introductory Statistics*. New York: John Wiley and Sons, Inc.

Martinson, D.G, Pisias, N.G., Hays, J.D., Imbrie, J., Moore, T.C. and Shackleton, N.J. (1987) Age dating and the orbital theory of the Ice Ages: development of a high-resolution 0 to 300,000-year chronostratigraphy. *Quaternary Research* **27**, 1–29.

McCarroll, D. (2002) Amino-acid geochronology and the British Pleistocene: Secure stratigraphical framework or a case of circular reasoning? *Journal of Quaternary Science* **17**, 647–651.

McCoy, W.D. (1987) The precision of amino acid geochronology and paleothermometry. *Quaternary Science Reviews* **6**, 43–54.

McGlone, M.S. and Wilmshurst, J.M. (1999) Dating initial Maori environmental impact in New Zealand. *Quaternary International* **59**, 5–16.

McGlone, M., Wilmshurst, J.M. and Wiser, S.K. (2000) Lateglacial and Holocene vegetation and climatic change on Auckland Island, Subantarctic New Zealand. *The Holocene* **10**, 719–728.

Meltzer, D.J. (1997) Monte Verde and the Pleistocene peopling of the Americas. *Science* **276**, 754–755.

Mercier, N., Valladas, H., Joron, J.-L., Reyss, J.-L., Lévêque, F. and Vandermeersch, B. (1991) Thermoluminescence dating of the late Neanderthal remains from Saint-Césaire. *Nature* **351**, 737–739.

Merkt, J., Müller, H., Knabe, W., Müller, P. and Weiser, T. (1993) The early Holocene Saksunarvatn tephra found in lake sediments in NW Germany. *Boreas* **22**, 93–100.

Michab, M., Feathers, J.K., Joron, J.-L., Mercier, N., Selos, M., Valladas, H., Valladas, G., Reyss, J.-L. and Roosevelt, A.C. (1998) Luminescence dates for the Paleoindian site of Pedra Pintada, Brazil. *Quaternary Science Reviews (Quaternary Geochronology)* **17**, 1041–1046.

Miller, G.H. and Brigham-Grette, J. (1989) Amino acid geochronology: Resolution and precision in carbonate fossils. *Quaternary International* **1**, 111–128.

Miller, G.H., Beaumont, P.B., Deacon, H.J., Brooks, A.S., Hare, P.E. and Jull, A.J.T. (1999a) Earliest modern humans in southern Africa dated by isoleucine epimerization in ostrich eggshell. *Quaternary Science Reviews* **18**, 1537–1548.

Miller, G.H., Magee, J.W., Johnson, B.J., Fogel, M.L., Spooner, N.A., McCulloch, M.T. and Ayliffe, L.K. (1999b) Pleistocene extinction of *Genyornis newtoni*: human impact on Australian megafauna. *Science* **283**, 205–208.

Moar, N.T. and Suggate, R.P. (1996) Vegetation history from the Kaihinu (last) interglacial to the present, west coast, South Island, New Zealand. *Quaternary Science Reviews* **15**, 521–547.

Montelius, O. (1885) *Dating the Bronze Age with special reference to Scandinavia*. K. Vitterhets Historie och Antikvitetsakademien.

Moore, P.D., Webb, J.A. and Collinson, M.E. (1991) *Pollen Analysis*. Cambridge: Blackwell Science.

Moss, P.T. and Kershaw, A.P. (2000) The last glacial cycle from the humid tropics of northeastern Australia: comparison of a terrestrial and a marine record. *Palaeogeography, Palaeoclimatology, Palaeoecology* **155**, 155–176.

Murray, A.S. and Roberts, R.G. (1997) Determining the burial time of single grains of quartz using optically stimulated luminescence. *Earth and Planetary Science Letters* **152**, 163–180.

Murray-Wallace, C.V. (1993) A review of the application of the amino acid racemisation reaction to archaeological dating. *The Artefact* **16**, 19–26.

Nelson, D.E., Korteling, R.G. and Stott, W.R. (1977) Carbon-14: direct detection at natural concentrations. *Science* **198**, 507–509.

Nelson, N.C. (1916) Chronology of the Tano ruins, New Mexico. *American Anthropologist* **18**, 159–180.

Newnham, R.M., Lowe, D.J., McGlone, M.S., Wilmshurst, J.M. and Higham, T.F.G. (1998) The Kaharoa Tephra as a critical datum for earliest human impact in northern New Zealand. *Journal of Archaeological Science* **25**, 533–544.

Nijampurkar, V.N., Rao, D.K., Oldfield, F. and Renberg, I. (1998) The half-life of ^{32}Si: a new estimate based on varved lake sediments. *Earth and Planetary Science Letters* **163**, 191–196.

Noller, J.S., Sowers, J.M. and Lettis, W.R. (2000) *Quaternary Geochronology: Methods and Applications*. Washington: American Geophysical Union.

O'Connell, J.F. and Allen, J. (2004) Dating the colonization of Sahul (Pleistocene Australia-New Guinea): A review of recent research. *Journal of Archaeological Science* **31**, 835–853.

O'Connor, S. (1995) Carpenter's Gap Rockshelter 1: 40,000 years of Aboriginal occupation in the Napier Ranges, Kimberley, WA. *Australian Archaeology* **40**, 58–59.

O'Sullivan, P.E. (1983) Annually-laminated lake sediments and the study of Quaternary environmental changes – a review. *Quaternary Science Reviews* **1**, 245–313.

Ojala, A.E.K. and Saarnisto, M. (1999) Comparative varve counting and magnetic properties of the 8400-yr sequence of an annually laminated sediment in Lake Valkiajärvi, central Finland. *Journal of Paleolimnology* **22**, 335–348.

Oldfield, F., Appleby, P.G. and Thompson, R. (1980) Palaeoecological studies of lakes in the Highlands of Papua New Guinea. *Journal of Ecology* **68**, 457–477.

Oliver, W.R.B. (1931) An ancient maori oven on Mount Egmont. *Journal of the Polynesian Society* **40**, 73–80.

Oms, O., Parés, J.M., Martinez-Navarro, B., Agusti, J., Toro, I., Martinez-Fernández, G. and Turq, A. (2000) Early human occupation of Western Europe: paleomagnetic dates for two paleolithic sites in Spain. *Proceedings of the National Academy of Sciences* **97**, 10666–10670.

Parenti, F., Fontugue, M. and Guérin, C. (1996) Pedra Furada in Brazil and its 'presumed' evidence: limitations and potential of the available data. *Antiquity* **70**, 416–421.

Partridge, T.C., Shaw, J., Heslop, D. and Clarke, R.J. (1999) The new hominid skeleton from Sterkfontein, South Africa: age and preliminary assessment. *Journal of Quaternary Science* **14**, 293–298.

Pearce, N.J.G., Westgate, J.A., Perkins, W.T., Eastwood, W.J. and Shane, P. (1999) The application of laser ablation ICP-MS to the analysis of volcanic glass shards from tephra deposits: bulk glass and single shard analysis. *Global and Planetary Change* **21**, 151–171.

Pearce, R.H. and Barbetti, M. (1981) A 38,000-year-old archaeological site at Upper Swan, Western Australia. *Archaeology in Oceania* **16**, 168–172.

Peglar, S.M. (1993) The mid-Holocene *Ulmus* decline at Diss Mere, Norfolk, UK: a year-by-year pollen stratigraphy from annual laminations. *The Holocene* **3**, 1–13.

Peglar, S.M., Fritz, S.C., Alapieti, T., Saarnisto, M. and Birks, H.J.B. (1984) Composition and formation of laminated sediments in Diss Mere, Norfolk, England. *Boreas* **13**, 13–28.

Petrie, W.F. (1899) Sequences in prehistoric remains. *Journal of the Royal Anthropological Institute* **29**, 295–301.

Pilcher, J.R. and Hall, V.A. (1992) Towards a tephrochronology for the Holocene of the north of Ireland. *The Holocene* **2**, 255–259.

Pilcher, J.R. and Hall, V.A. (1996) Tephrochronological studies in northern England. *The Holocene* **6**, 100–105.

Pilcher, J.R., Baillie, M.G.L., Schmidt, B. and Becker, B. (1984) A 7,272-year tree-ring chronology for western Europe. *Nature* **312**, 150–152.

Pilcher, J.R., Hall, V.A. and McCormac, F.G. (1996) An outline tephrochronology for the Holocene of the north of Ireland. *Journal of Quaternary Science* **11**, 485–494.

Piperno, D.R. and Stothert, K.E. (2003) Phytolith evidence for early Holocene Cucurbita domestication in southwest Ecuador. *Science* **299**, 1054–1057.

Powell, E.N., Logan, A., Stanton, R.J. Jr., Davies, D.J. and Hare, P.E. (1989) Estimating time-since-death from the free amino acid content of the mollusc shell: A measure of time-averaging in modern death assemblages? Description of the technique. *Palaios* **4**, 16–31.

Prescott, J.R. and Hutton, J.T. (1994) Cosmic ray contributions to dose rates for luminescence and ESR dating: Large depths and long-term variations. *Radiation Measurements* **23**, 497–500.

Prescott, J.R., Huntley, D.J. and Hutton, J.T. (1993) Estimation of equivalent dose in thermoluminescence dating – the Australian slide method. *Ancient TL* **11**, 1–5.

Pringle, H. (1998) The slow birth of agriculture. *Science* **282**, 1446–1450.

Pyle, D.M. (1989) The thickness, volume and grainsize of tephra fall deposits. *Bulletin of Volcanology* **51**, 1–15.

Raban, A. (2000) Three-hole composite stone anchors

from a medieval context at Caesarea Maritima, Israel. *The International Journal of Nautical Archaeology* **29**, 260–272.

Radtke, U., Janotta, A., Hilgers, A. and Murray, A.S. (2001) The potential of OSL and TL for dating Lateglacial and Holocene dune sands tested with independent age control of the Laacher See tephra (12 880 a) at the Section 'Mainz-Gosenheim'. *Quaternary Science Reviews* **20**, 719–724.

Raisbeck, G.M., Yiou, F., Bourles, D., Lorius, C., Jouzel., J. and Barkov, N.I. (1987) Evidence for two intervals of enhanced ^{10}Be deposition in Antarctic ice during the last glacial period. *Nature* **326**, 273–277.

Ramsey, C.B. (1998) Probability and dating. *Radiocarbon* **40**, 461–474.

Reimer, P.J. and Reimer, R.W. (2001) A marine reservoir correction database and on-line interface. *Radiocarbon* **43**, 461–463.

Reimer, P.J., McCormac, F.G., Moore, J., McCormick, F. and Murray, E.V. (2002) Marine radiocarbon reservoir corrections for the mid- to late Holocene in the eastern subpolar North Atlantic. *The Holocene* **12**, 129–135.

Rendell, H.M. (1992) A comparison of TL age estimates from different mineral fractions of sands. *Quaternary Science Reviews* **11**, 79–83.

Roberts, R.G. (1997) Luminescence dating in archaeology: from origins to optical. *Radiation Measurements* **27**, 819–892.

Roberts, R.G., Flannery, T.F., Ayliffe, L.K., Yoshida, H., Olley, J.M., Prideaux, G.J., Laslett, G.M., Baynes, A., Smith, M.A., Jones, R. and Smith, B.L. (2001) New ages for the last Australian megafauna: continent-wide extinction about 46,000 years ago. *Science* **292**, 1888–1892.

Roberts, R., Bird, M., Olley, J., Galbraith, R., Lawson, E., Laslett, G., Yoshida, H., Jones, R., Fullagar, R., Jacobsen, G. and Hua, Q. (1998a) Optical and radiocarbon dating at Jinmium rock shelter in northern Australia. *Nature* **393**, 358–362.

Roberts, R., Yoshida, H., Galbraith, R., Laslett, G., Jones, R. and Smith, M. (1998b) Single-aliquot and single-grain optical dating confirm thermoluminescence age estimates at Malakunanja II rock shelter in northern Australia. *Ancient TL* **16**, 19–24.

Roberts, R.G., Jones, R., Spooner, N.A., Head, M.J., Murray, A.S. and Smith, M.A. (1994) The human colonisation of Australia: optical dates of 53,000 and 60,000 years bracket human arrival at Deaf Adder Gorge, Northern Territory. *Quaternary Science Reviews* **13**, 575–583.

Roberts, R.G., Jones, R. and Smith, M.A. (1990) Thermoluminescence dating of a 50,000-year-old human occupation site in northern Australia. *Nature* **345**, 153–156.

Robinson, W.S. (1951) A method for chronologically ordering archaeological deposits. *American Antiquity* **16**, 293–301.

Roosevelt, A.C., da Costa, M.L., Machado, C.L., Michab, M., Mercier, N., Valladas, H., Feathers, J., Barnett, W., da Silveira, M.I., Henderson, A., Sliva, J., Chernoff, B., Reese, D.S., Holman, J.A., Toth, N. and Schick, K. (1996) Paleoindian cave dwellers in the Amazon: the peopling of the Americas. *Science* **272**, 373–384.

Rose, N.L., Golding, P.N.E. and Battarbee, R.W. (1996) Selective concentration and enumeration of tephra shards from lake sediment cores. *The Holocene* **6**, 243–246.

Rose, W.I. and Chesner, C.A. (1987) Dispersal of ash in the great Toba eruption, 75 ka. *Geology* **15**, 913–917.

Saarinen, T. (1999) Palaeomagnetic dating of late Holocene sediments in Fennoscandia. *Quaternary Science Reviews* **18**, 889–897.

Sander, M., Bengtsson, L., Holmquist, B. and Wohlfarth, B. (2002) The relationship between annual varve thickness and maximum annual discharge (1909–1971). *Journal of Hydrology* **236**, 23–35.

Sanders, D.H. (1995) *Statistics: A First Course*. McGraw-Hill, Inc., New York, 624 pp.

Sandweiss, D.H., McInnis, H., Burger, R.L., Cano, A., Ojeda, B., Paredes, R., del Carmen Sandweiss, M. and Glascock, M.D. (1998) Quebrada Jaguay: early South American maritime adaptions. *Science* **281**, 1830–1832.

Santos, G.M., Bird, M.I., Parenti, F., Fifield, L.K., Guidon, N. and Hausladen, P.A. (2003) A revised chronology of the latest occupation layer of Pedra Furada Rock Shelter, Piauí, Brazil: the Pleistocene peopling of the Americas. *Quaternary Science Reviews* **22**, 2303–2310.

Sarna-Wojcicki, A. (2000) Tephrochronology. In (J.S. Noller, J.M. Sowers and W.R. Lettis, eds.) *Quaternary Geochronology: Methods and Applications*, 357–377. AGU Reference Shelf 4.

Sarnthein, M., Kennett, J.P., Allen, J.R.M., Berr, J., Grootes, P., Laj, C., McManus, J., Ramesh, R., SCOR-IMAGES Working Group 117 (2002) Deacadal-to-millenial-scale climate variability – chronology and mechanisms: summary and recommendations. *Quaternary Science Reviews* **21**, 1121–1128.

Schramm, A., Stein, M. and Goldstein, S.L. (2000) Calibration of the ^{14}C time scale to >40 ka by ^{234}U-^{230}Th dating of Lake Lisan sediments (last glacial Dead Sea). *Earth and Planetary Science Letters* **175**, 27–40.

Schulz, M. and Stattegger, K. (1997) SPECTRUM: Spectral analysis of unevenly spaced paleoclimatic time series. *Computational Geosciences* **23**, 929–945.

Schulz, M., Berger, W.H., Baillie, M., Luterbacher, J., Meincke, J., Negendank, J.F.W., Paul, A. and Ramseier, R.O. (2002) Tracing climate-variability: The search for climate dynamics on decadal to millennial time-scales. In (G. Wefer, W. Berger, K.-E. Behre and E. Jansen, eds.) *Climate Development and History of the North Atlantic Realm*, 125–148. Berlin: Springer Verlag.

Schwarcz, H.P. and Lee, H.-K. (2000) Electron spin resonance dating of fault rocks. In (J.S. Noller, J.M. Sowers and W.R. Lettis, eds.) *Quaternary Geochronology: Methods and Applications*, 177–186. AGU Reference Shelf 4.

Schwarcz, H.P., Buhay, W.M., Grün, R., Valladas, H., Tchernov, E., Bar-Yosef, O. and Vandermeersch, B. (1989) ESR dating of the Neanderthal site, Kebara Cave, Israel. *Journal of Archaeological Science* **16**, 653–659.

Schwarcz, H.P., Grün, R., Vandermeersch, B., Bar-Yosef, O., Valladas, H. and Tchernov, E. (1988) ESR dates for the hominid burial site of Qafzeh in Israel. *Journal of Human Evolution* **17**, 733–737.

Scott, E.M., Harkness, D.D. and Cook, G.T. (1998) Interlaboratory comparisons: lessons learned. *Radiocarbon* **40**, 331–340.

Shane, P. (2000) Tephrochronology: a New Zealand case study. *Earth Science Reviews* **49**, 223–259.

Shelkoplyas, V.N. and Morozov, G.V. (1965) Determination of the relative age of the Quaternary deposits of the middle Dneiper by thermoluminescent methods. In: Osnovie Problemi Izucheniia Chetvertichogo Perioda. Moscow: Nauka.

Siani, G., Paterne, M., Michel, E., Sulpizio, R., Sbrana, A., Arnold, M. and Haddad, G. (2001) Mediterranean sea surface radiocarbon age changes since the Last Glacial Maximum. *Science* **294**, 1917–1920.

Sikes, E.L., Samson, C.R., Guilderson, T.P. and Howard, W.R. (2000) Old radiocarbon ages in the southwest Pacific Ocean during the last glacial period and deglaciation. *Nature* **405**, 555–559.

Simola, H.L.K., Coard, M.A. and O'Sullivan, P.E. (1981) Annual laminations in the sediments of Loe Pool, Cornwall. *Nature* **290**, 238–241.

Singhvi, A.K., Sharma, Y.P. and Agrawal, D.P. (1982) Thermoluminescence dating of sand dunes in Rajasthan, India. *Nature* **295**, 313–315.

Smith, B.D. (1997) The initial domestication of Cucurbita pepo in the Americas 10,000 years ago. *Science* **276**, 932–934.

Smith, C. (1992) *Late Stone Age Hunters of the British Isles*. London: Routledge.

Smith, F.H., Trinkaus, E., Pettitt, P.B., Karavanic, I. and Paunovic, M. (1999) Direct radiocarbon dates from Vindija G_1 and Velika Pécina late Pleistocene hominid remains. *Proceedings of the National Academy of Sciences* **96**, 12281–12286.

Smith, M.A. (1987) Pleistocene occupation in arid central Australia. *Nature* **328**, 710–711.

Snowball, I. and Sandgren, P. (2002) Geomagnetic field variations in northern Sweden during the Holocene quantified from varved lake sediments and their implications for cosmogenic nuclide production rates. *The Holocene* **12**, 517–530.

Sparks, R.S.J., Bursik, M.I., Ablay, G.J., Thomas, R.M.E. and Carey, S.N. (1992) Sedimentation of tephra by volcanic plumes. Part 2: controls on thickness and grain-size variations of tephra fall deposits. *Bulletin of Volcanology* **54**, 685–695.

Sparks, R.S.J., Bursik, M.I., Carey, S.N., Gilbert, J.S., Glaze, L.S., Sigurdsson, H. and Woods, A.W. (1997) *Volcanic Plumes*. Chichester: John Wiley and Sons.

Stihler, S.D., Stone, D.B. and Begét, J.E. (1992) 'Varve' counting vs. tephrochronology and ^{137}Cs and ^{210}Pb dating: A comparative test at Skilak Lake, Alaska. *Geology* **20**, 1019–1022.

Stokes, S. and Lowe, D.J. (1988) Discriminant function analysis of late Quaternary tephras from five volcanoes in New Zealand using glass shard major element chemistry. *Quaternary Research* **30**, 270–283.

Stokes, S., Lowe, D.J. and Froggatt, P.C. (1992) Discriminant function analysis and correlation of late Quaternary rhyolitic tephra deposits from Taupo and Okataina volcanoes, New Zealand, using glass shard major element composition. *Quaternary International* **13/14**, 103–117.

Stringer, C. (2002) Modern human origins: Progress and prospects. *Philosophical Transactions of the Royal Society of London* **B357**, 563–579.

Stringer, C.B. and Grün, R. (1991) Time for the last Neanderthals. *Nature* **351**, 701–702.

Stringer, C.B., Grün, R., Schwarcz, H.P. and Goldberg, P. (1989) ESR dates for the hominid site of Es Skhul in Israel. *Nature* **338**, 756–758.

Stuiver, M., Braziunas, T.F., Becker, B. and Kromer, B. (1991) Climatic, solar, oceanic and geomagnetic influences on Late-Glacial and Holocene atmospheric $^{14}C/^{12}C$ change. *Quaternary Research* **35**, 1–24.

Stuiver, M., Reimer, P.J., Bard, E., Beck, J.W., Burr, G.S., Hughen, K.A., Kromer, B., McCormac, F.G., van der Plicht, J. and Spurk, M. (1998) INTCAL98 radiocarbon age calibration 24,000–0 cal BP. *Radiocarbon* **40**, 1041–1083.

Swanson, S.E. and Begét, J.E. (1994) Melting properties of volcanic ash. Volcanic Ash and Aviation Safety: Proceedings of the First International Symposium on Volcanic Ash and Aviation Safety. *U.S. Geological Survey Bulletin* **2047**, 87–92.

Swisher III, C.C., Rink, W.J., Antón, S.C., Schwarcz, H.P., Curtis, G.H., Suprijo, A. and Widiasmoro (1996) Latest *Homo erectus* of Java: potential contemporaneity with Homo sapiens in southeast Asia. *Science* **274**, 1870–1874.

Tarling, D.H. (1991) Archaeomagnetism and palaeomagnetism. In (H.Y. Göksu, M. Oberhofer and D. Regulla, eds.) *Scientific Dating Methods*, 217–250. Dorderecht: Kluwer Academic Publishers.

ter Braak, C.J.F. (1987) Ordination. In (R.H.G. Jongman,

C.J.F. ter Braak and O.F.R. van Tongeren, eds.) *Data Analysis in Community and Landscape Ecology*, 91–173. Wageningen: Pudoc.

ter Braak, C.J.F. and Smilauer, P. (1998) *CANOCO reference manual and user's guide to Canoco for Windows: Software for Canonical Community Ordination (version 4).* Ithaca: Microcomputer Power.

Tercier, J., Orcel, A. and Orcel, C. (1996) Dendrochronological study of prehistoric archaeological sites in Switzerland. In (J.S. Dean, D.M. Meko and T.W. Swetnam, eds.) *Tree rings, Environment and Humanity.* Proceedings of the International Conference, Tucson, Arizona, 17–21 May 1994. *Radiocarbon* **(1996)**, 7–582.

Thellier, E. (1981) Sur la direction du champ magnétique terrestre, en France, durant les deux derniers millenaires. *Physics of the Earth and Planetary Interiors* **24**, 89–132.

Thompson, R. and Oldfield, F. (1986) *Environmental Magnetism*. London: Allen and Unwin.

Thorarinsson, S. (1944) Tefrokronologiska studier på Island. *Geografiska Annaler* **26**, 1–217.

Thorne, A., Grün, R., Mortimer, G., Spooner, N.A., Simpson, J.J., McCulloch, M., Taylor, L. and Curnoe, D. (1999) Australia's oldest human remains: age of the Lake Mungo 3 skeleton. *Journal of Human Evolution* **36**, 591–612.

Tiljander, M., Ojala, A., Saarinen, T. and Snowball, I. (2002) Documentation of the physical properties of annually laminated (varved) sediments at a sub-annual to decadal resolution for environmental interpretation. *Quaternary International* **88**, 5–12.

Topham, J. and McCormick, D. (2000) A dendrochronological investigation of stringed instruments of the Cremonese School (1666–1757) including 'The Messiah' violin attributed to Antonia Stradivari. *Journal of Archaeological Science* **27**, 183–192.

Törnqvist, T.E., de Jong, A.F.M. and van der Borg, K. (1990) Comparison of AMS ^{14}C ages of organic deposits and macrofossils: a progress report. *Nuclear Instruments and Methods in Physics Research* **B52**, 442–445.

Torrence, C. and Compo, G.P. (1998) A practical guide to wavelet analysis. *Bulletin of the American Meteorological Society* **79**, 61–78.

Towner, R.H. (2002) Archaeological dendrochronology in the southwestern United States. *Evolutionary Anthropology* **11**, 68–84.

Turney, C.S.M. (1998) Extraction of rhyolitic component of Vedde microtephra from minerogenic lake sediments. *Journal of Paleolimnology* **19**, 199–206.

Turney, C.S.M. (1999) Lacustrine bulk organic $\delta^{13}C$ in the British Isles during the Last Glacial-Holocene transition (14–9 ka ^{14}C BP). *Arctic, Antarctic and Alpine Research* **31**, 71–81.

Turney, C.S.M., Lowe, J.J., Davies, S.M., Hall, V., Lowe, D.J., Wastegård, S., Hoek, W.Z., Alloway, B., SCOTAV and INTIMATE members (2004) Tephrochronology of Last Termination sequences in Europe: A protocol for improved analytical precision and robust correlation procedures (a joint SCOTAV-INTIMATE proposal). *Journal of Quaternary Science* **19**, 111–120.

Turney, C.S.M. and Lowe, J.J. (2001) Tephrochronology. In (W.M. Last and J.P. Smol, eds.) *Tracking Environmental Changes in Lake Sediments: Physical and Chemical Techniques*, 451–471. Dordrecht, The Netherlands: Kluwer Academic Publishers.

Turney, C.S.M., Bird, M.I., Fifield, L.K., Roberts, R.G., Smith, M.A., Dortch, C.E., Grün, R., Lawson, E., Ayliffe, L.K., Miller, G.H., Dortch, J. and Cresswell, R.G. (2001a) Early human occupation at Devil's Lair, southwestern Australia 50,000 years ago. *Quaternary Research* **55**, 3–13.

Turney, C.S.M., Bird, M.I., Fifield, L.K., Kershaw, A.P., Cresswell, R.G., Santos, G.M., di Tada, M.L., Hausladen, P.A. and Zhou, Y. (2001b) Development of a robust ^{14}C chronology for Lynch's Crater (North Queensland, Australia) using different pretreatment strategies. *Radiocarbon* **43**, 45–54.

Turney, C.S.M., Bird, M.I. and Roberts, R.G. (2001c) Elemental $\delta^{13}C$ at Allen's Cave, Nullarbor Plain, Australia: Assessing post-depositional disturbance and reconstructing past environments. *Journal of Quaternary Science* **16**, 779–784.

Turney, C.S.M., Coope, G.R., Harkness, D.D., Lowe, J.J. and Walker, M.J.C. (2000) Implications for the dating of Wisconsinan (Weichselian) Lateglacial events of systematic radiocarbon age differences between terrestrial plant macrofossils from a site in SW Ireland. *Quaternary Research* **53**, 114–121.

Turney, C.S.M. and Bird, M.I. (2002) Determining the timing and pattern of human colonisation in Australia: Proposals for radiocarbon dating 'early' sequences. *Australian Archaeology* **54**, 1–5.

Turney, C.S.M., Harkness, D.D. and Lowe, J.J. (1997) The use of micro-tephra horizons to correlate Lateglacial lake sediment successions in Scotland. *Journal of Quaternary Science* **12**, 525–531.

Tyldesley, J.B. (1973) Long-range transmission of tree pollen to Shetland. I. Sampling and trajectories. *New Phytologist* **72**, 175–181.

Ugumori, T. and Ikeya, M. (1980) Luminescence of $CaCO_3$ under N_2 laser excitation and application to archaeological dating. *Japanese Journal of Applied Physics* **19**, 459–465.

Valladas, H., Joron, J.L., Valladas, G., Arensburg, B., Bar-Yosef, O., Belfer-Cohen, A., Goldberg, P., Laville, H., Meignen, L., Rak, Y., Tchernov, E., Tiller, A.M. and Vandermeersch, B. (1987) Thermoluminescence dates for the Neanderthal burial site at Kebara in Israel. *Nature* **330**, 159–160.

Valladas, H., Reyss, J.L., Joron, J.L., Valladas, G., Bar-Yosef, O. and Vandermeersch, B. (1988) Thermoluminescence dating of Mousterian 'Proto-Cro-Magnon' remains from Israel and the origin of modern man. *Nature* **331**, 614–616.

van der Knaap, W.O., van Leeuwen, J.F.N., Fankhauser, A. and Ammann, B. (2000) Palynostratigraphy of the last centuries in Switzerland based on 23 lake and mire deposits: chronostratigraphic pollen markers, regional patterns, and local histories. *Review of Palaeobotany and Palynology* **108**, 85–142.

Vermeersch, P.M., Paulissen, E., Stokes, S., Charlier, C., van Peer, P., Stringer, C. and Lindsay, W. (1998) A Middle Palaeolithic burial of a modern human at Taramsa Hill, Egypt. *Antiquity* **72**, 475–484.

Verosub, K.L. (2000) Varve dating. In (J.S. Noller, J.M. Sowers and W.R. Lettis, eds.) *Quaternary Geochronology: Methods and Applications*, 21–24. AGU Reference Shelf 4.

Voelker, A.H.L., Grootes, P.M., Nadeau, M.-J. and Sarnthein, M. (2000) Radiocarbon levels in the Iceland Sea from 25–53 kyr and their link to the Earth's magnetic field intensity. *Radiocarbon* **42**, 437–452.

Wagner, G.A. (1998) *Age determination of young rocks and artifacts*. Berlin: Springer Verlag.

Wehmiller, J.F. and Miller, G.H. (2000) Aminostratigraphic dating methods in Quaternary geology. In (J.S. Noller, J.M. Sowers and W.R. Lettis, eds.) *Quaternary Geochronology: Methods and Applications*, 187–222. AGU Reference Shelf 4.

Westgate, J.A. and Gorton, M.P. (1981) Tephra Studies. In (S. Self and R.S.J. Sparks, eds.) *Correlation techniques in tephra studies*, 73–94. Dordrecht: Reidel.

Westgate, J.A., Perkins, W.T., Fuge, R., Pearce, N.J.G. and Wintle, A.G. (1994) Trace-element analysis of volcanic glass shards by laser ablation inductively coupled plasma mass spectrometry: application to tephrochronological studies. *Applied Geochemistry* **9**, 323–335.

Wick, L. (2000) Vegetation response to climatic changes recorded in Swiss Late Glacial lake sediments. *Palaeogeography, Palaeoclimatology, Palaeoecology* **159**, 231–250.

Wintle, A.G. (1993) Luminescence dating of aeolian sands: an overview. In (K. Pye, ed.) *The dynamics and environmental context of aeolian sedimentary systems*, 49–58. Geological Society Special Publications 72.

Wintle, A.G. (1997) Luminescence dating: Laboratory procedures and protocols. *Radiation Measurements* **27**, 769–817.

Wintle, A.G. and Huntley, D.J. (1980) Thermoluminescence dating of ocean sediments. *Canadian Journal of Earth Science* **17**, 348–360.

Wohlfarth, B. (1996) The chronology of the last termination: a review of radiocarbon-dated, high resolution terrestrial stratigraphies. *Quaternary Science Reviews* **15**, 267–284.

Yamaguchi, D.K., Atwater, B.F., Bunker, D.E., Benson, B.E. and Reid, M.S. (1997) Tree-ring dating the 1700 Cascadia earthquake. *Nature* **389**, 922–923.

Yamazaki, T. and Oda, H. (2002) Orbital influence on Earth's magnetic field: 100,000-year periodicity in inclination. *Science* **295**, 2435–2438.

Yasuda, Y. (2002) Origins of pottery and agriculture in East Asia. In Y. Yasuda (ed.) *The Origins of Pottery and Agriculture*, 119–142. New Delhi: Roli Books/Lustre Press.

Yiou, F., Raisbeck, G.M., Baumgartner, S., Beer, J., Hammer, C., Johnsen, S., Jouzel., J., Kubik, P.W., Lestringuez, J., Stiévenard, M., Suter, M. and Yiou, P. (1997) Beryllium 10 in the Greenland ice core project at Summit, Greenland. *Journal of Geophysical Research* **102**, 26783–26794.

Yokoyama, Y., Esat, T.M., Lambeck, K. and Fifield, L.K. (2000) Last ice age millennial scale climate changes recorded in Huon Peninsula corals. *Radiocarbon* **42**, 383–401.

Zeller, E.J., Levy, P.W. and Mattern, P.L. (1967) Geologic dating by electron spin resonance. *Symposium on Radioactive Dating and Low Level Counting. I.A.E.A., Wien, Proceedings*, 531–540.

Zhu, R.X., Hoffman, K.A., Potts, R., Deng, C.L., Pan, Y.X., Guo, B., Shi, C.D., Guo, Z.T., Yuan, B.Y., Hou, Y.M. and Huang, W.W. (2001) Earliest presence of humans in northeast Asia. *Nature* **413**, 413–417.

Zillén, L.M., Wastegård, S. and Snowball, I.F. (2002) Calendar year ages of three mid-Holocene tephra layers identified in varved lake sediments in west central Sweden. *Quaternary Science Reviews* **21**, 1583–1591.

Zolitschka, B. (1996) Recent sedimentation in a high arctic lake, northern Ellesmere Island, Canada. *Journal of Paleolimnology* **16**, 169–186.

5

Integrated studies in environmental archaeology

5.0 Chapter summary

The preceding chapters have provided a conceptual and practical framework for the sub-discipline of environmental archaeology. In doing so they have provided the reader with overviews or details of the contextual, geoarchaeological, bioarchaeological and geochronological tools necessary to investigate the relationships between humans and their environment over a range of spatial and temporal scales. This final chapter demonstrates the wealth of information that may be obtained by applying these tools to provide fully integrated archaeological and environmental archaeological studies conducted over a range of spatial scales. This will be achieved using four case studies.

First, we present the results of the Dover Bronze Age boat project (UK), a spectacular archaeological discovery in 1992 that presented a unique opportunity to study the archaeology and environmental context of an internationally important structure using a range of scientific methods. Our second case study – the London Thames valley – demonstrates how the investigation of a wide range of archaeological and geological archives, using a range of environmental archaeological techniques, can be used to formulate a detailed model of the human environment at a sub-regional scale. Thirdly, we present the various lines of archaeological and scientific evidence that have engendered a fascinating debate on the relationships between irrigation, salinity and cultural change in ancient Mesopotamia. Finally, we outline and discuss the various archaeological and environmental archaeological hypotheses put forward to explain worldwide megafaunal extinction.

5.1 Microscale – the Dover Bronze Age boat

In September 1992, buried 6 m below the modern streets of Dover in south-east England and about 200 m inland from the present shore, the hull of a beautifully preserved sewn-plank boat of middle Bronze Age date was discovered by a small team of archaeologists working on a development site. The unexpected nature of the find and the very short time available (a few days only) meant that the excavation and sampling methodology was necessarily reactive. However, one end of the boat (of a complex and unique design) was safely recovered. The other end lay underneath existing buildings, and proved impossible to recover.

The excavation of the Dover boat was a dramatic and exhausting exercise. From a standing start, with few resources in place and in the teeth of formidable practical difficulties, the team recovered what has been described as one of 'the most spectacular examples of prehistoric

FIGURE 5.1 Working conditions at the base of the 6 m deep shaft during the excavation of the Dover Bronze Age boat (Photo: Andrew Savage © Canterbury Archaeological Trust).

woodwork ever found in Europe' in just 15 days. When word first got out about the discovery, English archaeology was galvanized into action; offers of assistance poured in from around the country, and overnight the initial small team was joined by what seemed like a small army of archaeological experts, all eager to help in whatever way they could. The success of the project is a testament to the skills of the huge multidisciplinary team (some 150 individuals) that finally worked on various aspects of the project from the initial excavation, and analysis (Clark, 1993a, 1993b, 1994; Clark and Keeley, 1997), through to the award-winning presentation at Dover museum and the final monograph (Clark, 2004).

A length of about 9.5 m of the vessel was recovered, the northern end of the boat lying outside the area of excavation. The hull was approximately 2.3 m wide at its widest point, and consisted of four sculpted oak planks, joined together by a system of twisted yew (*Taxus* sp.) withy stitches, wedges and transverse timbers (Marsden, 2004). It was made watertight by placing pads of moss along the seams between the planks, which were compressed and held in place by thin wooden laths. The holes cut through the planks for the stitches were sealed by some form of stopping material. The Y-shaped end of the boat, presumably the bow, was an extremely complex piece of woodworking that originally held an end board, giving the boat a punt-like appearance. (See Fig. 5.1.)

The boat was cut into 32 pieces, each placed on plywood boards and stored in fresh water. Though the inboard surfaces of the boat had been cleaned during excavation, sediment still adhered to the outboard surfaces of the timbers. Apart from the boat timbers themselves, the data recovered included around 550 bulk samples, 545 flint artefacts, 3 sherds of pottery, wood fragments, hand collected animal and fish bone, a fragment of shale, 9 monoliths, 7 Kubiena tins and thirteen 1 m drill cores. The analysis project was initially designed to address a wide range of research questions (Clark, 1994) requiring different approaches for different datasets.

Samples were taken of the constructional materials from various points on the boat and the wood samples thin sectioned for observation under transmitted light microscopy (Hather and Wales, 1993). Nearly all of the timbers of the boat were of oak (*Quercus* sp.); two samples, however were of yew (*Taxus baccata*), from a lath used to repair the hull where it had split, and one sample was of hazel (*Corylus avellana*) from a constructional lath. The moss (or bryophyte) samples were also studied microscopically and five taxa identified: *Plagiothecium cf denticulatum*, *Isopterygium cf elegans*, *Thamnobryum* sp., *Hypnum* sp. and *Eurhynchium* sp. – all common types in north-west Europe. There did not appear to be any selection of particular taxa for different purposes. The stopping material used to make the stitch holes watertight was analysed using a variety of techniques, including infrared spectrometry, thin-layer chromatography, gas chromatography and mass spectrometry. The results suggested that the stopping material consisted of a mix of beeswax and resin (probably from *Betula* sp.), sometimes mixed with animal fat.

Recording consisted of a series of 1:1 drawings of each of the 32 pieces the boat had been cut into, then bringing the results together so that an accurate drawing of the entire boat was produced (Marsden, 2004). The fact that the boat had been cut into pieces caused some problems, but also had several benefits. One of these was the series of cross sections through the timber planks that facilitated examination of both the medullary rays and annual rings of the trees from which the planks had been hewn, as well as providing evidence that the timbers had been severely compressed and distorted during their long burial under six metres of overburden. Moreover, it proved possible by calculation to correct this compression, and ascertain the original thickness and shape of the boat timbers when it was built (Darrah, 2004a) (see Fig. 5.2).

FIGURE 5.2 The Dover Bronze Age boat *in situ* (Photo: Andrew Savage © Canterbury Archaeological Trust).

Unfortunately, the boat was incomplete; not only was one end unretrievable, but there was clear evidence for a pair of upper side planks that had been removed in antiquity, leading to a number of hypotheses about the missing parts of the boat (Roberts, 2004). Three possible candidates for the original hull form were selected for further analysis: one barely longer than the recovered remains, with a flat transom ('Dover 1'); one with equal ends ('Dover 2'); and one similar to Dover 2, but having a length based on a 15 m one-length bottom planking, thought to be the maximum size possible ('Dover 3'). Each reconstruction was subjected to a series of mathematical analyses to calculate the amount of longitudinal bending, the stresses on the hull when afloat, the strength of the stitched seams, etc. in order to assess its seaworthiness. The results showed that as length increased, given the relative shallowness of the hulls, the vessels would flex excessively in even modest sea conditions, damaging the plank fastenings and ultimately the structure of the hull itself. Dover 2 and 3 could only have been calm water craft so, assuming that the boat was a sea-going vessel (there is no large river at Dover), the most likely candidate was thought to be Dover 1. The preferred solution was then subjected to further mathematical analysis; its dry weight was calculated (2.26 tonnes), as was its displacement and stability at various loadings, and the power needed to propel it at various speeds with different loads. This gave a range of possibilities, but the results indicated that the original Dover boat could withstand reasonable sea conditions produced by Force 3 winds, travelling between 4 and 5 knots (according to the number of crew), and carry a cargo of up to 3 tonnes. It was also clear that at least one crew member must have been employed in bailing out water, as the boat would have leaked!

The function of the boat proved difficult to determine. Its hull coefficients suggest it was a good load carrier, but not capable of high speeds (Roberts, 2004: table 10.12). Unfortunately there was no cargo *in situ*.

It is also difficult to assess the boat's operative range: analysis of the hull form suggests that, on a good day, a distance of around 30 nautical miles would be a reasonable aim, and of course a voyage could take many days. A fragment of shale found on the inboard surface of the boat was analysed by scanning electron microscope, X-ray fluorescence and infrared analysis prior to geological examination of a simple smear slide and nannofossil analysis (Bown *et al.*, 2004). The results of these studies (in particular the nannofossil (coccolith) analysis) demonstrated that the fragment originated in Kimmeridge Bay on the southern British coast, about 260 km west of Dover. Just 25 km to the north of Dover, on the Isle of Thanet, a large Trevisker urn has been excavated, dated to around 1600–1320 cal BC (Gibson *et al.*, 1997). It had been brought to Thanet from Cornwall, in the south-west of Britain some 465 km to the west. These finds, and others like them, may give some indication of the distances travelled in the middle Bronze Age.

The boat had a long life: its outboard surface was worn, presumably by periodic beaching, and the side had split and been repaired on several occasions. The method of repair was quite ingenious: stitch holes were cut above and below the split, moss placed over the split, covered with a thin lath (usually of oak, with one of yew), and the lath lashed to the boat side with yew withies, compressing the moss and making the split watertight. Also one of the main laths sealing the boat seams was of hazel (*Corylus* sp.), rather than oak, suggesting that the original had been replaced during the boat's working life.

It was clear, even at the time of excavation, that the remains of the boat did not represent a shipwreck: there was evidence that it had been deliberately dismantled at the time of its abandonment. Along the top edge of each side were a number of withy stitches, indicating that there was originally an upper side plank. These stitches had been severed with a blade of some kind, in order to purposefully remove the planks. Similarly, the end board had been deliberately removed; the wedges holding it in place had clearly been cut through with an axe.

The boat was found entirely enclosed in a series of tufa/silts, tufa-rich sand and gravels that appeared to have accumulated relatively rapidly, perhaps explaining the remarkable preservation of the boat timbers (Green, 1998). Study of the biological remains from these deposits demonstrated that the boat was abandoned in a freshwater environment, in the shallows of a braided stream, perhaps in a backwater subject to gentle flooding (Keeley *et al.*, 2004). The diatom assemblage was dominated by freshwater taxa (in particular *Fragilaria* taxa), almost all non-planktonic, i.e. associated with or attached to submerged surfaces, indicating a shallow water environment (Cameron and Dobson, 1998). The fish bones, mollusc shells, plant macrofossils, insect remains and pollen were overwhelmingly of freshwater species (Allison, 1998; Fairbairn, 1998; Locker, 1998; Lowe *et al.*, 1998; Wilkinson, 1998), as was the ostracod assemblage, which included specimens typical of temporary pools and flooded water meadows (e.g. *Priococypris serrata* and *Potamocypris fulva*; Robinson, 1998).

This environmental evidence is crucial in understanding the nature of the boat and the circumstances of its abandonment. First, the data is unequivocal that the Dour valley at the time of the boat's disposal did not possess a major river that would have required a large boat to cross it. Indeed, the shallow braided stream suggested by the environmental evidence, probably with exposed mud banks and shallow pools subject to drying out, may have necessitated the boat being manhandled to the place where it was eventually dismantled. This seems counter-intuitive. Why such effort to move the heavy vessel upstream simply to break it up? Similarly, it is hard to understand why the boat was dismantled at all. Although there is some evidence of boat parts being recycled in the construction of new vessels in antiquity (e.g. the Galilee boat; Wachsmann, 1995), partly seasoned oak is extremely difficult to work, particularly with bronze tools, and the suggestion that the boat was taken apart to re-use some of its elements seems highly unlikely. There were also some peculiar aspects of the damage done to the boat when it was dismantled. The cut withies on the upper edges of each side showed that two upper planks had been removed; the wedges originally holding the end board in place at the bow had been carefully cut through, probably with an axe, in order to remove the timber. The condition of the remaining wedges shows that the board itself would have had to have been broken in order to extract it. The central rails running along the inboard centre line of the vessel had also been carefully cut through, exposing one of the transverse timbers that held the bottom of the boat together. There seems no rational explanation for this.

So the boat was not an accidental loss, nor is it likely it was dismantled in order to re-use its constituent timbers. It was not abandoned at a place where it could have been used effectively, but rather manhandled upstream to a liminal location at the boundary between earth and water, where it was deliberately broken up and selectively and consciously damaged. In the context of our knowledge of Bronze Age deposition practices from other sites, perhaps the best explanation for these factors is that the boat was deliberately deposited in a ritual act; symbolically 'killed' and returned to the waters at the end of its working life, or as part of a rite of passage of an individual closely associated with it (Champion, 2004).

It was initially hoped that the sections through the timbers would facilitate dendrochronological dating of the boat, but the compression and distortion of the annual rings in the timber, exacerbated by the conservation process, made this very difficult. One boat piece, kept back from conservation specifically for wiggle-matching by high-precision radiocarbon dating, failed to produce an absolute date (Bayliss *et al.*, 2004). A small subsection of the ring sequence of this piece did produce consistent results against a series of reference chronologies from the Irish Midlands spanning the period 1742–1589 BC, but this was not considered an acceptable dendrochronological date. A group of five bidecadal blocks was therefore selected from this sequence for high-precision radiocarbon dating. This produced a date of 1685–1595 cal BC for the outer ring of the sequence, which correlated reasonably well with the date of

FIGURE 5.3 Some of the 32 pieces the Dover Bronze Age boat were cut into during its excavation. Here we can see part of the complex end of the boat (presumably the bow) prior to conservation (Photo: Andrew Savage © Canterbury Archaeological Trust).

1589 BC derived from tree ring dating. However, it could not be taken as the date of the felling of the trees used for the Dover boat, as an allowance had to be made for the wood removed during its construction (no sapwood was present on any of the boat planks). Taking into account the results of radiocarbon dating of samples of the yew withies and moss waterproofing, together with an estimate of the number of missing annual rings and sapwood, it can be suggested that the Dover boat was constructed in 1575–1520 cal BC (at 95 per cent confidence).

Analysis of the medullary rays and annual rings of the boat planks suggested that the trees used in their construction were at least 11 m long and 1 m in diameter. The absence of knots or any evidence of side branches suggests that the trees grew in dense oak woodland, and were about 350 years old when felled (Darrah, 2004b). Study of the yew withies showed they were of varying age (between 6 and 16 years) and had been cut in the early spring. The absence of compression wood suggests the branches grew vertically, perhaps indicating they were coppiced, though this is not conclusive (Hather and Wales, 1993). (See Fig. 5.3.)

The boat timbers were very fragile, and it was clear that they would need conservation from the moment of their discovery. Subsequent analysis showed that the water content of the main planks varied from around 70–90 per cent of the wet weight, with a specific gravity of 0.13–0.26 (as opposed to fresh oak which is 0.55–0.64; Watson, 2004). If allowed to dry naturally, the timbers were likely to shrink 10–15 per cent longitudinally, 14–16 per cent radially, and a significant 44–60 per cent tangentially. The fact that the boat had been cut into pieces fortuitously meant that they were small enough for treatment by polyethylene glycol (PEG) replacement and freeze-drying. The boat pieces were then returned to Dover, where they were reassembled. This process was complicated by the fact that the

FIGURE 5.4 Display of the Dover boat (© Canterbury Archaeological Trust).

exact shape of the outboard surface was unknown, and that the conserved pieces had differentially shrunk in relationship to each other, albeit by a small amount (Clark *et al.*, 2004). The boat now rests in a steel case 10.5 m long, fitted with 18 toughened-glass panels, maintained at a temperature of 18°C ±2°C and a relative humidity of 55 per cent ±4 per cent. It is the centrepiece of a multi-award-winning gallery focused on life in the Bronze Age, and a testament to the benefits of an integrated approach to archaeological research. (See Fig. 5.4.)

5.2 Mesoscale – the prehistoric human environment of the London Thames

The prehistoric environmental history (c. 11,500–2000 BP) of the London Thames has been the subject of intensive data gathering (Fig. 5.5), both proactive (archaeological and palaeoenvironmental research projects) and reactive (archaeological investigations as a consequence of urban development), resulting in the formulation of a comprehensive model (e.g. Sidell *et al.*, 2002: Fig. 5.6). The model has provided the basis for a better understanding of prehistoric human-environment interactions, and therefore represents an excellent example of how different lines of archaeological evidence (spatial archaeology, geoarchaeology,

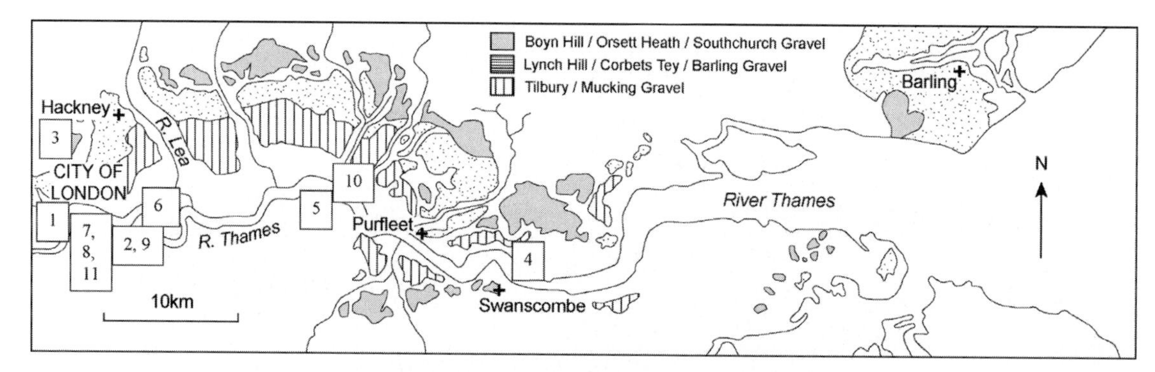

FIGURE 5.5 London basin showing the location of the gravel terraces, Pleistocene sites (+) and Holocene sites numbered as follows: (1) Westminster, (2) Bramcote Green, (3) Hampstead Heath, (4) Tilbury, (5) Crossness, (6) Bryan Road, (7) Joan Street, (8) Union Street, (9) Bricklayers Arms, (10) Hornchurch Marshes, (11) Phoenix Wharf.

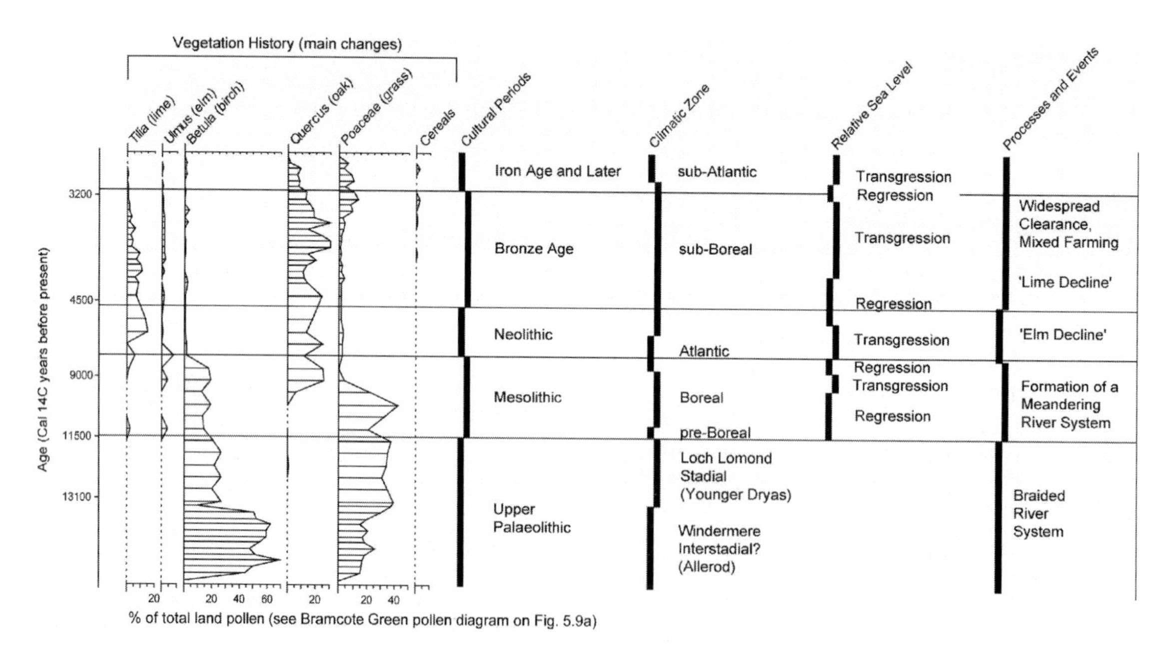

FIGURE 5.6 Model of environmental changes in the London Thames during the late Devensian and Holocene.

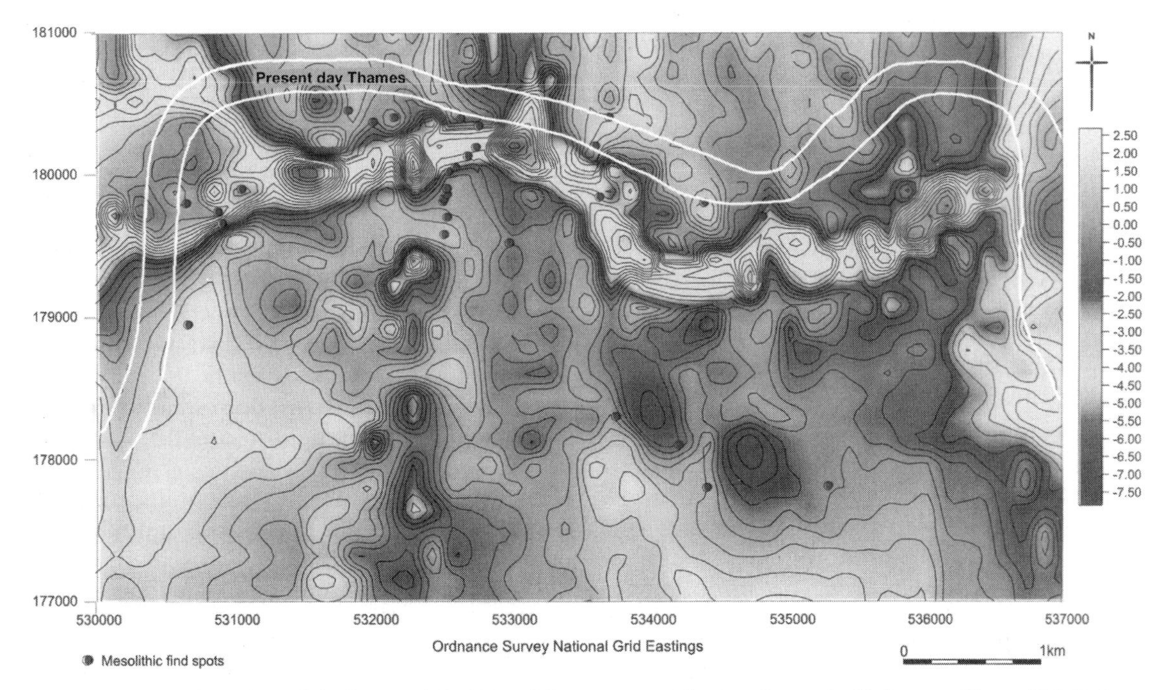

FIGURE 5.7 Predicted model of part of the River Thames route during the early Holocene. (From Sidell *et al.*, 2002)

bioarchaeology and geochronology) can be fully integrated.

The geomorphological and hydrological context for much of the human activity was shaped during the earlier climatic oscillations of the Pleistocene (Marine Isotope Stages 12 to 2), which led to the migration of the River Thames into its present position (MIS 12), and the formation of a highly complex series of river terraces and major tributary channels (Gibbard, 1994, 1995, 1999; Bridgland, 1995; Bridgland *et al.*, 1995). During the last cold stage (MIS 2, Devensian in Britain, c. 70,000–10,000 years ago), sands and gravels which today form the valley floor of the River Thames were deposited in a braided river system (the 'Shepperton Gravel'), the surface of which was differentiated by upstanding bars (eyots) and intervening channels. The climatic amelioration of the early Holocene (c. 11,500 BP) resulted in highly significant changes in the energy of the river system, and the formation of a relatively shallow meandering river with low sinuosity. Sedimentological data from central London suggest that during this stable period the position and form of the River Thames was therefore slightly different from the present day, with little evidence for flood plain sedimentation until approximately 6400 cal BP (Wilkinson *et al.*, 2000). Indeed, it is highly likely that the route of the River Thames was south of its present position (Fig. 5.7; Sidell *et al.*, 2002). This valley floor topography has been progressively buried, however, during the Holocene by mainly fine-grained, and sometimes organic, estuarine and fluvial sediments (e.g. Devoy, 1979, 1982). These interbedded mineral sediment and peat deposits are thought to have been the result of one or more of the following processes: (a) lateral channel migration; (b) flood plain formation; (c) estuarine contraction/expansion; (d) downwarping (subsidence); (e) changes in relative sea level; and (f) human impact within the River Thames catchment.

Of these natural and human processes that affected the environment, changes in the height of relative sea level following the end of the last glaciation had a considerable influence on the type of environment available for humans to exploit (particularly to the east of central London) and the timing of its availability.

During the early Holocene, these low lying areas were greatly affected by expansion of the estuary during the rapid rise in relative sea level by over 15 m between c. 11,500 and 6850 cal BP. This resulted in the formation of peat land (before c. 8000 cal BP and again between c. 7000 and 6500 cal BP) and estuarine conditions (c. 8000–7000 cal BP). Similar trends have been recorded during the middle Holocene, with estuarine conditions between c. 6500 and 5000 cal BP, followed by a prolonged period of contraction (reduction in channel width), and peat land formation (c. 5000–4000 cal BP), before being inundated once again after c. 3500 cal BP – although timing of this event within the lower Thames valley varied considerably. Against this background of widespread changes in estuary morphology and sedimentation, peat land formation was undoubtedly initiated in some areas as a consequence of local factors (e.g. tributary channel lateral movement or abandonment), or continued in others irrespective of relative sea level fluctuations. Nevertheless, certainly after c. 3500 cal BP, the effects of estuarine expansion were being registered further west (e.g. Westminster) with the deposition of brackish water sediments (Long *et al.*, 2000; Wilkinson *et al.*, 2000).

During the periods in which these changes in Holocene geomorphology and hydrology were taking place, there were several distinctive changes in vegetation cover in the London Thames which were due to the factors listed above, as well as northern hemispheric climate change, soil development and natural vegetation succession. Prior to the start of the Holocene, the ice-free tundra environment of London and south-east England undoubtedly witnessed a period of climatic oscillations for which there is considerable evidence across north-west Europe (Catt, 1979; Roberts, 1989). These changes may be summarized as follows:

FIGURE 5.8 Excavations at Bramcote Green. (Photo: James Rackham)

1 Approximately 16,000 years ago, warmer and more arid conditions resulted in a global retreat of ice sheets, and a rise in sea level.
2 Deglaciation in Britain between 14,000 and 11,500 cal BP oscillated between warm and cold conditions, with profound effects on soils and vegetation.
3 During the Windermere Interstadial, between 14,000 and 13,000 cal BP, summer and winter temperatures increased to approximately 17°C and 0–1°C (respectively), and soil development (e.g. raw humus or rendzina-type) provided conditions suitable for vegetation succession from open grassland (Poaceae, e.g. *Festuca rubra*) to shrubland (*Juniperus*, e.g. *J. communis*) and finally woodland (*Betula*, e.g. *B. pendula*).
4 A climatic deterioration between 13,000 and 11,500 BP (Loch Lomond Stadial) caused a return to periglacial conditions in London (Fig. 5.5; e.g. Bramcote Green and Silvertown), formation of arctic structure soils, colonization by shrub tundra vegetation (*Betula nana*; *Salix herbacea*; *Artemisia*, e.g. *A. glacialis*), and a reduction in summer and winter temperatures to 10°C and −20°C (respectively).

Following this transitional period, the climatic amelioration of the early Holocene (c. 11,500–8000 cal BP) and the return of full interglacial conditions had a significant effect on vegetation cover and soil development. The climatic amelioration led to the development of base-rich brown-earth soils (Catt, 1979) and a closed vegetation cover. Pollen analysis (Figs. 5.8 and 5.9; lacustrine sequence at Bramcote Green) suggests that *Betula* (e.g. *B. pendula*) and *Pinus* (e.g. *P. sylvestris*) woodland colonized the London Thames, together with *Alnus* (e.g. *A. glutinosa*), *Salix* (e.g. *S. alba*), *Corylus* (*C. avellana*), *Quercus* (e.g. *Q. pubescens*), *Tilia* (e.g. *Tilia cordata*) and a range of herbaceous taxa. This mosaic of dry land and wetland vegetation bordering a major river and nearby lake would have been seemingly ideal for occupation by human groups (Mesolithic cultural period: c. 11,500–6300 cal BP) and for the exploitation of wild plants and animals, including hazel nuts, acorns, *Cervus elaphus* and *Bos primigenius*. Archaeological evidence found on the southern shores of the Bramcote Green lake indicate the presence of a full range of tool preparation activities within flint assemblages: 'tested nodules, complete and fragmentary cores and preparation flakes (e.g. core tablets, core rejuvenators and crested pieces), large numbers of unmodified flakes and blades, and a range of retouched tools (principally microliths, scrapers and burins) and tool waste (microburins, sharpening flakes and burin spalls)' (Sidell *et al.*, 2002: 14). These tools, which were undoubtedly manufactured as well as repaired on the site, would have been used for a wide range of activities, including piercing, working and scraping dry hide, working antler, and cutting hide, meat or plant materials. These activities were associated with localized burning, and clearly suggest hunting and possible seasonal reoccupation of the site.

During the middle Holocene (c. 8000–3000 cal BP), two extremely important environmental and cultural changes occur. First, the annual temperature increased by approximately 1–2°C between c. 9000 and 5000 cal BP ('climatic optimum'), and secondly the hunter-gatherer subsistence strategies of Mesolithic human populations were replaced by farming communities (Neolithic cultural period: 6300–4100 cal BP). During this prolonged period of 'stable' climatic conditions, according to the ice core and glaciological records the climate was actually highly variable:

- c. 9000–7500 cal BP, maximum temperatures during the Holocene;
- dramatic short-term lowering of temperatures at c. 8200 cal BP;
- c. 7500–5000 cal BP, oscillating temperatures.

The consequent changes in geomorphology, sedimentology and hydrology (noted above), as well as soils and vegetation cover, associated with these events may have caused substantial environmental stress and affected the availability of natural resources. On dry land, there is evidence to suggest that after c. 7500 BP, warmer-drier (c. 7500–6300 cal BP) and colder-wetter (6300–5000 cal. BP) conditions led to a

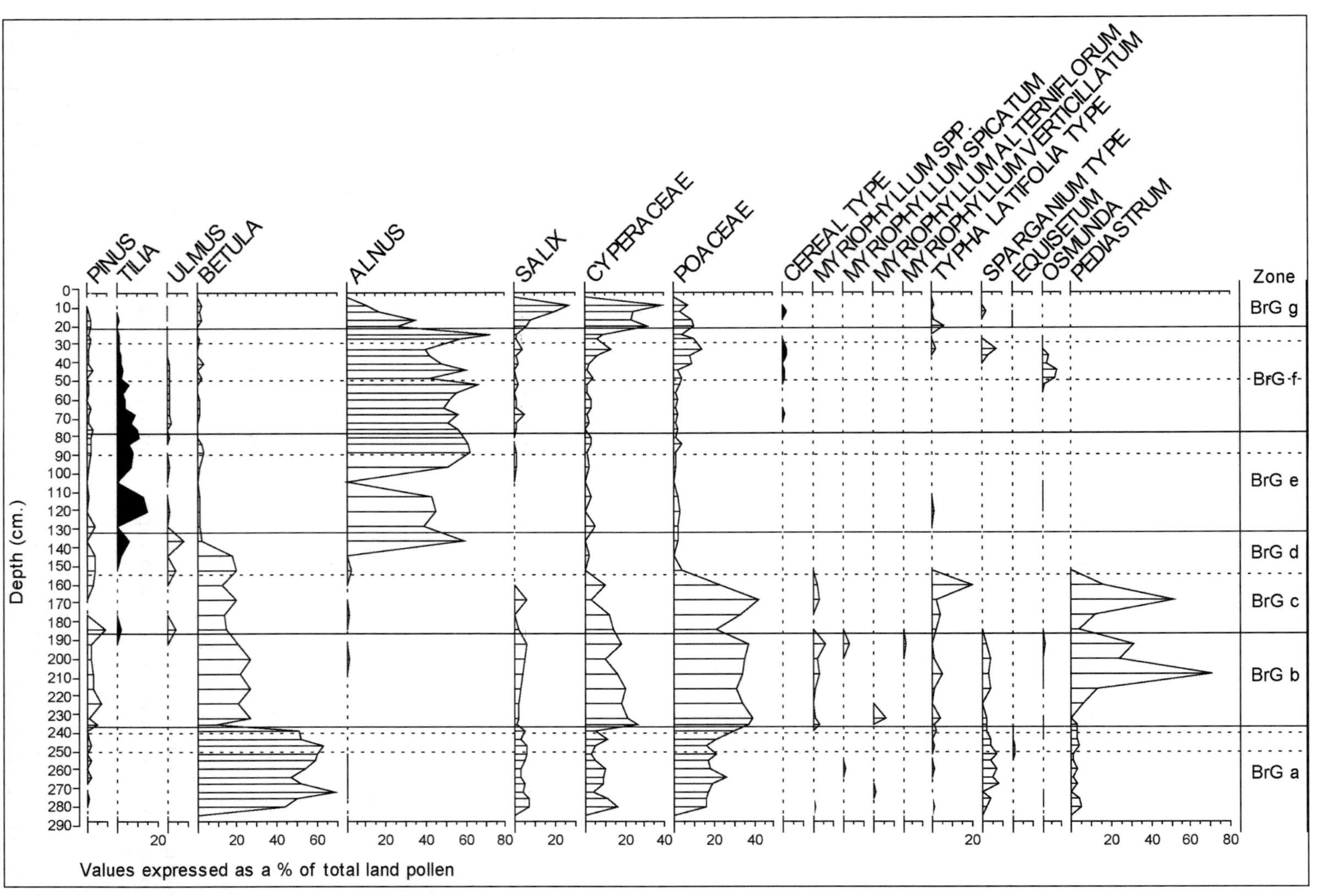

FIGURE 5.9 Summary pollen diagram from Bramcote Green, London. (From Branch and Lowe, in Thomas and Rackham, 1996)

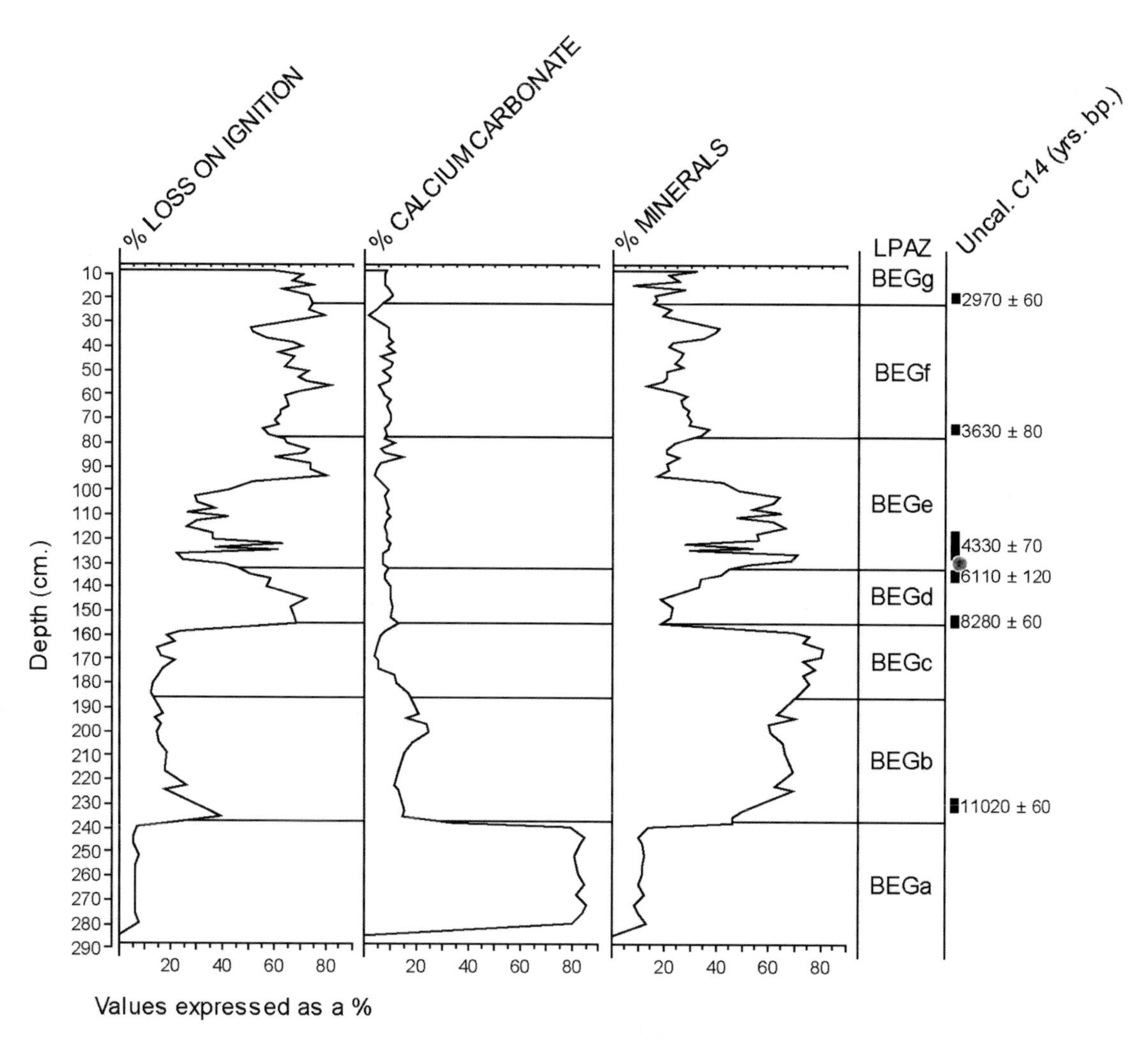

FIGURE 5.9 *(cont.)*
Sedimentary data and radiocarbon determinations to accompany the percentage pollen diagram from Bramcote Green.

deterioration in soil status, and in lowland areas this was characterized by a continuum from 'argillic brown earth to brown podzolic soil to podzol' (Bell and Walker, 1992: 100).

Following the transition to the Neolithic, there was a dramatic change in the landscape brought about by incoming domesticated plants (e.g. *Hordeum* sp. and *Triticum* sp.) and animals (e.g. cattle, sheep and goats). Agricultural activities may have been based on a system of shifting cultivation, involving temporary clearance of woodland by burning and felling prior to cultivation, abandonment and finally woodland regeneration (Fig. 5.10). The impact of early farming was felt mainly in those areas with better quality soils suitable for a mixed agricultural economy, and resulted in a modification of the environment to supply pasture, meadow and cropland. For example, evidence suggests that in the hinterland of London (Fig. 5.5; e.g. Hampstead Heath) cereal cultivation during the earliest Neolithic was

FIGURE 5.10 Clearing woodland. (After Merriman, in Sidell *et al.*, 2002; Courtesy of the Museum of London.)

accompanied by woodland clearance while on the margins of the River Thames, several archaeological sites have provided evidence for localized changes in woodland composition and structure. Against this background of human interference in natural vegetation succession, there is unequivocal evidence for soil erosion/degradation (Catt, 1979) and further changes in climate (warmer-drier) between c. 5000 and 3500 cal BP, however their relationships with vegetation change remain unclear.

The most significant change is the decline in *Ulmus* woodland, a macroscale event which occurred between about 6400 and 5300 cal BP (see Section 3.4.2). Unfortunately, in the London Thames there is a paucity of information due to very few sites having sedimentary sequences spanning the period of interest, imprecise dating and also possible pollen taphonomic problems detecting the *Ulmus* decline in low lying, wetland sites surrounded by closed woodland. The *Ulmus* decline is well represented on the eastern margins of London, although the radiocarbon dates from Tilbury (after c. 7075 cal BP), Crossness (after c. 6465 cal BP), Stonemarsh (c. 5625 cal BP) and Broadness marsh (c. 5965 cal BP) have a wide age range. Only three other sites have provided unequivocal evidence for the *Ulmus* decline: Bryan Road (c. 6000 cal BP), Silvertown (c. 5760 cal BP) and Bramcote Green (c. 7000–5000 cal BP). However, two sites, Joan Street and Union Street, provide important evidence suggesting that following the 'primary' *Ulmus* decline, woodland regenerated prior to a further decline sometime after 5500 cal BP.

The possible relationship between farming and the *Ulmus* decline ('primary' and 'secondary') was discussed in Chapter 3, where it was noted that there is a substantial body of wood macrofossil evidence throughout Europe for *Ulmus* exploitation for animal fodder. However, it is likely that many pollen diagrams will provide only equivocal evidence for human activity being the cause of the decline, since it is extremely difficult to confidently identify human activities (other than cereal cultivation) using this method. Indeed, this may explain why in many studies there is no evidence for human activity prior to, during or immediately after the 'primary' *Ulmus* decline, with the exception of Hampstead Heath (c. 6300–5800 cal BP; Greig, 1992). It may also highlight the significant spatial variation in Neolithic farming practices, with some areas probably more suitable for pasture/meadow, while others supported cultivated fields. Although these arguments do not dismiss other possible causes for the 'primary' *Ulmus* decline (see Chapter 3), they may provide a more secure case for explaining the 'secondary' *Ulmus* decline during the later Neolithic, especially where there is better bioarchaeological evidence (Fig. 5.5; e.g. Joan Street, Union Street and Bryan Road).

The removal of localized areas of woodland would have created a mosaic of pasture, cultivated fields, meadow and natural vegetation communities in the London Thames. Does archaeological evidence support this interpretation? In the London Thames there are only a few Late Mesolithic sites, perhaps as a consequence of rising river levels, linked to changes in relative sea level, resulting in the inundation of traditional activity areas (Sidell *et al.*, 2002). During the Neolithic, there are also a few sites with struck flint and pottery. One of the best examples, from Bricklayers Arms (Fig. 5.5), was located on the edge of the former lake mentioned above. Associated with two timber platforms displaying cut marks and made of *Salix*, *Alnus* and *Betula*, two flint axes and numerous flakes may represent either a working environment for exploitation of the lake and/or a ritual link between dry land and the open water body (Sidell *et al.*, 2002).

The possible correlation between the decline of *Ulmus* and a period of climatic deterioration between c. 6300 and 5000 cal BP has largely been disregarded because of the evidence, albeit limited, for disease and human activity. However, at several sites there are pronounced changes in peat stratigraphy and vegetation cover that might have been triggered by climate change (Fig. 5.5; e.g. Bramcote Green and Hornchurch Marshes). For example, the colonization by *Taxus* (e.g. *Taxus baccata*) woodland of peat land from c. 5200 cal BP, and possibly earlier, in the London Thames provides the most convincing evidence yet of the response by vegetation to Holocene climate change. At the present day, *Taxus baccata* prefers calcareous substrates in the United Kingdom and tends to avoid wet soils such as acidic peat land. The discovery of *Taxus* wood macrofossils and pollen preserved in peat at Hornchurch Marshes (Fig. 5.11) suggests, therefore, that environmental conditions were more suitable in the past for its growth. Perhaps the most important factor would have been the reduction in moisture content of the peat after c. 5200 cal BP that is reflected in the enhanced levels of peat humification. These relatively stable conditions on the peat surface may have facilitated suitable growing conditions for *Taxus*. Why did the peat surface become drier? The answer to this question lies in the ice core and glaciological records which show a synchronous period of drier climatic conditions from c. 5000 cal BP. The decline of *Taxus* at Hornchurch Marshes occurs sometime after c. 4100 cal BP, and most probably coincides with worsening climatic conditions after c. 3500 cal BP. This event is recorded at Hornchurch Marshes by the onset of estuarine sedimentation, an interpretation supported by other records of hydrological change in the London area.

At the beginning of the Late Neolithic (c. 4500 cal BP) and during the Bronze Age (c. 4100–3000 cal BP), a second major event causes further reduction in woodland cover: the *Tilia* decline. The phenomenon is well recorded across England, with radiocarbon dates indicating a clearly diachronous decline (c. 5000–3000 cal BP, or later). Three primary causes for the *Tilia* decline have been proposed:

1 Climate change inhibiting the growth of warmth-loving tree taxa, such as *Tilia*.
2 Paludification resulting in a decline in dry land tree taxa due to deteriorating soil conditions (Waller, 1994).
3 Human interference in natural vegetation succession.

Climate change to cooler and wetter conditions has been rejected as a possible cause due

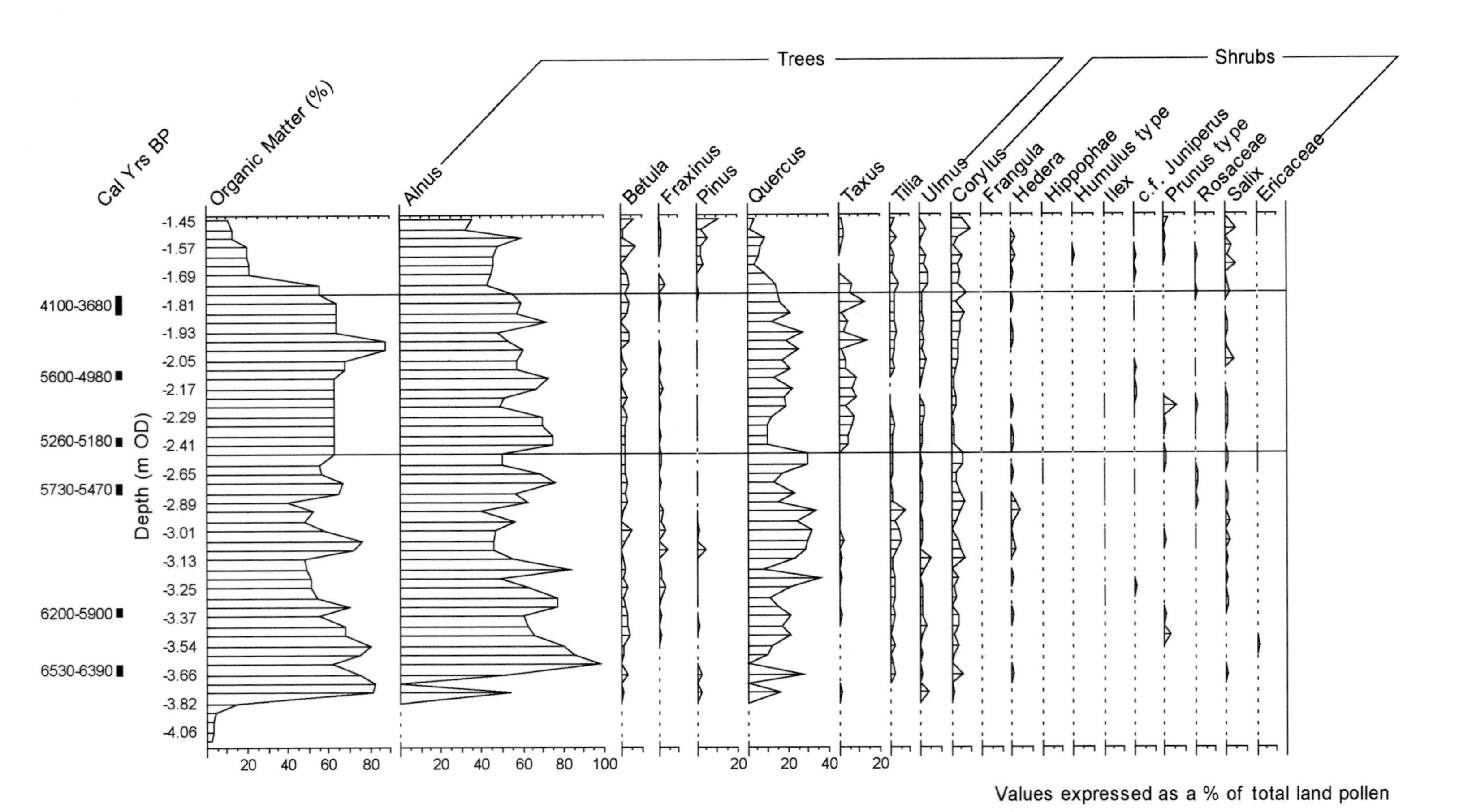

FIGURE 5.11 Percentage pollen diagram from Hornchurch Marshes, Dagenham, London. (From Branch *et al.*, forthcoming)

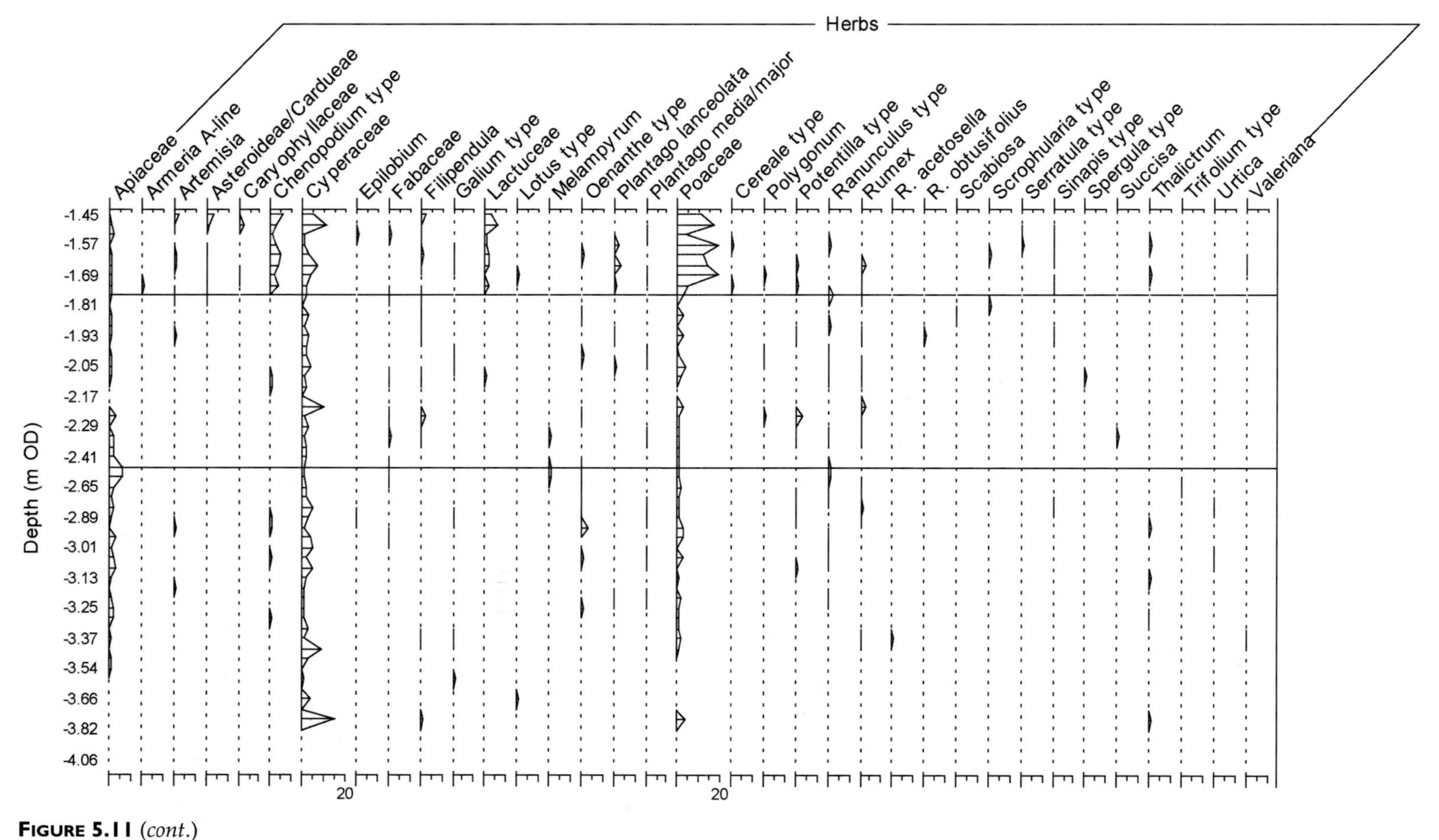

FIGURE 5.11 *(cont.)*

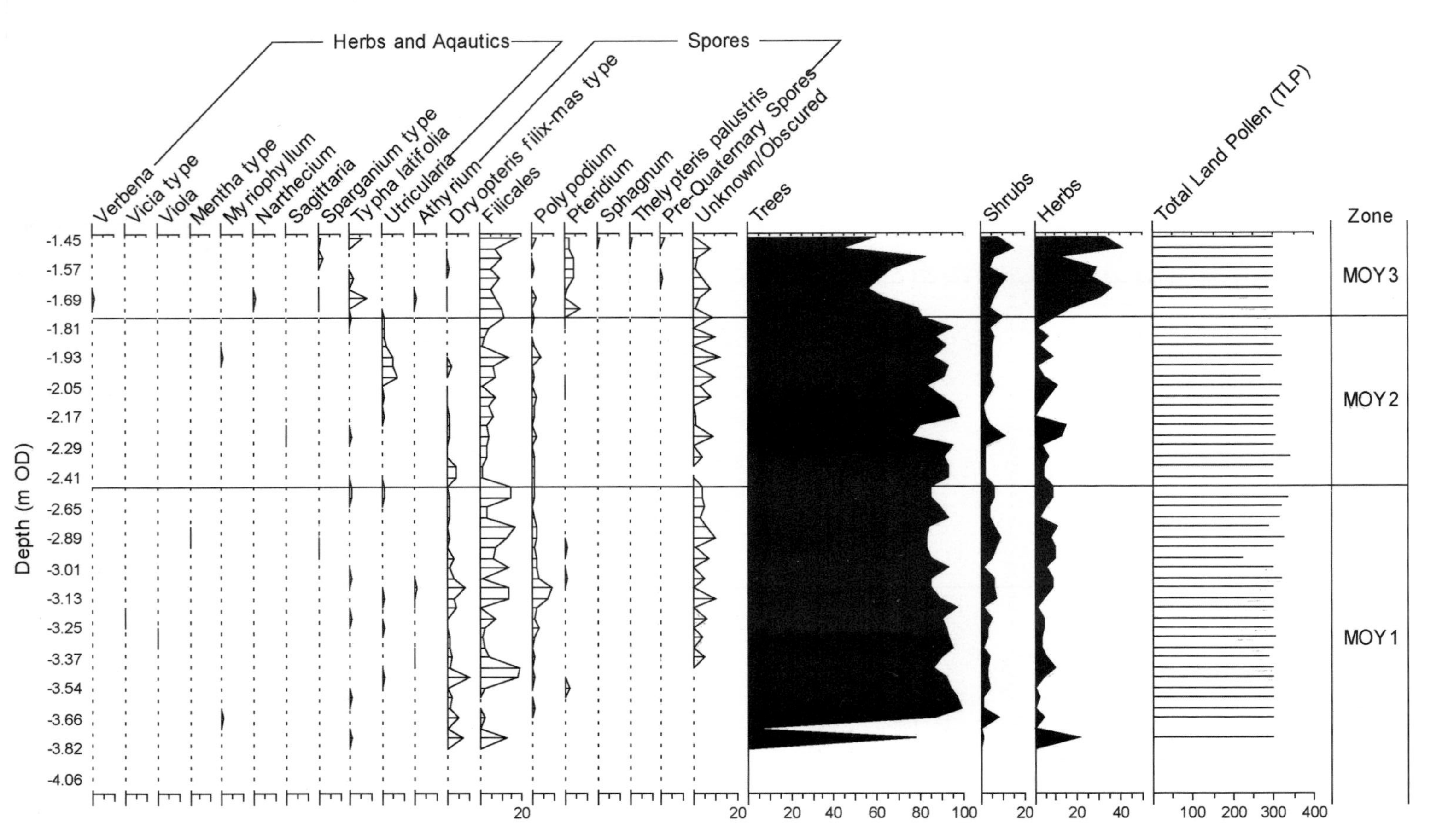

FIGURE 5.11 *(cont.)*

to the wide range of radiocarbon dates, and therefore the absence of a precise correlation between the *Tilia* decline and deteriorating climate that commenced c. 3500 cal BP. Although there is unequivocal evidence for paludification during this period, the precise temporal and spatial relationships between soil waterlogging, peat formation and the *Tilia* decline are unclear. The relationships between human activity and the *Tilia* decline are much clearer, and were first recorded in the 1960s in a seminal piece of high-resolution pollen analysis (Turner, 1962). The interpretation was based upon combining several lines of pollen evidence, most notably an increase in heliophilous (light-loving, e.g. *Fraxinus*), herbaceous (e.g. Poaceae) and disturbed ground (e.g. *Plantago lanceolata*) taxa, and the presence of cultivars (e.g. cereals).

The earliest record for the 'primary' *Tilia* decline from the London Thames is at c. 4500 cal BP (Fig. 5.5: Bramcote Green and Union Street), with other sites displaying significant spatial variations in the timing of the event. Similar to the *Ulmus* decline, these sites provide only circumstantial evidence for a link with human-induced woodland clearance, either for cultivation or pasture. A 'secondary' decline occurs at several sites, with the best record from Bramcote Green (c. 4000–3200 cal BP) clearly showing a strong correlation with cereal cultivation for approximately 800 years. These records provide evidence for a decline in woodland cover during this period, probably as a response to human interference from c. 4000 cal BP, and suggest that by the end of the Bronze Age substantial parts of the London Thames were cleared of woodland. The archaeology strongly supports this model, with evidence for many activities occurring on the River Thames flood plain and sand/gravel eyots (natural islands). These include records of ritual activities (e.g. cremations at Fennings Wharf), food preparation (e.g. burnt mounds possibly for boiling or dry roasting of meat at Phoenix Wharf), access to the river (e.g. a wooden jetty at Vauxhall), cereal cultivation (see Fig. 5.12: plough marks) and access across the marshland (e.g. wooden trackway at Bramcote Green and Erith; Thomas and Rackham, 1996; Sidell and Wilkinson, 2000).

Following the decline in *Ulmus* and *Tilia*, woodland cover was significantly depleted in the London Thames. There is evidence for a continuation of *Quercus* and *Corylus avellana*, although the extent of woodland cover during the last 1000 years of prehistory is unclear (e.g. Bramcote Green, Union Street, Joan Street). In addition, at several sites there is unequivocal evidence for cereal cultivation (*Triticum*/*Hordeum* and *Secale cereale*). These changes occur against a background of greater estuarine influence in the London Thames and a general deterioration in climate to cooler and wetter conditions after c. 3500 BP, with a reduction in temperature of 1–2°C.

FIGURE 5.12 Prehistoric plough marks at Phoenix Wharf, London. (Courtesy of the Museum of London.)

5.3 Macroscale – irrigation, salinity and culture in ancient Mesopotamia

Mesopotamia means literally 'between the rivers', and was the name used by the ancient Greeks to describe that area of the Near East (in modern day Iraq) controlled by the Tigris and Euphrates rivers (see Fig. 5.13). A unique set of geographical circumstances operate in the region, dominated by the influence of two subtly different water courses mostly between 100 and 200 miles apart, but closer in places and flowing within 20 miles of each other near Baghdad. The rivers rise in the Taurus mountains of Armenia and flow south-east, eventually joining up at Basra to become the Shatt-el-'Arab waterway which continues on down to the Persian Gulf. The northern part of the Tigris-Euphrates area tends to be rockier and the rivers are constrained by cliffs, while the southern part consists of broad flat alluvial plains capable of large amounts of food production (Roux, 1964).

Although there is considerable evidence for various lithic periods in Mesopotamia up to about 4500 BC (see Yoffee and Clark, 1993), the main cultures representing the chronology for this discussion occurred after that time, namely the Ubaid (4000 BC–3500 BC), Uruk (3500 BC–2900 BC) and Early Dynastic (2900 BC–2370 BC) periods. These divisions are based on the cul-

FIGURE 5.13 Map of Mesopotamia. (Adapted from Mallowan, 1965)

tures found in excavations at various cities in the region, and from the Uruk period onwards it is the city-state that becomes the main focus of political activity in the region. Cities such as Lagash, Nippur, Uruk itself and Kish all grew to considerable sophistication in architecture, and excavations produce evidence of early writing and record keeping. From these remains we learn that the people developed considerable skill in mathematics, surveying, ceramics, metalwork and medicine (Kramer, 1957). Transport along the rivers was well developed, with clear links existing between Uruk, Mari and Brak before 3000 BC despite these cities being up to 800 miles apart. Trade was also well developed; for example Sumerian grain was traded to the east with semi-precious stones from Iran.

Wheat, barley, meat and fish were produced abundantly on the southern plains from early times. This wealth led to population growth by the Ubaid period and a city at Eridu had grown to several thousand people even before 4000 BC (Mallowan, 1965). Irrigation must have been a critical part of this agricultural regime. Modern rainfall measurements from the first half of the last century range from 115 to 135 mm per annum which is at least 50 mm too little to sustain crop growth and there is no evidence for significant climate change during the intervening period (Charles, 1988). With minimal technology for irrigation, the settlement prior to this urban growth in the region had to follow the patterns of the smaller tributary channels, and agriculture took place in the depressions and levee backslopes where irrigation consisted primarily of simple local systems. Adams (1974) argued that this period would not have seen irrigation systems more sophisticated than the occasional brushwood weir feeding a few canals.

During the Uruk period, however, we start being able to access written sources and there are numerous references to canal and bund construction, as well as evidence for an administration concerned with de-silting and reservoir construction (Adams, 1974). Silt levels in the two rivers are variable but can be extremely high, reaching 25,000 ppm at peak flood times in the Tigris. Although this can be beneficial in some respects, depositing 'irrigation sediment' to great thicknesses on the plain, it causes significant problems in canals, which have to be regularly cleaned out (Charles, 1988). By Early Dynastic times (Pre-Sargonic), there is a written record of a regulator being constructed requiring 648,000 baked bricks and 265,000 litres of bitumen. However, it is only after the Pre-Sargonic period that evidence for large-scale works appears. Postgate (1992) quotes a self-proclaiming royal inscription describing canal construction from Larsa:

> *I fashioned the (canal's) two banks like awe-inspiring mountains. I established abundance at its mouth and its tail I extended. I made the fresh grass thrive on its banks. I called the canal Tuqmat-Erra, and thus restored the eternal waters of the Tigris and Euphrates.*

The different scales of civil engineering represented by these examples have fascinated archaeologists because of the indications they give about the degree of cohesion in the society and the extent to which they demonstrate a larger authority imposing its will on the different city states. Kramer (1963: 5), for example, described the need for irrigation and canal building as the reason for the rise of the Sumerian state. Since much of the evidence is textual, details of translation become critical. For instance, if the word 'eg' is redefined from 'a canal and bank' to 'a bank' only, significant changes immediately ensue for the apparent levels of supra-local organization implicit in particular texts (Pemberton *et al.*, 1988).

Archaeological evidence for the necessary engineering can be found, but it mostly dates to much later. One exception is the extraordinary brick structure at Tello (Parrot, 1948) interpreted by many as a regulator, and dating to the middle of the third millennium BC. It consists of a massive, roughly 12 m long, brick corridor with a 15–20 m entrance and exit funnels (Fig. 5.14). Four long, deep (c. 3–4 m) chambers (Fig. 5.15) extend out at right angles from the line of the brick corridor, two on each side. As pointed out by Pemberton *et al.* (1988) it is still not clear how it would actually work as a regulator. The whole structure gives the appearance of producing a constriction tapering to 3 m wide at the narrowest point, and the

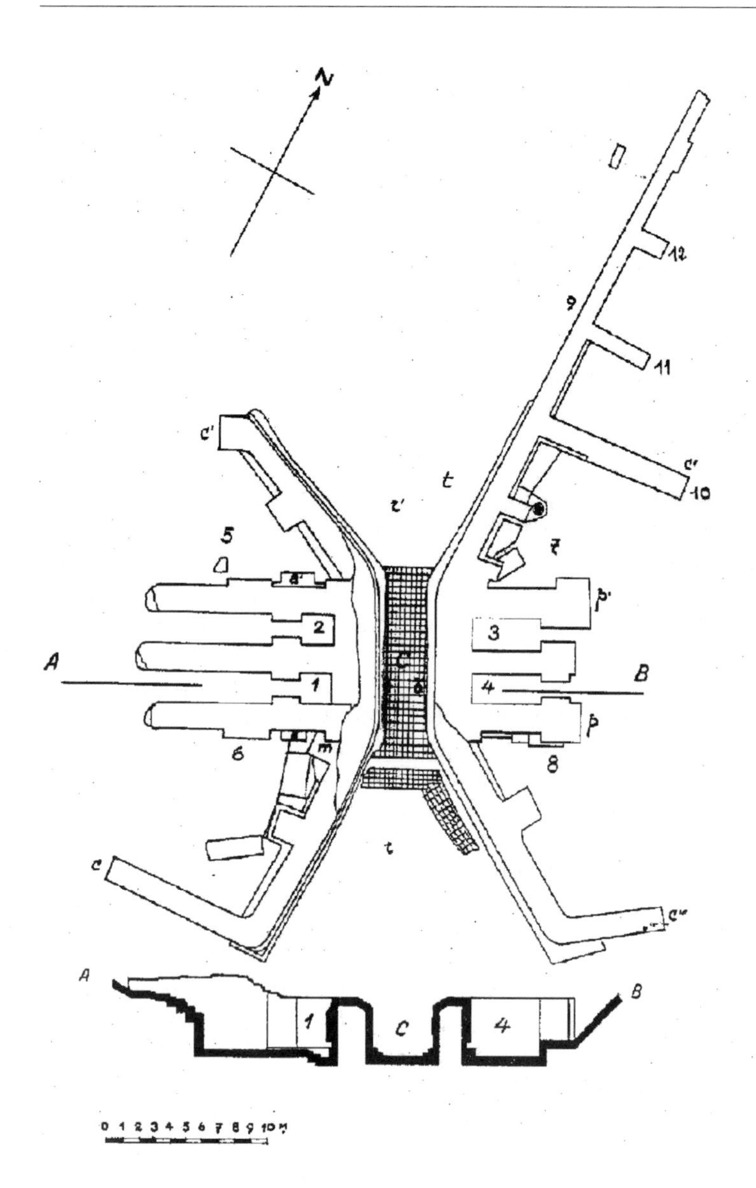

FIGURE 5.14 Plan of the possible regulator at Tello. (From Parrot, 1948)

possibility exists that timber needles were raised or lowered to alter the flow in this constriction.

With an increase in the scale of irrigation projects comes the danger of an increase in soil salinity. Tigris and Euphrates waters have low to medium salt contents, but some local waterways develop high levels (up to 900 ppm) resulting from evaporation or saline groundwater seepage (Charles, 1988). Problems in soils build up where the saline groundwater level gets too high, allowing it to evaporate and deposit its soluble salts at the soil surface. This can be overcome by improving the drainage (and thus removing the saline water ultimately to the sea) or by allowing periods of fallow in which a combination of wild plant growth and rainwater leaching gradually reduce the salt content of the topsoil rendering it suitable for crops again. The fallowing system is both widely understood and insisted on by modern rural villagers (Jacobsen in Gibson, 1974). Ultimately, whatever the method used, a critical

FIGURE 5.15 One of the deep chambers extending out from the brick corridor at Tello. (From Parrot, 1948)

balance has to be achieved between the salinity of the irrigation water, the amount used and the depth to the water table.

By Early Dynastic times, written records exist of considerable salinity problems in fields owned by the temple at Girsu. Similar records are found through the Akkadian and Ur III periods. The problem was clearly well known by the general populace, being used in curses against trespassers on boundary stones, and forming part of the Atrakhasis myth which recounts divine intervention in the form of a famine:

> *the field decreased its yield,*
> *repulsed the grain,*
> *(from being) black, the tilth turned white,*
> *the broad plain gave birth to wet salt,*
> *the womb of the earth revolted,*
> *no plants came up, the ewe conceived not.*
> (Jacobsen, 1982)

Jacobsen and Adams (1958) put forward the view that this salinization affected southern Iraq in three phases, the most serious of these (around 2700–1700 BC) playing a part in the shift of power from southern to central Iraq in the early second millennium BC. They argued that the problem was exacerbated by a dispute over Euphrates water between the neighbouring cities of Girsu and Umma, which led to Girsu's ruler Entemenak constructing a large canal to bring his city an independent supply from the Tigris. The previously limited irrigation west of Girsu then intensified, and salinization ensued. Clearly, the salinity itself does not survive archaeologically, but the authors used the textual descriptions of saline ground, as well as agricultural accounts of crop yields, to argue for an overall decline in productivity from around 2400 BC as well as a shift away from cultivating wheat to the more salt-tolerant barley. The southern alluvial plain subsequently declined in political importance and leadership passed permanently out of the region with the rise of Babylon in the eighteenth century BC. Although the detailed evidence was not published until Jacobsen (1982), the concept of salinity affecting political power in this way was clearly a stimulating one for archaeologists, anthropologists and environmentalists alike.

Gibson (1974) developed the ideas into a political scheme to describe the relationship between irrigation and administration. As governance became more and more centralized, larger schemes of irrigation works would be introduced, taxation would increase, and the small-scale local producer would be gradually replaced by larger landholders, often absentees. This process would affect salinization by reducing the implementation of the vital fallow period in which saline soils could recover from the high water table induced by irrigation, and salts could be leached out.

Powell (1985) argued forcefully against the detail of Jacobsen's (1982) assumptions and deductions. The textual evidence could be shown to have different interpretations and there are translations that do not support the view that salinity was a large-scale environmental crisis. Jacobsen's use of a particular account as a benchmark yield against which production declined is singled out for criticism, especially as the subsequent reports are not necessarily derived from exactly the same piece of land and thus the harvests cannot automatically be compared. Even assumptions such as a widespread change from wheat to barley were not judged by Powell to be necessarily indicative of salinization, since the relative requirements of the two grain species cannot be reduced to salinity factors alone.

Evidence from elsewhere could point to less polarized explanations. The Deh Luran plain just inside modern Iran has many similarities to Mesopotamia, but with slightly higher rainfall and more saline river waters available from the Dawaraij and Mehmeh rivers (Neely, 1974). It appears to have been a favourable place for settlement during the period 7500–3000 BC, and 13 village mounds had been located by 1969 – the largest (Tepe Musiyan) being 1.5 km wide. In addition, scatters of pottery were found, but the 3 metres depth of alluvium in some areas suggests that much of the evidence, including whole villages, will be buried without trace (Hole *et al.*, 1969). By the end of the Bayat phase, around 3700 BC, two sites showed clear signs of only salt-tolerant plants in their uppermost layers (Helbaek, 1969), followed in one case by abandonment. Hole *et al.* (1969) believed that it took around 2000 years for salinity to develop, after which sites were abandoned in favour of others nearby. Salinization was thus a localized phenomenon, occurring on the poorly drained areas of alluvium which were balanced on the edge of salinity, while better drained areas on gravel could sustain longer-lived settlements. The culture changed from a wheat-goat economy to a barley-sheep economy as irrigation progressed with an attendant rise in population. The well-drained areas became complex societies, but the more marginal soils, such as those at Deh Luran, gradually became zones of shifting settlement barely keeping ahead of the environmental destruction (Hole *et al.*, 1969).

Can the different views of Mesopotamian salinity be reconciled? The grander scale picture of a civilization bringing about its own destruction by overuse or misuse of natural resources is a tempting interpretation to follow in the light of modern concerns with both local and global environmental instabilities resulting from human action. On the other hand, the clear evidence that salinity can be coped with at a local scale by individual farmers today immediately raises the simple question: why wouldn't similar measures have been equally applicable in the past? Revision of the archaeological and textual evidence will always provide a core debate among data analysts, but the wider interpretation ultimately falls into one or other camp as a result of instinctive beliefs about the nature of history. Although occasionally inviting re-analysis, the facts of salinization are not challenged. However, the way those facts are narrated cannot in any way be proscribed: there is a true past, but not a true history.

5.4 Megascale – megafaunal extinction and human settlement

A topical issue of environmental archaeology that regularly crops up in the scientific literature is megafaunal extinction. The questions raised by the extinctions, and the details of how and when they happened are commonly the cause of division in the archaeological science community (Diamond, 1989).

Various definitions of the term 'megafauna' exist, but most refer to animals with a mean adult body weight exceeding 40kg. It has long been observed that the world seems to be relatively impoverished in animals of this size (Wallace, 1876), but it is only recently that we have been able to determine the numbers that must have died in the past. This is a crucial point because although in many places around the world whole genera vanished, not all megafauna have become extinct. In South America, 46 of the 58 genera disappeared (80 per cent of the total); in Europe, 7 out of 24 (29 per cent); while intriguingly south of the Sahara, only 1 out of 44 (only 2 per cent) have been lost (Stuart, 1993). In addition to the numbers involved, a key part of the debate is the question of *when* the extinctions occurred. Megafaunal extinction is loosely thought of as a

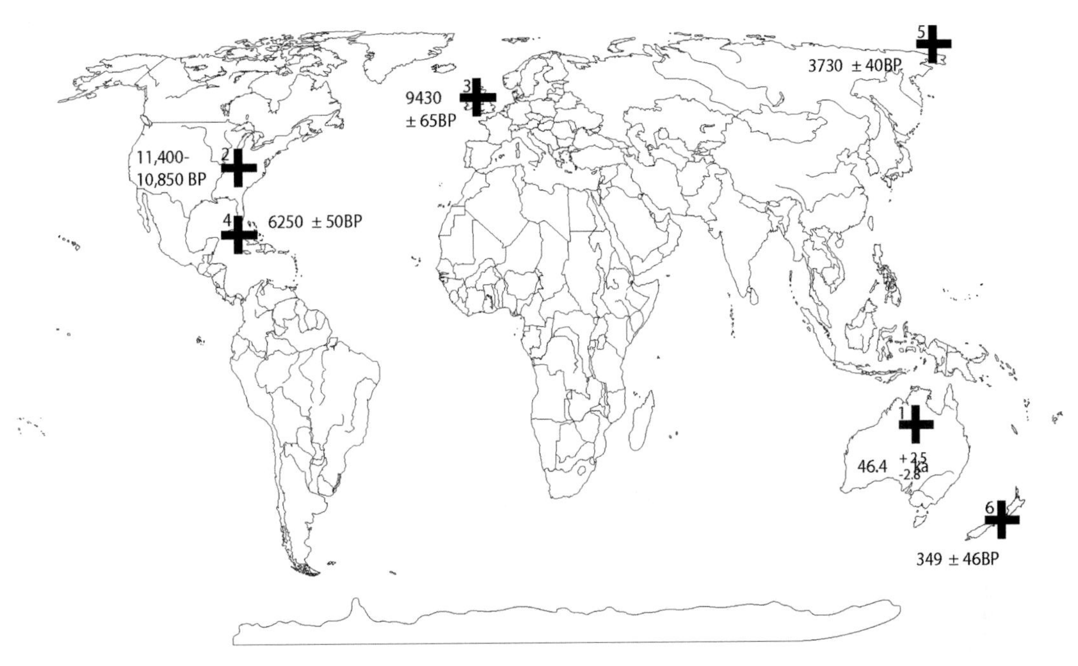

FIGURE 5.16 Most recent ages of megafaunal extinction in selected regions of the world, numbered according to age: 1. Australia (numerous species) (Roberts *et al.*, 2001); 2. North America (numerous species) (Elias, 1999); 3. Isle of Man, UK (Giant Irish Elk) (Gonzalez *et al.*, 2000); 4. Cuba (Ground Sloth) (MacPhee *et al.*, 1999); 5. Wrangel Island, Siberia (Dwarf Mammoth) (Vartanyan *et al.*, 1993); 6. New Zealand (Moa) (Anderson, 1989a).

late Quaternary event, but as we have discussed in Chapter 4, this covers a significant period of time. Detailed investigation shows a more complex pattern with extinctions happening at different times in different places.

The arguments put forward for megafaunal disappearance typically fall into one of two camps. The hypothesis that human overhunting is the principal cause of megafaunal extinction was first suggested by Owen (1846) and popularized by Martin (1973, 1984) and Flannery (1990). The idea is that the arrival of humans in different continents led to rapid overkill ('blitzkrieg') of the native populations and ultimately their extinction. Several arguments have been put forward against this view (Schuster and Schüle, 2000) including the observations that many megafauna are still present in Africa where there is a long record of human occupation, and that in Eurasia they coexisted with different hominin species over at least 100,000 years. Furthermore, it is assumed that most fauna would be shy of humans (though Darwin's (1839) reports from the Galapagos Islands show this is not true in all cases) and that hunter-gatherers have small population densities with only low environmental impact. The second main view put forward to explain the extinctions is that rapidly changing climate (and associated environmental gradients) led to the disappearance of the biomes on which the megafauna depended (Graham and Lundelius, 1984; Guthrie, 1984). A major argument against this hypothesis is that significant climatic change has happened repeatedly throughout the Quaternary without repeated megafaunal extinction.

This section will not attempt to provide the definitive cause of the extinctions but will review and critically assess the archaeological, bioarchaeological, stable isotopic and dating evidence for megafaunal disappearance in three parts of the world: Australia, North America and New Zealand. (See Fig. 5.16.)

5.4.1 Australia

Australia experienced extreme and rapid climate changes over glacial-interglacial cycles (Harrison and Dodson, 1993; Kershaw and Nanson, 1993), with significant fluctuations in precipitation and temperature (Miller *et al.*, 1997). Despite changes in sea level of up to 130 m associated with these cycles, Greater Australia (Papua New Guinea, Australia and Tasmania) remained an island land mass detached from Asia, and developed a unique flora and fauna as a remnant of the ancient super-continent of Gondwanaland (Flannery, 1997). In addition to natural climatic fluctuations, human arrival and impact have been invoked to explain significant changes in Australian landscapes during the late-Quaternary (Flannery, 1990; Jones, 1998; Webb, 1998), including modification of the Australian summer monsoon system by disruption of the vegetation cover through burning (Johnson *et al.*, 1999), though such views are fiercely contested (Bowman, 1998; Choquenot and Bowman, 1998; Horton, 2000). A significant element of the debate concerns the actual timing of human arrival in Australia. As we saw in Chapter 4, radiocarbon and luminescence dating imply human arrival by at least 45,000–55,000 cal BP (Roberts *et al.*, 1990, 1994, 1998; Turney *et al.*, 2001a; Bowler *et al.*, 2003). In addition, secondary evidence interpreted as resulting from human activity (principally increases in burning) is recorded at Lynch's Crater and ODP-820 at around 45,000 BP (based on ^{14}C and oxygen isotope stratigraphy respectively; Turney *et al.*, 2001b; Moss & Kershaw, 2000).

Interestingly, the megafauna appear to have been relatively small compared to those found in other parts of the world. For instance, the largest representative in Australia was the marsupial herbivore diprotodon (a fur-covered beast, similar to a wombat in shape), with adults only reaching 2 m in height and around 3.5 m in length (Flannery, 1997). The reason for the smaller size of the Australian megafauna is not clear but seems likely to be due to the relatively low nutrient levels (particularly iodine, cobalt and selenium) available in the landscape. Indeed, some have argued that these low levels may have contributed to their extinction (Milewski and Diamond, 2000). Overall, in Australia, 15 out of 16 genera (94 per cent) of megafauna disappeared sometime during the late-Quaternary.

Until recently, many of the megafaunal remains and their contexts were relatively poorly dated, and a key part of the debate has been taken up with the details of chronology. In many circumstances, recent ages had been reported for fossils that were subsequently found to be either reworked or contaminated (e.g. Lancefield Swamp, van Huet *et al.*, 1998). Recent years have therefore seen a concerted effort among the scientific community to improve site investigations and provide more robust dating for the extinction of the Australian megafauna.

Genyornis newtoni appears to be the largest (up to 200kg) Australian flightless bird to have become extinct. Miller *et al.* (1999) reported a comprehensive dating program for the eggshells of this species from three different Australian regions (particularly the Lake Eyre Basin), found within natural exposures or deflation hollows. The calcite matrices of the shells were measured for AAR, ^{14}C and Uranium-series while the associated sediments were dated by OSL. In addition to this, stable carbon isotopic analysis (δ^{13}C) of the eggshells was undertaken. An extinction event at 50,000 $\pm$5000 BP was reported for *G. newtoni*, contrasting with similarly dated eggshells of the similarly sized *Dromaius novaehollandiae* (emu) through to the present day. Although *G. newtoni* may have been selectively hunted compared to emu due to its larger body mass, its extinction cannot be directly attributed to human overkill, as there are no unambiguously accepted megafaunal kill sites in Australia (cf. Cuddie Springs – Field and Fullagar, 2001). The climatic explanation is also unsupported, since significant climate change is not believed to have taken place at this time. Stable isotopic analysis of the eggshells suggests that *G. newtoni* primarily fed on C3 plants (typical δ^{13}C values of -19 to -26‰) whereas emu appears to have browsed on both C3 and C4 plants (with a δ^{13}C range of -9 to -26‰), potentially making the latter less sensitive to changes in vegetation make-up as a result of burning (Miller *et al.*, 1999). To minimize the risk of dating reworked megafaunal remains, Roberts *et al.* (2001) concentrated on OSL and U-series dating of articulated megafaunal remains across Australia, and identified an apparently synchronous extinction event of all species at 46,400 (±2500–2800) BP, consistent with ecosystem disruption through fire, and statistically indistinguishable from the ages obtained from Miller *et al.* (1999).

5.4.2 North America

Unlike Australia, North America has periodically been connected to Asia via the Bering Strait, which sometimes became a land bridge depending on changing sea levels in glacial periods (Goebel, 1999). 'Beringia' is used as a collective term to include the Bering Strait and a region of around 3200 km from the Kolyma River in Siberia to the Mackenzie River in Canada. It has acted alternately as the continental crossroads between the Old and the New Worlds, and as the marine connection between the Pacific and Arctic Oceans. In North America, the existence of the Laurentide and Cordilleran ice caps effectively blocked a terrestrial migration of humans from Beringia into the American continent, although they almost certainly reached what is now Alaska sometime around 13,000 BP (Dixon, 2001). With the melting of the ice, the land bridge is traditionally believed to have opened, allowing humans to rapidly colonize the continent. The debate concerning the arrival of humans in the Americas is complex and will not be explored in detail here (see Section 3.4.4). Clearly, the Clovis complex of 12,000–11,000 cal BP (Taylor *et al.*, 1996) provides a minimum age of human arrival, and the recent unambiguous identification of human occupation sites in Quebrada Jaguay, Peru between 13,000 and 11000 cal BP (Sandweiss *et al.*, 1998) and Monte Verde, Chile between 13,000 and 12,500 BP (Dillehay, 1989) suggests that humans may have taken a coastal route prior to the retreat of the ice (Dixon, 2001).

North America lost 33 genera of megafauna out of 45 (73 per cent). These include the ground sloth (*Glossotherium harlani*), at least two species of horse (*Equus* spp.), the mastodon (*Mammut americanum*) and the Colombian mammoth (*Mammuthus columbi*), which reached up to 3.4 m in height. Comprehensive radiocarbon dating of megafaunal remains and their evaluation indicate that no species survived beyond 10,000 BP (Meltzer and Mead, 1983) and most had probably become extinct prior to 12,000 BP (Grayson, 1989). As in Australia, ideas on megafaunal extinction are very much centred on the two principal potential mechanisms of climate change or human arrival, creating considerable debate in the literature (Meltzer and Mead, 1983; Grayson, 1989; Steadman *et al.*, 1997; Flannery, 2002).

Pollen evidence from sequences across North America shows that the end of the last ice age was a period of significant environmental change (Shane, 1987; Elias, 1996; Maenza-Gmelch, 1997). Increasing temperatures from around 15000 BP encouraged the development of closed forests and the disappearance of the steppe grassland favoured by megafauna such as the mammoths that are believed to have closely co-evolved with unique flora through the previous glacial. Rapid warming might have led to a nutritional bottleneck for the fauna if they could not adapt sufficiently quickly to the changing conditions (Graham and Lundelius, 1984). Such a view is supported by $\delta^{13}C$ values obtained on mammoth and mastodon remains from Florida, USA (Koch *et al.*, 1998). Here, $\delta^{13}C$ from carbonate in mastodon tooth enamel apatite gives a relatively small range of values between −10 and −12‰ (consistent with a predominantly C3 diet) whereas mammoth remains showed values of −6 to 0‰ (consistent with a predominantly C4 diet), supporting a view that these megafauna were specialized feeders. More recently, using a modified soil DNA extraction technique and specific polymerase chain reaction primers, Willerslev *et al.* (2003) documented dramatic changes in the taxonomic diversity and composition of the Siberian biota over the last 400,000 years. They obtained ancient DNA sequences from at least 19 different plant and 8 different vertebrate taxa (both extinct and extant). The results indicated the steady decline in the proportion of herbs to shrubs over the last 400,000 years, with a rapid reduction in herbs and an increase of mosses during the Holocene. The evidence from sedimentary DNA that Siberia experienced a profound ecological shift at the Pleistocene/Holocene boundary may have implications for the extinction of the megafauna at this time (Willerslev *et al.*, 2003).

Such a view is not supported by all workers, however. Martin (1973) first suggested that the megafauna had become extinct due to a 'blitzkreig' created by overkill as humans rapidly colonized North America. Unlike Australia, North America has kill sites where humans clearly interacted with the megafauna. There are at least 12 such sites, including Naco in Arizona (Haury, 1986), where an adult mammoth was identified with eight Clovis points embedded in the skeleton, and one at the base of the skull. Interestingly, recent modelling studies indicate that even with slow human population growth and relatively low hunting densities, significant megafaunal extinction (of a similar magnitude to that observed in North America) can take place (Alroy, 2001). Furthermore, the recent dating of the now extinct ground sloth (*Megalocnus rodens*) in neighbouring Cuba at 6250±50 BP (MacPhee *et al.*, 1999) is contemporaneous with the arrival of humans on the island and raises fascinating questions within the North American context if climatic and environmental changes were the primary cause of megafaunal extinction (Flannery, 2002).

5.4.3 New Zealand

The moas (Dinornithiformes) of New Zealand comprise 11 species of large flightless birds that as adults reached between 20 and 250kg. Unlike Australia and North America, however, their extinction appears to have been relatively recent and the cause of their disappearance can be unambiguously attributed to the arrival of humans (Anderson, 1989a). No species of moa became extinct at the same time as the Australian megafauna (45,000–50,000 BP) despite the relatively close proximity between the two land masses and similar climatic controls on their environments.

New Zealand was the last major land mass on the planet to be colonized by humans. Considerable debate has focused on when the first humans arrived, with the dating of associated Pacific/Polynesian rat (*Rattus exulans*) remains suggesting a date as early as 2155±130 BP (Holdaway, 1996). Although this supports the early arrival view (Sutton, 1987), it is not backed up by direct archaeological evidence. Indeed, exhaustive studies of pollen and stratigraphic records indicate that evidence previously taken to be of early human impact in the environment resulted from inwashed old carbon and hard water effects, biasing radiocarbon samples to older ages (McGlone and Wilmshurst, 1999). The first unambiguous human occupation occurs around 600 BP when the moa appear to have rapidly become a major part of the human diet, with numerous locations showing evidence of hunting and butchering (Anderson, 1991). Indeed, the link is so clear that the identification of moa remains in archaeological contexts has often been used as an indication of early human occupation (the 'Moa-hunter' period; Anderson, 1989b).

Estimates of the duration of moa hunting suggest it may have been as short as a couple of hundred years (Holdaway and Jacomb, 2000), with the moa becoming scarce by 500 BP. Confirmed kill sites suggest rapid depletion of moa, e.g. <20 years at Wairau Bar, South Island (1288–1300 cal AD; Higham *et al.*, 1999), thereby quickly reducing food resources in the immediate area. The most recent accepted moa age of 349±46 BP (Anderson, 1989a) indicates that they were most probably extinct by 300 BP. Modelling studies point to long-lived birds such as moas being extremely susceptible to human predation and suggest that relatively low levels of hunting can cause an irreversible decline and extinction of the populations within 160 years (Holdaway and Jacomb, 2000). The effect was so devastating that by the time of European arrival no living moas were present in New Zealand.

REFERENCES

Adams, R. McC. (1974) Historic patterns of Mesopotamian Irrigation Agriculture. In (T.E. Downing and M. Gibson, eds.) *Irrigation's Impact on Society*, 1–6. Tucson: University of Arizona Press.

Allison, E. (1998) *The Insect Remains from the Dover Bronze Age Boat*. Unpublished archive report, Canterbury Archaeological Trust.

Alroy, J. (2001) A multispecies overkill simulation of the end-Pleistocene megafaunal mass extinction. *Science* **292**, 1893–1896.

Anderson, A. (1989a) Mechanics of overkill in the extinction of New Zealand moas. *Journal of Archaeological Science* **16**, 137–151.

Anderson, A. (1989b) *Prodigious Birds: Moas and Moa Hunting in Prehistoric New Zealand*. Cambridge: Cambridge University Press.

Anderson, A. (1991) The chronology of colonization in New Zealand. *Antiquity* **65**, 767–795.

Bayliss, A., Groves, C., McCormac, F.G., Bronk Ramsey, C., Baillie, M.G.L., Brown, D., Cook, G.T. and Switsur, R.V. (2004) Dating. In (P. Clark, ed.) *The Dover Bronze Age Boat*, 251–255. London: English Heritage.

Bell, M. and Walker, M.J.C. (1992) *Late Quaternary Environmental Change*. Harlow: Longman.

Bowler, J.M., Johnston, H., Olley, J.M., Prescott, J.R., Roberts, R.G., Shawcross, W. and Spooner, N.A. (2003) New ages for human occupation and climatic change at Lake Mungo, Australia. *Nature* **421**, 837–840.

Bowman, D.M.J.S. (1998) The impact of Aboriginal landscape burning on the Australian biota. *New Phytologist* **140**, 385–410.

Bown P., Bristow, C. and de Silva, N. (2004) The shale. In (P. Clark, ed.) *The Dover Bronze Age Boat*, 216. London: English Heritage.

Branch, N.P. and Lowe, J.J. (1996) Pollen analysis. In Thomas, C. and Rackham, J. (1996) Bramcote Green, Bermondsey: a Bronze Age trackway and palaeoenvironmental sequence. *Proceedings of the Prehistoric Society* **61**, 221–253.

Branch, N.P., Cameron, N.G., Coope, R., Densem, R., Gale, R., Green, C.P., Lowe, J.J., Palmer, A.P. and Williams, A.N. (forthcoming) Middle Holocene environmental changes at Hornchurch Marshes, Dagenham, and their implications for our understanding of the history of *Taxus* (L.) woodland in the lower Thames valley, London, UK. *Journal of Archaeological Science*.

Bridgland, D.R. (1995) The Quaternary sequence of the eastern Thames basin: problems of correlation. In (D.R. Bridgland, P. Allen and B.A. Haggart, eds.) *The Quaternary of the Lower Reaches of the Thames: Field Guide*. Durham: Quaternary Research Association.

Bridgland, D.R., Allen, P. and Haggart, B.A. (1995) (eds.) *The Quaternary of the Lower Reaches of the Thames: Field Guide*. Durham: Quaternary Research Association.

Cameron, N. and Dobson, S. (1998) *Diatom analysis of sediments associated with the Dover Bronze Age boat*. Unpublished archive report, Canterbury Archaeological Trust.

Catt, J.A. (1979) Soils and Quaternary geology in Britain. *Journal of Archaeological Science* **30**, 607–642.

Champion, T. (2004) The deposition of the boat, 276–281. In (P. Clark, ed.) *The Dover Bronze Age Boat*. London: English Heritage.

Charles, M.P. (1988) Irrigation in Lowland Mesopotamia. *Bulletin on Sumerian Agriculture* **4**, 1–39.

Choquenot, D. and Bowman, D.M.J.S. (1998) Marsupial megafauna, Aborigines and the overkill hypothesis: application of predator-prey models to the question of Pleistocene extinction in Australia. *Global Ecology and Biogeography Letters* **7**, 167–180.

Clark, P. (ed.) (2004) *The Dover Bronze Age Boat*. London: English Heritage.

Clark, P. (1993a) *The Dover Boat; Proposal for Assessment*. Canterbury Archaeological Trust, unpublished report.

Clark, P. (1993b) *The Dover Bronze Age Boat; Review of Assessment Phase*. Canterbury Archaeological Trust, unpublished report.

Clark, P. (1994) *The Dover Boat; An Assessment*. Canterbury Archaeological Trust, unpublished report.

Clark, P. and Keeley, H. (1997) *The Dover Bronze Age Boat – Palaeoenvironmental Assessment and Updated Project Design for Analysis, July 1997*. Canterbury Archaeological Trust, unpublished report.

Clark, P., Corke, B. and Waterman, C. (2004) Reassembly and display. In (P. Clark, ed.) *The Dover Bronze Age Boat*, 290–304. London: English Heritage.

Darrah, R. (2004a) Illuminating the original shape of the Dover boat timbers. In (P. Clark, ed.) *The Dover Bronze Age Boat*, 96–105. London: English Heritage.

Darrah, R. (2004b) Woodland management and timber conversion. In (P. Clark, ed.) *The Dover Bronze Age Boat*, 106–123. London: English Heritage.

Darwin, C. (1839) *The Voyage of the Beagle*. Republished by Penguin Classics, London, England.

Devoy, R.J. (1979) Flandrian sea-level changes and vegetational history of the lower Thames estuary. *Philosophical Transactions of the Royal Society of London* **B285**, 355–410.

Devoy, R.J. (1982) Analysis of the geological evidence for Holocene sea-level movements in southeast England. *Proceedings of the Geologists Association* **93**, 65–90.

Diamond, J.M. (1989) Quaternary megafaunal extinctions: Variations on a theme by Paginini. *Journal of Archaeological Science* **16**, 167–175.

Dillehay, T.D. (1989) *Monte Verde: A late Pleistocene settlement in Chile 2: Paleoenvironmental and site context.* Washington (D.C.): Smithsonian Institution Press.

Dixon, E.J. (2001) Human colonization of the Americas: timing, technology and process. *Quaternary Science Reviews* **20**, 277–299.

Elias, S.A. (1996) Late Pleistocene and Holocene seasonal temperature reconstructed from fossil beetle assemblages in the Rocky Mountains. *Quaternary Research* **46**, 311–318.

Elias, S.A. (1999) Quaternary biology update, debate continues over the cause of Pleistocene megafauna extinction. *Quaternary Times*, June, 11.

Fairbairn, A. (1998) *Dover Bronze Age Boat Plant Macrofossil Analysis*, unpublished archive report, Canterbury Archaeological Trust.

Field, J. and Fullagar, R. (2001) Archaeology and Australian megafauna. *Science* **294**, 7a.

Flannery, T.F. (1990) Pleistocene faunal loss: implications of the aftershock for Australia's past and future. *Archaeology in Oceania* **25**, 45–67.

Flannery, T. (1997) *The Future Eaters*. Sydney: Reed New Holland.

Flannery, T. (2002) *The Eternal Frontier*. London: Vintage.

Gibbard, P.L. (1994) Pleistocene History of the Lower Thames Valley. Cambridge: Cambridge University Press.

Gibbard, P.L. (1995) Palaeogeographical evolution of the Lower Thames. In (D.R. Bridgland, P. Allen and B.A. Haggart, eds.) *The Quaternary of the Lower Reaches of the Thames: Field Guide*, 5–34. Durham: Quaternary Research Association.

Gibbard, P.L. (1999) The Thames valley, its tributary, valleys and their former courses. In (D.Q. Bowen, ed.) *A Revised Chronology of Quaternary Deposits in the British Isles*, 45–58.

Gibson, A., Macpherson-Grant, N. and Stewart, I. (1997) A Cornish vessel from farthest Kent. *Antiquity* **71**, 438–41.

Gibson, M. (1974) Violations of fallow and engineered disaster in Mesopotamian civilisation. In (T.E. Downing and M. Gibson, eds.) *Irrigation's Impact on Society*, 7–20. Tucson: University of Arizona Press.

Goebel, T. (1999) Pleistocene human colonization of Siberia and peopling of the America: an ecological approach. *Evolutionary Anthropology* **8**, 208–227.

Gonzalez, S., Kitchener, A.C. and Lister, A.M. (2000) Survival of the Irish elk into the Holocene. *Nature* **405**, 753–754.

Graham, R.W. and Lundelius, E.L. (1984) Co-evolutionary disequilibrium and Pleistocene extinctions. In (P.S. Martin and R.G. Klein, eds.) *Quaternary Extinctions: A Prehistoric Revolution*, 211–222. Tucson: University of Arizona Press.

Grayson, D.K. (1989) The chronology of North American late Pleistocene extinctions. *Journal of Archaeological Science* **16**, 153–165.

Green, C. (1998) *The Dover Bronze Age Boat. Technical Report – Sedimentology*. Archaeoscape Consulting Report, **ASC 97/26**, unpublished.

Greig, J.R.A. (1992) The deforestation of London. *Reviews of Palaeobotany and Palynology* **73**, 71–86.

Guthrie, R.D. (1984) Mosaics, allochemicals and nutrients: An ecological theory of late Pleistocene megafaunal extinctions. In (P.S. Martin and R.G. Klein, eds.) *Quaternary Extinctions: A Prehistoric Revolution*, 259–298. Tucson: University of Arizona Press.

Harrison, S.P. and Dodson, J. (1993) Climates of Australia and New Guinea since 18,000 yr BP. In (H.E. Wright, J.E. Kutzbach, T. Webb III, W.F. Ruddiman, F.A. Street-Perrott and P.J. Bartlein, eds.) *Global climates since the Last Glacial Maximum*, 265–293. Minneapolis: University of Minnesota Press.

Hather, J. and Wales, S. (1993) *Dover Boat: Constructional Materials Analysis: Pre PEG Identification of Constructional Wood and Packing Materials.* Unpublished archive report, Canterbury Archaeological Trust.

Haury, E. (1986) Artifacts with Mammoth Remains, Naco, Arizona: Discovery of the Naco Mammoth and the Associated Projectile Points. In (J.J. Reid and D.E. Doyel, eds.) *Prehistory of the American Southwest*, 78–98. Tucson: University of Arizona Press.

Helbaek, H. (1969) Plant collecting, dry-farming and Irrigation in Prehistoric Deh Luran. In (F. Hole, K.V. Flannery and J.A. Neely, eds.) *Prehistory and Human Ecology of the Deh Luran Plain*, 383–426. Ann Arbor: Michigan Museum of Anthropology.

Higham, T., Anderson, A. and Jacomb, C. (1999) Dating the first New Zealanders: The chronology of Wairau Bar. *Antiquity* **73**, 420–427.

Holdaway, R.N. (1996) Arrival of rats in New Zealand. *Nature* **384**, 225–226.

Holdaway, R.N. and Jacomb, C. (2000) Rapid extinction of the moas (Aves: Dinornithiformes): Model, test and implications. *Science* **287**, 2250–2254.

Hole, F., Flannery, K.V. and Neely, J.A. (1969) *Prehistory and Human Ecology of the Deh Luran Plain*. Ann Arbor: Michigan Museum of Anthropology.

Horton, D.R. (2000) *The Pure State of Nature*. New South Wales: Allen and Unwin.

Jacobsen, T. (1982) *Salinity and Irrigation Agriculture in Antiquity*. Malibu: Undena Publications.

Jacobsen, T. and Adams, R.M. (1958) Salt and silt in

ancient Mesopotamian agriculture. *Science* **128**, 1251–1258.

Johnson, B.J., Miller, G.H., Fogel, M.L., Magee, J.W., Gagan, M.K. and Chivas, A.R. (1999) 65,000 years of vegetation change in central Australia and the Australian Summer monsoon. *Science* **284**, 1150–1152.

Jones, R. (1998) Dating the human colonization of Australia: radiocarbon and luminescence revolutions. *Proceedings of the British Academy* **99**, 37–65.

Keeley, H., Allison, E., Branch, N., Cameron, N., Dobinson, S., Ellis, I., Ellison, P., Fairbairn, A., Green, C., Hunter, R., Lee, J., Locker, A., Lowe, J., Palmer, A., Robinson, E., Stewart, J. and Wilkinson, K. (2004) The environmental evidence. In (P. Clark, ed.) *The Dover Bronze Age Boat*, 229–250. London: English Heritage.

Kershaw, A.P. and Nanson, G.C. (1993) The last full glacial cycle in the Australian region. *Global and Planetary Change* **7**, 1–9.

Koch, P.L., Hoppe, K.A. and Webb, S.D. (1998) The isotopic ecology of late Pleistocene mammals in North America. Part 1. Florida. *Chemical Geology* **152**, 119–138.

Kramer, S.N. (1957) The Sumerians. In *Old World Archaeology: foundations of civilisation*. Readings from *Scientific American*. San Francisco: W.H. Freeman and Co.

Kramer, S.N. (1963) *The Sumerians*. Chicago: University of Chicago Press.

Locker, A. (1998) *The fish bones from deposits associated with the Dover Boat*. Unpublished archive report, Canterbury Archaeological Trust.

Long, A.J., Scaife, R.G. and Edwards, R.J. (2000) Stratigraphic architecture, relative sea-level and models of estuary development in southern England: new data from Southampton Water. In (K. Pye and J. Allen, eds.) *Coastal and Estuary Environments: Sedimentology, Geomorphology and Geoarchaeology*, 253–290. Geological Society Special Publication 175.

Lowe, J., Branch, N. and Ellis, I. (1998) *The Dover Bronze Age Boat Technical Report – Pollen Stratigraphy*. Unpublished archive report, Canterbury Archaeological Trust.

MacPhee, R.D.E., Flemming, C. and Lunde, D.P. (1999) 'Last occurrence' of the Antillean insectivoran Nesophontes: New radiometric dates and their interpretation. *American Museum of Natural History* **3261**, 1–19.

Maenza-Gmelch, T.E. (1997) Late-glacial-early Holocene vegetation, climate, and fire at Sutherland Pond, Hudson Highlands, southeastern New York, U.S.A. *Canadian Journal of Botany* **75**, 431–439.

Mallowan, M.E.L. (1965) *Early Mesopotamia and Iran*. London: Thames and Hudson.

Marsden, P. (2004) Description of the boat. In (P. Clark, ed.) *The Dover Bronze Age Boat*, 32–95. London: English Heritage.

Martin, P. (1973) The discovery of America. *Science* **179**, 969–974.

Martin, P. (1984) Prehistoric overkill: A global model. In (P.S. Martin and R.G. Klein, eds.) *Quaternary Extinctions: A Prehistoric Revolution*, 354–403. Tucson: University of Arizona Press.

McGlone, M.S. and Wilmshurst, J.M. (1999) Dating initial Maori environmental impact in New Zealand. *Quaternary International* **59**, 5–16.

Meltzer, D.J. and Mead, J.I. (1983) The timing of North American late Pleistocene extinctions. *Journal of Archaeological Science* **19**, 130–135.

Milewski, A.V. and Diamond, R.E. (2000) Why are very large herbivores absent from Australia? A new theory of micronutrients. *Journal of Biogeography* **27**, 957–978.

Miller, G.H., Magee, J.W. and Jull, A.J.T. (1997) Low-latitude glacial cooling in the Southern Hemisphere from amino-acid racemization in emu eggshells. *Nature* **385**, 241–244.

Miller, G.H., Magee, J.W., Johnson, B.J., Fogel, M.L., Spooner, N.A., McCulloch, M.T. and Ayliffe, L.K. (1999) Pleistocene extinction of Genyornis newtoni: human impact on Australian megafauna. *Science* **283**, 205–208.

Moore, P.D. (1977) Ancient distribution of lime trees in Britain. *Nature* **268**, 13–14.

Moss, P.T. and Kershaw, A.P. (2000) The last glacial cycle from the humid tropics of northeastern Australia: comparison of a terrestrial and a marine record. *Palaeogeography, Palaeoclimatology, Palaeoecology* **155**, 155–176.

Neely, J.A. (1974) Sassian and early Islamic water control and irrigation systems on the Deh Luran Plain, Iran. In (T.E. Downing and M. Gibson, eds.) *Irrigation's Impact on Society*, 21–42. Tucson: University of Arizona Press.

Owen, R. (1846) *A History of British Fossil Mammals and Birds*. London: Van Voorst.

Parrot, A. (1948) *Tello. Vingt campagnes de fouilles (1877–1933)*. Paris: Editions Albin Michel.

Pemberton, W., Postgate, J.N. and Smyth, R.F. (1988) Canals and bunds, ancient and modern. *Bulletin on Sumerian Agriculture* **4**, 207–221.

Postgate, J.N. (1992) *Early Mesopotamia: society and economy at the dawn of history*. London: Routledge.

Powell, M.A. (1985) Salt, seed and yields in Sumerian agriculture: a critique of the theory of progressive salinization. *Zeitschrift fur Assyriologie* **75**, 7–38.

Roberts, R.G., Flannery, T.F., Ayliffe, L.K., Yoshida, H., Olley, J.M., Prideaux, G.J., Laslett, G.M., Baynes, A., Smith, M.A., Jones, R. and Smith, B.L. (2001) New ages for the last Australian megafauna: continent-wide extinction about 46,000 years ago. *Science* **292**, 1888–1892.

Roberts, N. (1989) *The Holocene: An Environmental History*. Oxford: Blackwell.

Roberts, R.G., Jones, R. and Smith, M.A. (1990) Thermoluminescence dating of a 50,000-year-old human occupation site in northern Australia. *Nature* **345**, 153–156.

Roberts, R.G., Jones, R., Spooner, N.A., Head, M.J., Murray, A.S. and Smith, M.A. (1994) The human colonisation of Australia: optical dates of 53,000 and 60,000 years bracket human arrival at Deaf Adder Gorge, Northern Territory. *Quaternary Science Reviews* **13**, 575–583.

Roberts, R., Yoshida, H., Galbraith, R., Laslett, G., Jones, R. and Smith, M. (1998) Single-aliquot and single-grain optical dating confirm thermoluminescence age estimates at Malakunanja II rock shelter in northern Australia. *Ancient TL* **16**, 19–24.

Roberts, O. (2004) Reconstruction and performance. In (P. Clark, ed.) *The Dover Bronze Age Boat*, 189–210. London: English Heritage.

Robinson, E. (1998) *Report on the Ostracod fauna associated with the Dover Boat.* Unpublished archive report, Canterbury Archaeological Trust.

Roux, G. (1964) *Ancient Iraq.* London: George Allen and Unwin.

Sandweiss, D.H., McInnis, H., Burger, R.L., Cano, A., Ojeda, B., Paredes, R., del Carmen Sandweiss, M. and Glascock, M.D. (1998) Quebrada Jaguay: early South American maritime adaptions. *Science* **281**, 1830–1832.

Schuster, S. and Schüle, W. (2000) Anthropogenic causes, mechanisms and effects of Upper Pliocene and Quaternary extinctions of large vertebrates. *Oxford Journal of Archaeology* **19**, 223–239.

Shane, L.C.K. (1987) Late-glacial vegetational and climatic history of the Allegheny Plateau and the Till Plains of Ohio and Indiana, USA. *Boreas* **16**, 1–20.

Sidell, J. and Wilkinson, K. (2000) The interaction of environmental change and human habitation. In (J. Sidell, K. Wilkinson, R. Scaife and N. Cameron, eds.) *The Holocene Evolution of the London Thames: Archaeological Excavations (1991–1998) for the London Underground Limited Jubilee Line Extension Project*, 118–124. Museum of London Monograph 5. London: Museum of London.

Sidell, J., Cotton, J., Rayner, L. and Wheeler, L. (2002) *The Prehistory and Topography of Southwark and Lambeth.* Museum of London Monograph 14. London: Museum of London.

Steadman, D.W., Stafford, T.W. and Funk, R.E. (1997) Nonassociation of paleoindians with AMS-dated late Pleistocene mammals from the Dutchess Quarry Caves, New York. *Quaternary Research* **47**, 105–116.

Stuart, A.J. (1993) Death of the megafauna: mass extinction in the Pleistocene. *Geoscientist* **2**, 17–20.

Sutton, D.G. (1987) A paradigmatic shift in Polynesian prehistory: implications for New Zealand. *New Zealand Journal of Archaeology* **9**, 135–155.

Taylor, R.E., Haynes, C.V. Jr. and Stuiver, M. (1996) Clovis and Folsom age estimates: stratigraphic context and radiocarbon calibration. *Antiquity* **70**, 515–525.

Thomas, C. and Rackham, J. (1996) Bramcote Green, Bermondsey: a Bronze Age trackway and palaeoenvironmental sequence. *Proceedings of the Prehistoric Society* **61**, 221–253.

Turner, J. (1962) The *Tilia* decline: an anthropogenic interpretation. *New Phytologist* **61**, 328–341.

Turney, C.S.M., Bird, M.I., Fifield, L.K., Roberts, R.G., Smith, M.A., Dortch, C.E., Grün, R., Lawson, E., Ayliffe, L.K., Miller, G.H., Dortch, J. and Cresswell, R.G. (2001a) Early human occupation at Devil's Lair, southwestern Australia 50,000 years ago. *Quaternary Research* **55**, 3–13.

Turney, C.S.M., Kershaw, A.P., Moss, P., Bird, M.I., Fifield, L.K., Cresswell, R.G., Santos, G.M., di Tada, M.L., Hausladen, P.A. and Zhou, Y. (2001b) Redating the onset of burning at Lynch's Crater (North Queensland): Implications for human settlement in Australia. *Journal of Quaternary Science* **16**, 767–771.

van Huet, S., Grün, R., Murray-Wallace, C.V., Redvers-Newton, N. and White, J.P. (1998) Age of the Lancefield megafauna: a reappraisal. *Australian Archaeology* **46**, 5–11.

Vartanyan, S.L., Garutt, V.E. and Sher, A.V. (1993) Holocene dwarf mammoths from Wrangel Island in the Siberian Arctic. *Nature* **362**, 337–340.

Wachsmann, S. (1995) *The Sea of Galilee Boat.* New York: Plenum Press.

Wallace, A.R. (1876) *The Geographical Distribution of Animals, with a Study of the Relations of Living and Extinct Faunas as Elucidating Past Changes of the Earth's Surface.* New York: Harper.

Waller, M. (1994) Paludification and pollen representation: the influence of wetland size on Tilia representation in pollen diagrams. *The Holocene* **4**, 430–434.

Watson, J. (2004) Conservation. In (P. Clark, ed.) *The Dover Bronze Age Boat*, 282–287. London: English Heritage.

Webb, R.E. (1998) Megamarsupial extinction: the carrying capacity argument. *Antiquity* **72**, 46–55.

Wilkinson, K. (1998) *An examination of non-marine Mollusca from the Dover Bronze Age boat: Analytical report.* Unpublished archive report, Canterbury Archaeological Trust.

Wilkinson, K., Scaife, R., Sidell, J. and Cameron, N. (2000) The palaeoenvironmental context of the JLE project. In (J. Sidell, K. Wilkinson, R. Scaife and N. Cameron, eds.) *The Holocene Evolution of the London Thames: Archaeological Excavations (1991–1998) for the London Underground Limited Jubilee Line Extension Project*, 11–19. Museum of London Monograph 5. London: Museum of London.

Willerslev, E., Hansen, A.J., Binladen, J., Brand, T.B.,

Gilbert, M.T.P., Shapiro, B., Bunce, M., Wiuf, C., Gilichinsky, D.A. and Cooper, A. (2003) Diverse plant and animal genetic records from Holocene and Pleistocene sediments. *Science* **300**, 791–795.

Yoffee, N. and Clark, J. (eds.) (1993) *Early Stages in the Evolution of Mesopotamian Civilization: Soviet Excavations in Northern Iraq*. Tucson: University of Arizona Press.

Index